Intermodal Freight Transport

Intermodal Freight Transport

David Lowe FCILT

ELSEVIER
BUTTERWORTH
HEINEMANN

AMSTERDAM • BOSTON • HEIDELBERG • LONDON • NEWYORK • OXFORD
PARIS • SAN DIEGO • SAN FRANCISCO • SINGAPORE • SYDNEY • TOKYO

Elsevier Butterworth-Heinemann
Linacre House, Jordan Hill, Oxford OX2 8DP
30 Corporate Drive, Burlington, MA 01803

First published 2005

British Library Cataloguing in Publication Data
A catalogue record for this book is available from the British Library

Library of Congress Cataloguing in Publication Data
A catalogue record for this book is available from the Library of Congress

ISBN 0 7506 5935 1

For information on all Elsevier Butterworth-Heinemann publications visit our website at http://books.elsevier.com

Typeset by Charon Tec Pvt. Ltd, Chennai, India
www.charontec.com
Printed and bound in Great Britain

Contents

Front cover captions

1. Distribution of Blue Circle (Lafarge) cement in the UK by piggyback road–rail system. (*Photo*: David Lowe.)
2. Geest North Sea Line's latest short sea vessel shipping containers in costal trade
3. Transferring a Maersk Line ISO shipping container from articulated road vehicle on to rail with the aid of a twist-lock equipped spreader on a gantry crane. (Publicity photo.)

To my wife
Patricia

Disclaimer

The legal explanations in this text are provided for general information purposes only and are not definitive interpretations of the law, which only the courts may make. Readers are advised to seek appropriate legal advice before making any decisions based on the legal information contained herein.

List of illustrations

The Author

David Lowe is a freelance writer with a lifetime of road haulage and freight industry experience. He is the author of the best-selling *Transport Manager's and Operator's Handbook*, published annually since 1970, and of numerous other transport-related titles and was a regular contributor to the UK transport press for many years. Besides having a detailed knowledge of UK and European Union (EU) road transport law and extensive 'hands-on' experience of freight transport operations, he has taken a particular interest in the development and growth of intermodal transport. He is a Fellow of the Chartered Institute of Logistics and Transport, a Liveryman of the Worshipful Company of Carmen and is a former winner of the Carmen Company's prestigious Herbert Crow Memorial Award for, 'over 30 years of written contributions towards the furtherance of transport industry knowledge'.

Foreword

One of the few joys of being a Transport Minister in the UK government is that it is a mercifully politics-free zone. I have never yet come across a convincing Marxist-Leninist view of traffic management and the Adam Smith version, which would presumably turn all traffic lights off and rely on market forces, seems equally unappealing. It would, however, be wrong to conclude that there is no science behind modern transport policy. Since the doctrine of 'predict and provide' rightly came under attack, ministers in both Conservative and Labour governments have emphasised the need to concentrate on sustainable integrated transport that offers practical alternatives to the way we have historically managed the movement of people and freight with the objective of protecting the environment, reducing air pollution and noise, and improving social mobility.

Much of the effort to create a new, more intelligent transport policy has understandably concentrated on ways of offering alternatives to the use of the private car. The inherent inefficiency of a single occupant in a large metal container consuming large quantities of fossil fuel compared to the same person using a train or bus is easy to appreciate. But until recently much less thought has been given as to how the movement of goods rather than people can change the way we procure the lifestyle consumers demand while impacting significantly less on the world around us. My suspicion is that this is because freight policy has been in the 'too difficult' file on transport planners' desks. Yet the challenges facing an advanced economy such as ours in which currently 420 000 trucks of varying degrees of environmental efficiency deliver virtually all our goods and services are actually very real.

In this excellent and extremely comprehensive book David Lowe builds on his established reputation as one of this country's leading transport writers to explain and illustrate the key concept of intermodal transport in detail. In the next decade much more attention will be paid to the use of rail, canals and short sea freighters as alternatives to trucks over part if not all of the freight journey. This will be a priority for politicians but also for consignors and consignees. It will be integral to the corporate social responsibility of any serious enterprise. More to the point, it will also be a driver toward more efficient and less costly transportation. This is a timely and readable volume which will appeal to policy-makers and practitioners at every level. I thoroughly commend it.

Steven Norris FCILT, FIHT
Minister for Transport 1992–1996
Director General,
The Road Haulage Association 1997–1999

Acknowledgements

Many organizations, both governmental and private, numerous commercial firms in the UK and throughout Europe and many individuals from the industry very kindly provided, by various means, much of the information used in compiling the text for this book both in its original form in 1996 and in its current expanded and updated form. The names of some of them are to be found where appropriate in the text. To all of these I would like to record my sincere appreciation for their valued help.

However, it would be remiss of me not to record my particular appreciation for the assistance provided for the original text by Mr. J. R. Fells, formerly of the Freight and Road Haulage Division, Department of Transport, London and Mr. J. Hugh Rees, formerly of the Directorate-General for Transport (DG VII), Commission of the EU, Brussels. Also, various staff members from the freight division of Eurotunnel both in Calais and Folkestone; Andrew Trasler, Through Transport Mutual Services (UK) Limited – the TT Club, London; and Ms Helen Berry, Peter Huggins and Ms Lynn Mentiply, Library staff at the former Chartered Institute of Transport library, London, and the Chartered Institute of Logistics and Transport library, Corby. I am also indebted to Gerhard Muller, noted United States transportation specialist from whose book *Intermodal Freight Transport* (1995) I have gleaned much and to which I have made reference.

Various relevant intermodal and freight organizations have been the source of much useful information, especially; the International Union of Combined Road–Rail Transport Companies (UIRR), and the European Intermodal Association (EIA); both of these are Brussels-based organizations. I am also grateful for the considerable knowledge that I gained about intermodalism from attending the series of intermodal exhibitions and conferences organized by Informa Maritime and Transport.

Acknowledgement is made to The Stationery Office (TSO) in respect of the reproduction of extracts from legislative material in accordance with HMSO guidelines; to the European Commission (EC) in respect of the reproduction of extracts from its reports as identified in the text (also legally permitted); and to the UK International Road Freight Office for the reproduction of notes from its explanatory literature for transport operators, and to its helpful staff who patiently answered many questions.

Additional help and support has been forthcoming from Lord (Tony) Berkeley, head of the Rail Freight Group; Paul Kneller of Eurotunnel's freight team; Duncan Buchanan of the Strategic Rail Authority (SRA); Roy Parker, Freight Marketing Manager, British Waterways; and Edmund Athayde of Herfurth Shipping (UK) Ltd, Hull. I also acknowledge the valuable information gained from material published on the Internet by British Waterways Board and the Association of Inland Navigation Authorities; by the UK Department for Transport; by the UK-based 'Freight on Rail' pressure group; and by the various sectoral branches of the EC within its Directorate-General for Energy and Transport on the Commission's extensive Europa web site, for which reproduction is permitted. Various industry journals have provided a constant source of reference as to what is happening in intermodal freighting, most notably; *International Freighting Weekly* (IFW), *Rail* and *Commercial Motor*. Finally, I would like to express my gratitude to my publishers Elsevier for their considerable forbearance while awaiting the manuscript for this book.

Despite all the kind assistance received, I accept that responsibility for the interpretation and presentation of all the information herein and its accuracy rests solely on my shoulders.

David Lowe
March 2005

Preface

The movement of freight is a vital ingredient in achieving economic sustainability and for enhancing our lives in many different ways. However, for all the benefits that it brings, such as stocking the shelves of our corner shops, local supermarkets, DIY stores and garden centres for instance, it does have its downside. Road haulage, in particular, which is the aspect of transport with which we, the general public, are likely to be most familiar because of the large lorries that we see on our streets, adds to the congestion in our town and city centres and on our motorways; these lorries pollute the air we breath with unhealthy exhaust fumes, but so too do buses and the thousands of cars on our roads; they cause visual intrusion, noise and vibration and they contribute, albeit only marginally compared to all the other vehicles and pedestrians, to the awful accident toll on our roads. But, no matter what the environmental penalties of lorry use amount to, we cannot do without road transport: the heavy lorry will never go away and can never be replaced by some magical form of conveyance that is noiseless, fumeless and takes up no road space – no such contrivance exists now and is never likely to.

However, all is not doom and gloom. There is a partial solution which is operationally feasible, economically viable and, most importantly, environmentally sustainable. It is the concept of intermodal freighting. This is a system in which two or more different modes of transport, such as road and rail, road and waterway or rail and shipping are combined, or integrated, to enable goods contained within a single loading unit, to be moved from their place of origin to their final destination. The loading unit, depending on the system used, may be a container, a swap body, a complete road vehicle or an unattached articulated semi-trailer. Importantly, at the interchange of the modes the goods remain undisturbed, only the loading unit is transferred from one mode to another, and in the case where a load is packed within a road vehicle or road semi-trailer, which thus constitutes the loading unit, this is either driven, or lifted, on to a rail wagon or a roll-on/roll-off ferry ship.

The intermodal alternative – or in contemporary terminology, intermodalism – currently holds sway over other, individual, transport modes, especially direct door-to-door delivery by road haulage, mainly for reasons of its so-called 'green credentials' because, in other words, it is less of a blight on the environment. It is a transport system where road vehicles are employed to do what they can do best; namely, the essential, short-haul, collections and deliveries in locations where trains and canal barges cannot go, but where the long-haul leg of the journey to the final delivery is carried out by the much more environmentally friendly freight train or waterway transport system which, for their part they are best suited. In fact, such is the current support for the intermodal transport concept that it comprises a major building block of the EC's Transport Policy White Paper for 2010: *Time to Decide*, for the reason that it contributes to the desirable objective of shifting the balance between modes.

The stated aim of the Commission's policy on intermodal freight transport is to support the efficient door-to-door movement of goods, using two or more modes of transport, in an integrated transport chain. As the Commission says:

> Each mode of transport has its own advantages; for example, potential capacity, high levels of safety, flexibility, low energy consumption, low environmental impact. Intermodal transport allows each mode to play its role in building transport chains, which overall, are more efficient, cost effective and sustainable.

Road haulage invariably features significantly in the intermodal equation, usually in combination with rail freighting and using the two modes together – commonly referred to as combined road–rail freighting – is the predominant intermodal solution. But inland waterway and/or short-sea and coastal shipping too, where appropriate, may also be combined with road transport to form an intermodal operation. Multimodal transport, on the other hand, involves the use of more than just two modes; for instance, a typical multimodal freight journey may involve a combination of road haulage, a rail freight journey leg and then either a short sea, coastal or even a trans-ocean crossing.

The inter- or multi-modal concept is simple to comprehend. But the whys and wherefores of effecting delivery of a particular freight consignment by either a single mode, direct, door-to-door road haulage operation or by utilizing a combination of different modes, is a much more complex issue. Broadly, the aim of intermodalism is to utilize the most operationally efficient and cost effective combination of transport modes to convey a load of goods to its final destination – using each individual mode to its best effect. However, of equal importance, is the need to switch freight traffic from our inadequate and overcrowded road network on to rail or on to the relatively under-utilized waterway systems of the UK and Europe, principally with the environmentally beneficial objectives of reducing air pollution, noise, vibration and the risk of road traffic accidents. While the commercial considerations are obviously important to shippers consigning the goods, achieving environmental harmony is a key objective of the sustainable transport policies of both the UK Government and the EC.

The general public depends heavily on the heavy lorry for carrying most of its goods traffic – without it our supermarket shelves would be virtually empty. However, while being extremely convenient and flexible with its direct door-to-door delivery capability, the lorry is nevertheless notorious for polluting the atmosphere with noxious exhaust fumes which damage human health, for using-up scarce fossil-based fuel resources, for creating undue noise and vibration – which is especially noticeable in urban areas – and for adding to the traffic congestion and accident risks on our largely inadequate road network. But, it is possible, through intermodalism, to harness the particular advantages of road transport to the freight carrying capabilities of other quieter, less polluting, and generally more environmentally friendly transport modes. And by so doing cost effective freight deliveries can be achieved while avoiding, or at least reducing, the environmental blight which massive and ever-growing volumes of heavy lorry traffic inflicts upon us. From the lorry operator's point of view, intermodalism offers the possibility of avoiding the proliferation of lorry bans in urban areas; congestion charging zones; motorway tolls; and the prospect of Lorry Road User Charges (currently existing in parts of Europe and to be with us in the UK in 2007/2008); the ever increasing costs of delay and disruption to operating schedules and in the UK, the extortionate price of diesel fuel inflated by excessive levels of excise duty imposed by government.

This book examines, in a practical manner, the whole concept of intermodalism as it relates specifically to freight transport: passenger inter-modalism being a wholly different genre is not covered. Some of the text herein follows on from a study on combined road–rail freighting written by the author in 1996. However, while that particular work focused mainly on a single aspect of freight intermodalism, namely combined road–rail transport, this new text, besides being significantly updated to take account of new Government and EU policy initiatives, legislative changes and operational developments in the interim period, has also been broadened to include the wider concept of intermodalism involving inland waterway freighting and short-sea and coastal shipping. Such has been the expansion of interest in this form of transportation in recent times that there has been almost an explosion of information about it published on the Internet, and the author has taken the opportunity to draw together and incorporate some of this accumulated wisdom to provide readers with a much greater insight into the structure of intermodalism, the policy-making activities of Brussels and national governments and to relevant legislative developments.

The text examines intermodalism from a number of perspectives; first of all from the political angle in regard to governmental policies and strategies and then from a geographical viewpoint where intermodal developments in the UK, in Europe and, collectively in North America, the Middle East, Asia and

Australia, are considered. Then, importantly, individual chapters are devoted to the key constituent elements of intermodalism, namely road haulage; rail freighting; inland waterway, short-sea and coastal shipping; networks and terminals; and hardware in the form of intermodal vehicles, loading, lifting and transfer equipment.

The book also covers many other essential factors involved in making intermodalism work in an efficient and cost-effective manner; for example, the legal aspects of carrier liability; vital safety regulations; governmental grant aid designed to encourage a switch of freight from road to other more sustainable modes of transport; environmental considerations in favour of intermodal transport; and documentation and authorization requirements. To help the reader, an extensive glossary of related terms is included and so to is a bibliography of useful further reading.

Overall, it has been the author's intention to provide, so far as is possible with such a fast-moving scenario, a study of intermodal freight transport that will be a valid and useful source of reference for freight shippers, intermodal road hauliers – and especially those road hauliers contemplating a switch of their long-haul traffic to other modes, rail service suppliers, terminal operators, equipment manufacturers and ancillary suppliers to the industry. The book will also be of value to students of transport and others who may wish to keep abreast of developing trends in transportation. Additionally, it is hoped that the text will be of general interest to industry at large as it endeavours to placate public concern about environmental issues and achieve such worthwhile objectives as reduced pollution, reduced road traffic congestion and fewer road traffic accidents, but without any diminution in the levels of service, speed and security with which goods – many of them being the vital ingredients of everyday life – are transported to their destination.

1
What Is Intermodal Freight Transport?

Intermodal freight transport, as previously outlined in the Preface, is the concept of utilizing two or more 'suitable' modes, in combination, to form an integrated transport chain aimed at achieving operationally efficient and cost-effective delivery of goods in an environmentally sustainable manner from their point of origin to their final destination.

While some freight movements may use, and justify the use of, a number of different transport modes, such as road, rail or inland waterway or either short- or deep-sea shipping, thus making them multimodal operations, in the majority of instances efficient movements are invariably achieved by the use of just two modes: most commonly road haulage collection and final delivery journeys combined with a rail-freight trunk-haul journey, what is known as a 'combined road–rail' operation. However, where operational circumstances dictate or a feasible alternative option is available, road haulage or rail freighting may be combined instead with an inland waterway journey via river or canal, or with a short-sea shipping (SSS) operation, typically a coastal or a cross-Channel sailing. Combined transport operations involving either road haulage or rail freight in conjunction with deep-sea container services or with an airfreight operation also feature in intermodal and multimodal scenarios, albeit the latter occurring in only a relatively small number of instances and small scale in terms of the freight volumes shipped.

The word 'suitable' in the context used above may have a number of alternative connotations. It is possible in a given set of circumstances that cost alone will determine the choice of mode or modes, but frequently other considerations are decisive. For instance, operational practicalities like frequency of service, speed of delivery, the availability of special handling facilities, the ability to meet particular packaging requirements, security considerations or the sheer volumes of freight to be moved may be the determining factors, or indeed it may be that a number of these various service 'pluses', in combination, produce the ideal solution. But other less tangible issues may also enter the equation such as the need to follow corporate environmental policies or to assuage a shipper's social conscience.

Freighting by one of the intermodal or multimodal combinations mentioned above is the alternative, of course, to consigning loads for the whole of their journey by a single mode, as is the case with some 62 per cent of domestic freight moved in Great Britain (according to the Department for Transports' (DfTs'), *Transport Statistics for Great Britain – 2004*, publication) for example, which is transported by road. It is a fact that the haulage of goods by road from their source direct to their final destination remains the preferred method in the majority of cases, and it is this preference which those individuals, corporate bodies, and government departments alike who champion the cause of intermodal transport are trying to break down.

Since by far the largest proportion of freight traffic commences and ends its journey on the back of a lorry, intermodalism is principally understood to mean the use of an alternative mode to undertake the middle, long-haul, or trunk, leg of the journey. Typically this involves trans-shipping the unitized load from a lorry at a railhead, inland waterway terminal, or at a seaport, for onward shipment by rail, inland waterway barge, sea-going ship, and then trans-shipping it back again onto a lorry for the final delivery of the leg to the customer; that is, the consignee. In some instances, as we shall see later, it is not just the unitized load that is trans-shipped, but the whole road vehicle, or at least its semi-trailer, which is loaded aboard a rail freight wagon, an inland waterway barge, a short-sea vessel, invariably a roll-on/roll-off (RO-RO) ferry ship, or an ocean-going ship for onward transportation.

No matter what the particular freighting arrangement is, the essence of the whole operation is to utilize the key characteristics of each individual transport mode to its best advantage. The lorry has the benefits of immediacy and flexibility in its favour plus the ability to affect collections and deliveries of goods from locations that have no rail sidings or waterway quays for loading. Rail freighting offers a lower-cost alternative for multiple loads carried over longer transits and a much less-polluting effect on the environment, as does barge traffic shipped on the navigable rivers, and canals of inland waterway networks. Thus road haulage in any combination with either rail freighting, inland waterway, short-sea, or coastal shipping may prove to be the most viable option, both economically and operationally. But in certain cases, particularly where no RO-RO vehicle ferry service, road or rail tunnel facility exists, shipping by container vessel may be necessary, and especially for trans-global freight movements.

As this book will show, there are politically motivated policy moves within the European Union (EU) to find ways of switching as much freight as possible from road onto the rail, and to a lesser extent onto the waterway, networks. This is seen as a beneficial antidote to the adverse impact of heavy lorries on traffic flows both on motorways and in urban areas, although the latter is something of a misconception since heavy lorries will still be required to serve many road–rail terminals, and the substantial numbers of freight originators and recipients located in urban areas who are not rail connected (e.g. most small businesses and High Street retail outlets). The European Commission (EC) talks of the 'complementary' qualities of road and rail transport, and it is this aspect of complementarity that it believes to be the key to transport policy for the future. However, we should not overlook the keen interest now being shown in Brussels (i.e. in the EC's Directorate-General for Energy and Transport – DG VII) in revitalizing the role of inland waterway and short-sea and coastal shipping, as we shall see in greater detail in Chapter 8.

Also to be gained from a modal switch are perceived social benefits, such as reduced air pollution, resulting from fewer heavy lorries, the theoretical saving in road accidents – according to the European Commission's *Community Road Accident Database* (CARE), there were some 40 000 road accident deaths in the 15 Member State-EU in 2002 costing 160 billion euros (£115 billion at the November 2004 currency exchange rate) – and the separation of freight movements from a 'people' environment, which roads substantially are and rail, largely, is not.

The economic concept of road–rail combined transport is that it keeps the expensive element of the road haulage operation, namely the operation of the road vehicle tractive unit and the driver, fully utilized on short-haul, road-borne collections and deliveries, for which it is ideally suited and sufficiently flexible to go anywhere at any time to suit individual requirements, while the less expensive part of the operation, namely the loaded semi-trailer, swap body, or container, is sent unaccompanied on the long-haul leg of the journey by rail. Given that the rail-haul leg is long enough to justify the switch from road (generally it needs to be at least 500 kilometres, although new thinking suggests that short-distance intermodalism becomes viable for distances in excess of only around 200 kilometres), this produces benefits by way of savings in journey time, reduced consumption of carbon fuel, less pollution from exhaust emissions, reduced heavy traffic flows on motorways, and, theoretically, fewer road accidents. The road haulier benefits from not having his truck and driver missing for hours, if not days, on end, pounding up and down the roads of the UK or across Europe to meet customer deadlines, and incurring wear and tear, damage, driver subsistence costs, plus the well-known risks of penalty for speeding, traffic, and other law infringements. Given the right combination of circumstances, the goods transported on the long haul by rail rather than by road will be at their destination sooner since trains, unlike truck drivers, are not obliged by law to park up en-route for a statutory night or weekend rest period.

The essence of efficient intermodal transport lies in the use of a unit-load system capable of transfer between road, rail, and other transport modes, and which allows for the collection of consignments by, for example, a road vehicle followed by a trunk-haul journey by rail or waterway and a final road-borne delivery without trans-shipment or repacking of the load itself. Standard loading units take the form of either road-going semi-trailers conforming to standard dimensions and designed to be piggybacked aboard rail wagons, or more commonly, swap bodies and shipping containers built to international (ISO) standards which are fully interchangeable between a variety of road vehicle combinations, rail wagons, river and canal barges, and sea-going ships. In all circumstances the load remains intact and secure within the loading unit which is lifted or transferred by purpose-built equipment onto a rail wagon, a canal barge, or into the hold of a ship and then back to a road vehicle at the end of the trunk-haul leg of the journey.

Such systems provide greater flexibility for the customer, who may be either the consignor or the consignee, by allowing goods to be loaded or unloaded at his premises in the conventional manner without changing the current practices applied to his domestic or local traffic. It also assures his piece of mind if, having seen his freight securely stowed and sealed in an intermodal-loading unit, he knows that it will not be disturbed again until it reaches its final destination, unless it is to comprise part of a groupage load. The principal benefits of unit-load intermodalism is that it can provide:

- lower transit costs over long journeys;
- potentially faster delivery times in certain circumstances (these obviously need to be individually assessed for particular cases);
- a reduction in road congestion (a major beneficial factor in these modern times);
- a more environmentally acceptable solution to congestion and related problems (such as the emission of noise and fumes, the damage caused to the built environment by vibration and so on);
- reduced consumption of fossil fuels since the long-haul section of the route is more fuel efficient;
- safer transit for some dangerous products.

1.1 The background to intermodalism

The practice of transferring road trailers and road-borne containers onto rail wagon for trunk haulage has existed since the earliest days of rail. The hardware has obviously changed over the years and today's domestic and international journeys are much longer than the domestic operations of yesteryear, but the basic principles remain the same. Simple wooden box containers, used even in the days of horse-drawn transport, have given way to the latest form of steel shipping container and swap body built to international (ISO) strength and dimensional standards, while road-hauled semi-trailers have developed from simple two-wheeled affairs with 'cart-like' springing – as drawn by the well-known railway 'turn-on-a-sixpence' three-wheeled Scammell mechanical horse – into high-capacity multiaxle, sophisticatedly-suspended units. In fact, present-day articulated semi-trailers are highly sophisticated pieces of equipment, cushioned with air suspension and equipped with airbrakes, and capable of safely carrying a 30-odd tonne laden ISO container or swap body within the current legal maximum vehicle gross weight limit of 44 tonnes and at the maximum permitted speed; namely 56 miles per hour (mph) for speed-limited heavy trucks. The parallel development of technically sophisticated lifting and transfer equipment enables these loading units and semi-trailers to be transferred rapidly and efficiently from road to rail or barge for long-haul transport, and back again for final delivery.

Intermodal road–rail transport as we know, today it has been widely and successfully employed in mainland Europe for many years; especially notable, for example, being the French Novatrans 'Kangaroo' system for the piggyback carriage on rail of unaccompanied road-going semi-trailers and the similar German Kombiverkehr system in which swap bodies, piggyback semi-trailers, and complete road vehicles are also carried by rail on what is known as a rolling motorway system.

1.2 The impact of the Channel Tunnel

It is useful to consider here the impact of the Channel Tunnel between the UK and France, which opened in May 1994, on the longer-haul potential of UK–Europe intermodal freighting. Eurotunnel, the Tunnel operator, originally estimated that it would carry around 400 000 heavy goods vehicles annually on its drive-on/off freight shuttle service, a target which, in 2003, it significantly exceeded with 1 284 875 trucks being carried – in fact, capturing some 40 per cent of the total cross-Channel driver-accompanied freight traffic market. By September 2003 Eurotunnel was operating up to seven freight shuttles each way every hour during peak periods. However, these carryings have somewhat diminished since these figures were published with a 4 per cent reduction in truck shuttles reported for the third-quarter of 2004 and a 2 per cent loss of market share.

Besides this lorry traffic, substantial volumes of rail-borne inter-Continental swap body and container traffic passes through the Tunnel from inland freight terminals in the UK to international destinations via Europe's 241 000-kilometre rail networks. In 2003 this amounted to 1 743 686 tonnes; albeit this tonnage was some 40 per cent less than its carryings in 1999 and lower than for both 1998 and 1997, largely due to its slow recovery from the widely publicized problems of 2002 caused by the influx of illegal immigrants into the UK who were stowing away on the freight trains from the French side of the Tunnel, resulting in many service cancellations. Encouragingly, however, by November 2004, Eurotunnel had reported a 7-percent increase in its rail-freight carryings through the Tunnel.

These statistics show the Tunnel to have been an important catalyst for increased interest in the development of intermodal services between the UK and Europe. Furthermore, additional encouragement was provided by legislative measures permitting heavier lorries for use in intermodal transport operations. From March 1994, 44-tonne lorries (compared with the previous domestic maximum weight limit of 38 tonnes) were allowed on UK roads, initially for use only in road–rail transport operations to and from rail terminals, subject to specific technical and administrative conditions, but since 1 February 2001 44-tonne vehicles have also been permitted, unconditionally, for general-freight carrying in the UK. This development proved to be a boon for Eurotunnel's heavy-vehicle carryings since it encouraged greater interest among UK road hauliers in undertaking trans-European operations via the Tunnel. Another boost for Eurotunnel's market potential share, although not yet, at the end of 2004, reflected in its carryings, was the addition of the 10 new Member States to the EU from May 2004; road hauliers from these countries being permitted to carry freight to and from the UK subject to meeting the necessary EU legal requirements relating to goods vehicle operation.

1.3 Freight transport growth

Continued development of intermodal transport between the UK and Continental Europe is, of course, dependent upon a growing freight transport market throughout the EU, the rest of Western Europe, and the former Eastern Bloc countries – now, largely, part of the EU. The EU expanded from its former 15 Member States in May 2004 with the admission of the 10 so-called 'accession' states; namely, the Czech Republic, Estonia, Cyprus, Latvia, Lithuania, Hungary, Malta, Poland, Slovenia, and Slovakia – increasing the potential market to some 480 million people – and is due to expand again in 2007 when Bulgaria and Romania expect to be admitted, with Turkey following at some point in the future. There is no doubt that

Year	1970	1980	1990	1995	2000
Pipeline	6.8	7.1	5.0	5.3	4.7
Inland water	10.9	8.9	7.6	7.4	6.9
Rail	30.2	24.2	18.2	14.2	13.8
Road	52.0	59.9	69.2	73.2	74.6

Fig. 1.1 EU Freight transport by mode statistics 1970–2000 (in tonne-kilometres %). (*Source*: EU Energy and Transport in Figure 2003, via Internet.)

with all these new Member States, the expanded EU will provide colossal opportunities for the development of intermodal freight transport – in October 2003, European Commissioner in charge of transport, Sna Loyola de Palacio, said that goods transport would increase by 36–40 per cent in the next decade and that besides the new infrastructures needed, alternative modes of transport would also have to be developed. Certainly, if she is right, which undoubtedly she is, it would be a frightening prospect if the resultant increase in intra-European trade and consequently its transportation needs were to be funnelled onto the EU's existing heavily congested road network.

We have, of course, already seen significant transport growth within the EU over past years, but it has not been shared equally between modes. While road transport has grown to account for roughly 75 per cent of all intra-Community goods transport activity, in the same period (i.e. 1970–2000) rail transport decreased in relative terms from 30.2 to 13.8 per cent. The inland waterways of Europe, as Figure 1.1 shows, also declined from carryings of 10.9 per cent of traffic to just 6.9 per cent, albeit since 2000 we have begun to see a reversing trend in the fortunes of this mode, while SSS now carries some 40 per cent of all trade within the EU.

The trend towards the growth of road freighting in favour of other modes as shown in the table above continues. In its 2001 White Paper, *European Transport Policy for 2010: Time to Decide*, the EC predicted that by 2010 heavy goods vehicle traffic will have increased by nearly 50 per cent over its 1998 level. And with the strong economic growth expected in the acceding countries (i.e. the 10 countries which joined in May 2004) and better links with outlying countries it suggests that there will be further increases in traffic flows, and in particular road haulage traffic. This inexorable level of road-traffic growth is unsustainable, and everybody knows that. The EC for its part is of the opinion that road transport alone will not cope with the projected expansion in traffic; it suggests that it will need the combined strengths of both road and rail services to meet the challenge. In fact, the EC's White Paper proposed some 60 specific measures to improve transport across the Community, including an action programme extending until 2010. In the context of intermodalism, the most interesting proposals are three-fold:

- revitalizing the railways;
- improving quality in the road transport sector;
- promoting transport by inland waterway (generally referred to in Europe as inland navigation) and by sea.

To relieve congestion on Europe's roads and to protect the environment, the Commission is desperately striving to direct pressure towards a switch of more freight from road to rail and to inland waterway – it talks of achieving 'modal shift'. Thus, through both government direction and commercial pressure, we shall see this developing trend towards intermodal transport continuing. And it has to be in everyone's interest for this to happen. Undoubtedly, the continuing development of combined road–rail transport, associating the economic and environmental advantages of rail and inland waterway freighting for long-distance inter-city or international trunk hauls with the practical advantages of road haulage for local collection and delivery is a strategy which holds much promise for the future.

However, to be a competitive alternative to direct 'door-to-door' lorry transport over long distances, intermodal transport must offer frequent schedules, fast transit times, a high degree of reliability, and all at a cost that fully meets the expectations of markets with the keenest service requirements. In practice, its development is most likely to take place mainly between large industrial conurbations where the problems of road infrastructure congestion are currently most acute and operating costs are thus higher, and where air pollution from road vehicles is at its worst, as in the UK and the Northern European industrial triangle.

Before venturing further into this exploration of what intermodal freight transport is about, it is useful in this opening chapter to consider a few definitions for the variety of terms used in connection with road–rail transport, and in intermodal transport in general, and the equipment needed for its operation. Often these terms are used quite indiscriminately, resulting in people saying one thing, but yet really meaning something quite different. For example, not all containers are built to the international (ISO) standards, or are of the type with which we are principally concerned in the context of this book, and demountable bodies used in domestic transport operations are not to be confused with ISO standard intermodal swap bodies, which often they are. To fully set the scene, we should also examine in this chapter the pressures that have brought about renewed interest in intermodal freight transport, and finally, in summary, what the future holds.

1.4 Definitions

Although this book contains a glossary of terms some of the key components and systems that are encountered in intermodal operations are described here.

1.4.1 Unit loads and loading units

A unit load is a consignment of freight – invariably, but not always, comprising a combination of small consignments, as in a groupage load, which is unitized to save trans-shipment and repacking time and cost at each individual stage of the journey, and also for ease of handling. Such loads are usually consolidated into an ISO container or a swap body built to internationally recognized and accepted standards or into an articulated lorry semi-trailer. Unitization of freight into standard loading units in this manner is a vital element of the intermodal transport concept providing speed and efficiency in handling, security for the load in transit and reduced risk of damage.

1.4.2 Intermodal transport

The term 'intermodal' and the practice of 'intermodalism' are relatively new, being absent, for example, from the *Concise Oxford English Dictionary* of 1980, although, as we have already seen, intermodalism was far from a new concept even at that time. However, by 1993 the terminology was included, being defined as:

> a vehicle/container system, etc. employing, suitable for, or able to adapt or be conveyed by two or more modes of transport.

By the 10th (1999) edition of the same Dictionary it was obviously felt unnecessary to credit the word 'intermodal' with a more detailed definition than:

> involving two or more different modes of transport.

The term 'intermodalism' may thus be taken to mean the practice or activity of conveying freight in unit loads by two or more transport modes such as, for example, by road and rail, by road and inland waterway, or by road and air. Invariably a road element is necessary to make the initial collection of the goods from the consignor's premises and to make final delivery to the consignee since in the majority of cases, there is no direct connection or access between the loading/unloading point and the rail, inland waterway, sea, or air transport system.

1.4.3 Multimodal transport

This term means much the same as intermodal transport, namely the use of a variety of different transport modes for the movement of unitized freight from its place of origin to the final destination. Its more specific use, however, should be confined solely to the situations where more than just two transport modes are employed to complete a freight journey.

1.4.4 Bimodal transport

Bi, meaning two, when used with the term 'modal', correctly implies the employment of just two modes of transport to complete a freight movement, for example the use of road and rail, or road and inland waterway. However, there is an increasingly individualized use of this term to refer specifically to an established system whereby specially strengthened road-going articulated semi-trailers are converted, by being directly attached to rail bogies at the terminal, to run directly on rail either in conjunction with other forms of intermodal rail freight traffic or in combination to form complete bimodal trains.

'RoadRailer' is a brand name for an interchangeable, bimodal road–rail trailer system of this type conceived in the USA by Wabash National Inc. of Indiana. It is based on the use of a road-going semi-trailer with a six-wheel undercarriage that is pneumatically raised allowing the front and rear ends of the trailer to be mounted on rail bogies and formed into a train for rail transit. Road-trailer manufacturer, Fruehauf, and rail-wagon builder, Talbot of Germany, jointly pioneered a similar concept, known as 'Kombirail'. Yet another system similar in concept is in use by Canadian National Railways.

1.4.5 Combined transport

The term combined transport, as with bimodal transport described above, invariably means the use of just two transport modes in combination, such as road and rail or road and inland waterway. Referring to intermodal transport generally as combined transport is not incorrect, although obviously not so precise in circumstances where more than two modes are involved in a multimodal operation and tradition that the term 'combined' transport means just road–rail transport rather than any two other modes in combination.

1.4.6 Rolling motorway systems

A rolling motorway system is one where complete road vehicles are driven onto specially built rail wagons for the rail transit. The system gets its name because, in effect, vehicles are driven straight off the motorway onto the rail wagon at one end of the journey, then off again and onto the motorway at the other end of the trunk leg, the driver travelling on the train accompanying his vehicle. Hence the rail link is seen as a continuation of the motorway journey. Eurotunnel's freight shuttle service through the Channel Tunnel between the UK and France is a rolling motorway system, having special transfer wagons enabling vehicles to be driven onto the train from the platform, and off again at the other end of the through-Tunnel journey. By this means loading and unloading are both easily and rapidly accomplished reducing journey delays to an absolute minimum; one of its main advantages over RO-RO ferry-ship operation where there are often delays in loading and unloading.

1.4.7 Piggyback transport

Unaccompanied articulated semi-trailers are carried on certain UK and European rail services by a method known as 'piggyback' transport. The semi-trailers are either lifted onto special low-height rail wagons by container crane or by straddle carriers fitted with grapple arms which locate into strengthened pockets in the trailer under frame or are shunted, by a terminal tractor, onto special rail wagons with a swivelling, tilting, load-bed, which rotates and lowers to form a ramp. Neither the tractive unit nor the driver travels on the train, the unaccompanied semi-trailer being picked up from a rail terminal at the other end of the journey by another tractive unit and driver for final delivery to its destination.

Piggyback was at one time seen in many circles to be one of the most promising methods of switching long-haul freight from road to rail, and a number of major studies have been carried out in connection with its potential development. For example, the European Conference of Ministers of Transport (ECMT) issued a report in 1992 on improvements in so-called piggyback links across Europe, while in the UK a consortium of consultants and operators, among others, studied and for a while vigorously promoted this particular concept of road–rail transport (as discussed further in Chapter 3). However, for its development to progress in the UK, substantial and very costly infrastructure works are needed to increase the rail loading gauge to provide both top corner and platform clearance for the passage of normal height semi-trailers and 9-foot 6-inch-high ISO containers.

Two new developments gaining currency in 2003 that could have led to further development of the piggyback concept were first, the Modalohr system (described later in the book, but currently still in its experimental stages), and second, the 'Trailers on Trains' study project undertaken jointly by the Dutch Rail Users Platform, the Netherlands Ministry of Transport and the Port of Rotterdam authority. The purpose of this latter study being to find a suitable solution to overcome the traffic congestion and environmental problems caused by some 2.2 million unaccompanied semi-trailers passing through Rotterdam annually (700 000 of them incidentally being en-route to and from the UK) of which, at present, only some 2000 are carried on trains. It is interesting to note, in this connection, that the study initially identified some 80 different techniques that have so far been developed to put trailers on trains.

Overall, however, it has to be said that current (2004) thinking is veering away from further development of the piggyback concept on the basis that it is just not economical to keep lifting and carrying wheeled units when shipping containers and swap bodies can do the same job just as effectively. And it is not just the weight of the wheels that is the problem, heavy though they are; it is the case that piggyback-type semi-trailers need to be purpose-built with strengthened under frames and lifting pockets.

1.4.8 Articulated vehicles

For readers unfamiliar with the technicalities of road vehicles, an articulated vehicle is one comprising a towing vehicle, correctly called the *tractive* unit, but often referred to as the cab or towing unit or incorrectly as the *tractor* unit, and a load-carrying trailer, the *semi-trailer*. They are hitched together with the front end of the semi-trailer superimposed on the rear of the tractive unit (transferring at least 20 per cent of the weight of the load carried by the semi-trailer onto the drawing vehicle, to meet legal requirements), being attached by means of a kingpin on the semi-trailer engaging in a 'fifth-wheel' turntable on the tractive unit. When detached from the tractive unit a semi-trailer is supported on forward-mounted landing legs (or landing gear) that are raised or folded away for travel.

Articulated semi-trailers used in intermodal transport have two (tandem) or three (tri-) axles depending on legal gross weight requirements and tractive unit/semi-trailer axle configuration choices (e.g. 2 axles + 3 axles or 3 axles + 2 axles) – mainly determined by weight distribution and traction requirements. A minimum of five axles is needed for 38 tonne and six axles (i.e. 3 + 3 configurations) for 44-tonne operation in the UK. Most semi-trailers are built to the current maximum permitted overall length of 13.6 metres (to operate within the European 16.5 metres maximum overall length limit for complete articulated vehicles). They are usually fitted with either general-purpose platform bodies or, if used exclusively in container transport, they have skeletal frames (which offer reduced tare weight and increased payload), and are equipped with twist-lock attachments designed for securing standard length containers and swap bodies. Normally such vehicles carry one 40- or 45-foot or two 20-foot ISO containers, or a single, standard, 13.6-metre-long swap body.

1.4.9 Lorry and trailer (Drawbar) combinations

These vehicles comprise a combination of a conventional freight-carrying rigid vehicle towing a separate load-carrying trailer, known as a 'road train' in European terminology. The combination may also have a four-, five-, or six-axle configuration (i.e. 2 + 2, 2 + 3 or 3 + 3), again depending on weight requirements. In the case of the trailer axles, one axle at the front is steerable, and a single- or double-axle (tandem) bogie is fitted at the rear so that when separated from the towing vehicle the trailer stands alone on its wheels rather than being supported at the front by landing legs as with an articulated semi-trailer. An alternative design has two or three closely spaced (non-steerable) axles located centrally along the length of the trailer.

A conventional articulated semi-trailer, as described above, may be converted into a 'composite' trailer suitable for towing by a rigid goods vehicle by the use of what is known as a 'towing dolly'. This is a single independent vehicle axle surmounted by a fifth-wheel turntable, which connects into the coupling of the semi-trailer and supports the semi-trailer.

Drawbar combinations, as these vehicles are commonly called in the UK, when used in international operations are invariably designed to carry either a standard 20-foot ISO container or a standard, 7.15- or 7.45-metre swap body on the drawing vehicle and another on the trailer. This is accomplished within the current legal maximum overall length of 18.75 metres, which allows a total available load space between the drawing vehicle and the trailer of 15.65 metres. Vehicle bodywork is, similarly to articulated vehicles, either a conventional platform or a skeletal frame equipped with twist locks for securing containers or swap bodies.

1.4.10 Rail freight

In the context of this book, rail freight relies on the use of a variety of intermodal wagons specifically designed to carry ISO containers, standard swap bodies, or whole vehicle combinations (e.g. piggyback-style). The former are usually skeletal-framed flat wagons, often built to provide a low-loading height to accommodate 9-foot 6-inch-tall ISO containers and are fitted with twist locks at 20- and 40-foot centres. Vehicular traffic (i.e. semi-trailers and whole vehicle combinations) is carried on special pocket wagons or on spine wagons which provide safe and secure accommodation for vehicles in transit, their road wheels sitting low on the wagon to provide adequate overhead clearance for bridges and tunnels (known as 'loading-gauge' clearance). Depending on the system being used, the vehicles are either lifted on and off the rail wagon or are driven on/off via special loading ramps built into the wagon (e.g. as with the Modalohr system described in Chapter 12). The Channel Tunnel freight shuttle also uses a direct drive-on system.

1.4.11 Inland waterways

Intermodal freight traffic on the inland waterways is carried on barges equipped to accommodate either complete road vehicles, but more usually ISO containers stacked both in the cargo hold and on deck. Rarely are such vessels seen on UK inland waterways, but they are a common sight on European rivers, such as the Rhine, the Elbe, and the Danube which, along with other inland navigations, carry a great deal of freight traffic, a fair proportion of which being intermodal traffic. Two other forms of inland waterway vessels are those employed in the LASH (Lighter Aboard Ship) system and in the BACAT (Barge Aboard Catamaran) system. These are described in greater detail in Chapter 8.

1.4.12 Short-sea shipping

Increasingly, SSS is being seen as a key alternative to road freighting and a solution to the burdens caused by traffic congestion and air pollution across the whole of Europe (this is not just a UK problem).

The ships used in these types of operation are familiar enough to anybody making a port visit. They are generally small freighters (or coasters in common terminology to differentiate them from ocean-going vessels), typically of some 1000–5000 deadweight tonnes (dwt).

1.4.13 Containers

Freight, or shipping containers as they are usually termed are generally constructed of steel for strength and to meet the requirements of legislation and to standard dimensions established by the International Standards Organisation (ISO), hence the term ISO container. They are capable of being lifted from the top corner twist-lock castings by purpose-built container cranes, straddle carriers or stackers, or from the bottom by heavy-duty forklift trucks, for which purpose many containers also have strengthened fork pockets in the underside. These containers are sufficiently rigid and strong to be stacked 8, or even in some cases 9 or 10 units high in container terminals or within specially built cellular container ships where they are subjected to considerable stress. Standard ISO containers are 20-, 30-, 40- and 45-foot long by 8-foot wide (2.44 metres) and 8 feet 6 inches, and increasingly, 9 feet 6 inches, high.

Container ship capacity is measured and container industry statistics are compiled in relation to a single 20-foot ISO container defined as 1 teu (i.e. a 20-foot equivalent unit). Thus, a 40-foot container is counted as 2 teu. The largest of the current generation of sea-going containerships carry 9000 (or even more) 20-foot equivalent containers (i.e. 9000 teu).

Twenty- and forty-foot ISO units are those most commonly used since they provide greater flexibility in loading and more effectively match the legal dimensions for road vehicles – 30 footers being an odd size and square-ended 45 footers being too long for legal carriage on European road vehicles. However, 45-foot containers with the Geest-patented type of twist-lock corner casting are now becoming increasingly common and these overcome the difficulties of meeting the legally specified 'swept circle' dimension for maximum length semi-trailers thus allowing the 45 footer to be carried legally. A variety of different container designs are used for special purposes including half-height types for carrying particularly heavy, low-capacity freight, such as machinery or sheet steel, refrigerated units with built-in 'reefer' units, and bulk liquid or powder tanks mounted within standard-dimension steel container frames.

All these containers are fitted with standard design corner castings that enable them to be quickly and safely secured and released from matching twist-lock equipment mounted on road vehicle trailers, rail wagons, and lifting and loading equipment. Twist locks on vehicles and rail wagons may be recessed within load platforms to allow a clear deck for general-freight carrying, or mounted on special purpose-built skeletal frames, hence the term skeletal trailer. These can be used for no other purpose than container/swap body carrying, but by their nature they weigh less than a standard platform body and therefore provide a lower unladen vehicle weight which in turn allows a greater load-carrying capacity within legal gross weight limits. Other non-standard, non-stackable containers, and demountable bodies used in domestic transport and builders' skips are not included in this definition, and are not found in intermodal transport operations.

A new generation of ISO pallet-wide containers is in use now, patented by GESeaCo and brand named 'SeaCell'. These units have a revolutionary sidewall construction permitting sideways stowage of two, 1200 millimetres, standard Euro-pallets within the standard ISO exterior dimensions (45 feet by 8 feet). This configuration increases load capacity from 27 Euro-pallets (i.e. 1200 millimetres by 800 millimetres) to 33 pallets.

1.4.14 Swap bodies

Swap bodies (in French, *caisses mobiles*, a term frequently encountered) are loading units sufficiently strong for lifting, when loaded, to or from road vehicles and rail wagons with ease. Many of the shorter (i.e. 7.15 and 7.45 metres) units have fold-down legs so they can be free standing. All swap bodies are equipped with specially strengthened lifting pockets, located at standard dimensions, in the underside of the base frame for lifting by stacker or by overhead crane with grapple arms. They are built to a lighter construction than an ISO container thus saving on tare weight and increasing payload potential, but in consequence they are not normally sufficiently strong for top lifting or stacking like a container, although some more recent versions allow for stacking when empty.

A variety of bodywork types are encountered: the solid closed type, or those with curtain sides, or open with drop-down sideboards and tailboard for ease of loading and unloading. When used exclusively in intermodal operations, they are built to standard dimensions of 7.15, 7.45, 7.82, and 13.6 metres in length; 2.5, 2.55, or 2.6 metres wide; and up to 2.77 metres high allowing them to be carried within legal limits on European road vehicles and providing sufficient gauge (i.e. bridge and tunnel) clearance for rail freighting.

1.4.15 Initial and final journey legs

In combined road–rail transport operations, the relevant legislation as described in Chapter 2 refers to the road elements of journeys by this means as being the 'initial' and 'final' road legs. These are defined as being from the point of loading of the goods to a vehicle to the 'nearest suitable rail loading station', and from the 'nearest suitable rail unloading station' to the point of unloading the goods.

1.4.16 Operators and shippers

Operators are firms that operate intermodal transport services by road and rail. Shippers are firms that consign freight by a combination of transport modes. In many cases freight shippers initially contract with a freight forwarder or a road haulier who arranges the onward movement by combined modes, sending the loaded semi-trailer, swap body, or container to a local freight terminal for transfer to rail or barge for the long-haul journey to its destination, with final delivery being undertaken by yet another road-haul operation.

1.5 Why intermodalism now?

Given that so many benefits are proclaimed for intermodal transport in general and combined road–rail transport in particular, one could be forgiven for asking why it is only now, at around the end of the twentieth century, that it is being promoted with such enthusiasm. Certainly, the concept of the system has been in existence for a very long time. In the UK, the railways, since their very earliest days, have carried containers on long-haul journeys and delivered them locally by road. By today's standards, of course, these were small and clumsy contraptions (being made of wood and far too heavy in relationship to the loads they carried), but it was an intermodal, or if you like, a combined road–rail system, and it worked.

Then we had two transport revolutions. First was the advent of the RO-RO ferry, basically developed as trade with Europe re-started in the early 1950s following the end of the second World War, which provided a system for combining road and sea freighting whereby both wheeled freight (i.e. lorries and unaccompanied trailers) and unit-load freight was shipped across short-sea routes and hauled off at the other end to continue its journey to its final destination by road. In this particular context, it is useful not to lose sight of the fact that until the Channel Tunnel opened in 1994 a substantial volume of rail-hauled freight also used the RO-RO concept via the, now defunct, cross-Channel rail ferries.

Second was the container revolution which came over to Europe from the USA in the early 1960s, with ships initially crossing the Atlantic loaded with containers that had been delivered to the USA docksides on wheels (which were detached for shipping) and were hauled off into the UK and Europe by road vehicles. Initially standard flat-platform semi-trailers were used, in many cases with containers being precariously, dangerously even, secured by ropes and chains, until the subsequent development of the skeletal system with built-in twist locks, which ensure safe carriage of ISO-type containers.

The UK Freightliner, domestic container, system in the UK – part of the state-owned British Rail prior to rail privatization in 1996 – is a prime example of a combined road–rail transport operation although concentrating solely on the carriage of ISO shipping containers, rather than a mix of containers and swap bodies, principally between the UK ports and inland destinations. In fact, until much more recent times, intermodal swap bodies have not been a popular means of road–rail freighting in the UK, albeit demountable body systems are used extensively in domestic distribution operations.

In Europe, intermodalism has grown significantly; typically with such operations as the French and German road–rail systems of Novatrans and Kombiverkehr, respectively, which carry containers, swap bodies, and road vehicles on long-haul, trans-European, rail journeys. Among the many catalysts, which prompted such developments, was the fact that (pre-Single European Market – 1 January 1993) international road hauliers wishing to cross France or Germany needed road haulage permits authorizing transit through these countries. These were available only on a very limited allocation basis, but hauliers willing to load their vehicles onto rail, piggyback-style (i.e. on the 'Kangourou' system in France and the 'Kombi-Trans' system in Germany), from one side of the country to the other, were granted transit rights without the need for one of these scarce permits.

So, if there is nothing new in this concept, where does the renewed interest come from? It can largely be put down to three key factors:

1. First, the opening of the Single European Market 1993, as a result of which more freight is in transit throughout the EU (and between the UK and the rest of Europe). With customs and other cross-border formalities abolished the quest is for rapid, efficient, and uninterrupted transport over long-haul inter-Continental journeys.
2. Second, increasing awareness by the EC, national governments, commercial firms, and the public at large that the growing environmental impact of road freighting, especially by way of undue levels of traffic congestion and air pollution, must be constrained by achieving modal shift to rail, inland waterway, and SSS wherever it is practicably and economically feasible to do so.
3. Third, and most specifically from the UK point of view, the inauguration of the Channel Tunnel fixed rail link between Britain and France in 1994 which exposed the reality (and sensibility) of being able to effectively consign long-distance international freight other than just by 'door-to-door' road haulage via the Channel ferries.

1.6 The potential market for intermodal transport

While there is considerable interest for many reasons in increasing the use of intermodal transport as a means of diverting freight from our overcrowded roads network, as outlined above, it is important to recognize that not all freight currently carried by road is suitable for switching to other modes. Generally, for intermodal transport to offer a practicable or economic solution to road haulage there is a need for significant traffic flows loaded in semi-trailer/swap body/container-sized consignments, originating near to, and for delivery within close proximity of, road–rail freight transfer terminals.

Figures for the amount of freight that could be transferred from direct door-to-door delivery by road vehicle to a more environmentally favourable combined transport system vary quite widely, and no organization has made firm predictions. Anything between the existing European estimate for intermodal freight of between 4 and 15 per cent of the total freight market is suggested as possible. Professor Michael Brown of the Polytechnic of Central London has said that it would be very difficult to judge exactly what proportion of freight could transfer to a combined transport-based system, even after having carried out a survey of potential users across six major industry sectors.

However, the fact is that combined road–rail transport, for example, is increasing its share of the freight market, indicating greater acceptance by shippers of this mode. The International Union of Combined Transport Companies (UIRR) – an association of 19 national combined road–rail operators across Europe, headquartered in Brussels – shows in its annual report for 2002 that combined road–rail traffic (measured in tonne-kilometres) carried by its member companies had increased by around 3.7 per cent in 2002 over 2001, but when compared to the 1993 figure this shows a growth rate of over some 55 per cent. Most of this traffic (in fact 76 per cent) is international freight and by far the greater proportion of it is carried in containers and swap bodies (around 1 367 000 consignments), with rolling road and semi-trailer traffic falling far behind in statistical terms. These figures also show that the average distance for an international journey is around 760 kilometres internationally and 550 kilometres nationally, and that the average weight of consignments is 25 and 21 tonnes, respectively.

By contrast, over the 30 years from 1970 inland waterway traffic (in tonne-kilometres) increased by around 22 per cent while international intra-EU SSS traffic increased by some 150 per cent between 1970 and 1996 (the latest EU figure available). As one might expect, the leading inland waterway nations, Germany (66.5 million tonne-kilometres) and the Netherlands (41.3 million tonne-kilometres), saw the highest growth levels (at around 33 and 35 per cent, respectively) whereas UK inland waterway tonne-kilometres fell by around one-third in the same period from 0.3 to 0.2 million tonne-kilometres. Conversely, and quite surprisingly, the UK was the leading proponent of intra-European SSS (at 159.6 million tonne-kilometres in 1996) showing a growth rate of over 123 per cent from 1970 to 1996 (i.e. from 71.4 to 159.6 million tonne-kilometres).

1.7 The future for intermodal freighting

Intermodal transport has received so many accolades over recent years that it is difficult to select a phrase that most appropriately sums up the likely future impact of this mode of transport. The UK weekly road transport journal, *Commercial Motor*, has suggested that intermodal transport is the 'biggest development in international transport since the advent of the RO-RO ferry,' and that was quite some transport revolution. Logistics experts predict that purpose-built, high-speed, unit-load freight trains carrying swap bodies and containers will be the most cost-effective way to make trans-European deliveries in the twenty-first century. These systems are substantially in place; the legislation permits it to happen, the trains are operating now; and all it needs is for shippers – especially the road haulage industry – to overcome their prejudices and misconceptions. There is need for a change of mind-set that persuades them, and others with the power or authority to influence modal choice, of the real-time benefits of switching more of their freight from a single mode operation (i.e. largely road haulage) to an intermodal alternative.

From another direction intermodal transport is being spurred on by ever-restrictive EU legislative measures (both existing and threatened), and by the ability of intermodal options to beat road congestion, lorry traffic bans, and goods vehicle drivers' hours restrictions. Europe's road network is becoming more and more congested, weekend, and town-centre lorry bans are being extended, goods vehicle drivers' working hours are subject to regulatory restriction, and diesel fuel prices are soaring. Added to which is the prospect of the significant financial impact of new road-toll systems for heavy trucks. Germany's LKW Maut system finally commenced in 2005 and the UK is expected to follow suit from 2008. All this makes it sensible to encourage a switch from road haulage to rail and/or waterway for long-haul freighting.

Gone for ever are the days of stand-alone individualism, especially for the blinkered truck operator who still believes in the merits of pounding a heavy truck across the Continent for the driver to arrive breathless – 'I'm here!' – on the customer's doorstep like some intrepid pioneer. Does he not know that the load, sent, for example, by a combined road–rail operation, would probably have been delivered sooner, with significantly less wear and tear on the truck and on the roads it uses, less stress on the driver, less environmental harm to the populace of every town and village through which it passed, less damage to the ozone layer, and less consumption of our limited and much threatened fossil fuel reserves?

There is increasing concern to promote the benefits of intermodal transport across a wide spectrum of interests. On the political front there is considerable pressure by the EC and by national governments for greater investment in the further development of intermodalism as a means of removing freight from the roads. It is worth remembering that anything that encourages more use of rail, in particular, is a vote-winning issue for politicians, especially in the UK. Major corporate freight shippers too are becoming increasingly concerned on a strategic level to direct their operations towards the most environmentally favourable transport modes (to mollify their socially aware shareholders for one thing), given that costs are kept within acceptable bounds (which on long-haul operations they should be) and that services are sufficiently frequent, fast and, reliable to meet demanding delivery schedules.

2

UK and EU Policies for Intermodal Transport

The development of intermodal freight transport, as indeed with all other forms of transport, is significantly influenced by the official policies of both individual national governments and, collectively, of the European Union (EU) via the European Commission (EC). Without such a driving force keeping the inertia moving forward it seems likely that the progress we are currently seeing in intermodalism, as opposed to single-mode transport operation, would hardly be noticeable. Commercial arguments alone appear insufficient to propel growth in this sector, so it is down to 'official' persuasion on account of the environmental and humanitarian (i.e. anti-pollution, anti-noise, anti-traffic congestion, and road accident reduction) benefits of modal switch, plus a certain amount of encouragement by way of financial grants to aid the development of suitable terminal facilities for modal transfer, to provide the incentive for change. An understanding of the policies set out in the various documents outlined in this chapter will provide a useful background to the way in which intermodalism has developed, and to the technical and operational characteristics of this form of transport and its constituent individual modal ingredients.

2.1 UK Government policy

For many years transport policy in the UK was obsessed first with nationalization and then with privatization, and the issues of deregulation and fair competition. There was little in the way of practical effort to secure a better transport system – better for users of transport services and better for the population that has to live with the consequences of an inadequate transport system. Since 1997, however, the Government has been strong on the issue of meaningful policy documents. A comprehensive list of these is too numerous to describe fully here (and in any case they are not all completely relevant to our study), but the following publications provide key insights into current policy thinking on freight and intermodal transport.

In 1998 a so-called 'new approach' was adopted with the publication by the Department for Transport (DfT) of the White Paper, *A New Deal for Transport: Better for Everyone*, signalling that the Government was looking for a radical change in transport – in particular, seeking to achieve a more integrated transport system to tackle the growing problems of congestion and pollution. Prior to publication of this White Paper, a number of other key documents set out the Government's strategies for overcoming the adverse effects of the motor vehicle, especially the heavy lorry, on our lives. While not specifically aimed at increasing inter- and multimodalism rather than single-mode (i.e. road) freighting, they tend to show how the thinking in this direction was developing.

2.1.1 Royal Commission Report: 'Transport and the Environment'

The 1994 Report of the Royal Commission on Environmental Pollution, *Transport and the Environment*, determined that an important objective for a sustainable transport policy must be founded on the concept of increasing:

> the proportions of personal travel and freight transport by environmentally less damaging modes and to make the best use of existing infrastructure.

It effectively warned that the likely consequences of taking more and more land to provide new infrastructure to cope with the forecast massive growth in road use could not be sustainable, so alternatives had to be devised and developed. The report highlighted the fact that the projected growth of heavy-goods vehicle traffic between 1989 (when the National Road Traffic Forecasts were made) and 2025 was no less than 140 per cent. The Report went on to discuss the scope for mode switching and described in detail the attributes of individual modes, particularly in terms of their energy efficiencies, emissions, noise levels, and accident rates. It concluded that even with a large-scale shift to rail, road transport would remain the dominant mode for freight transport and, therefore, it was important to explore all possible ways of limiting its adverse environmental impact.

2.1.2 Planning guidance

Hard on the heels of the Royal Commission Report came the Government's *Planning Policy Guidance Note 13* (i.e. PPG 13, HMSO 1995) which set out far-reaching aims among which was recognition of the need to encourage the use of alternative transport modes to divert freight away from road and onto rail or the waterways. It noted the fact that many other European countries achieve greater use of rail for freight transport than the UK. Importantly, the Report said that planning should identify and allocate appropriate sites for distribution and warehousing, and that consideration should be given to safeguarding rail and waterside sites for future industrial use and to encourage current users of rail and water not to switch to road.

2.1.3 White Paper: 'A New Deal for Transport'

The Government's White Paper, *A New Deal for Transport: Better for Everyone* (DETR 1998), mentioned above, is rightly concerned primarily with improving public transport so as to provide a more acceptable alternative to private car use, the cause of much of the congestion and pollution. However, freight transport features importantly with the recognition that the nation wants and needs a 'reliable and efficient transport system that supports prosperity, to provide the jobs and wealth we all want', but that the growth in freight transport risks being met at the expense of our environment. 'This is why we want to reduce the extent to which a healthier economy results in high levels of road-traffic growth. We want to see a real increase in the use of rail freight, inland waterways and coastal shipping.'

In other words the Government wants to see a clear modal switch away from road freight, which obviously means a need for more intermodal freighting rather than ever increasing use of the heavy lorry. In consequence of this, a number of key proposals of interest to the intermodal sector were made in the White Paper as follows:

- The establishment of a new rail authority to promote rail freight and its infrastructure (i.e. the Strategic Rail Authority (SRA).
- Encouraging greater use of six-axle lorries (to enable the carriage of heavy intermodal loading units (ILUs)).
- Improving best practice in the freight industry.

- Introducing legal powers to impound illegally operated lorries.
- Facilitating shipping as an efficient and environmentally friendly means of carrying trade.
- Extending the pre-existing freight grants scheme for developing rail terminal facilities to include coastal and short-sea shipping activities.

The White Paper stated categorically that:

> we can move more freight by rail, relieving pressure on the road network and bringing environmental benefits.

In fact, it cited the ambitious target of English, Welsh, and Scottish Railway (EWS – the UK's leading rail-freight operator) to double its traffic, measured in tonne-kilometres, over 5 years and triple it over 10 years (i.e. by 2003 and 2008, respectively). Similarly, Freightliner (the former state-owned deep-sea shipping container carrier) stated its aim to increase the number of containers it carries between the deep-sea ports and inland terminals by 50 per cent over 5 years (i.e. from 1998).

The Government endorsed these targets saying that reaching them could mean that in 2010 the share of freight by road would be 10 per cent lower than in 1998. It added that for every percentage point reduction in road freight, some 1000–2000 heavy lorries would be taken off the roads. To achieve this target, a number of important promises were made:

- To ensure (through the SRA) that freight is given proper consideration in the operation and planning of the rail network.
- That proper attention would be given to such obstacles as rail loading gauge, track capacity constraints, and access to additional land.
- That revised planning guidance would be issued to facilitate the carriage of more freight by rail (i.e. via PPG 13, mentioned previously).

As part of its strategy to encourage more efficient distribution of goods, the White Paper stated the Government's intention to increase lorry weights (initially to 41 tonnes from 1 January 1999 and subsequently to 44 tonnes from 1 February 2001) in line with EU law to facilitate the legal carriage of heavy ISO container and international swap body traffic.

Another aspect of the Government's intentions for developing freight modes concerns coastal shipping and inland waterways. Quoting from a 1993 research paper, *Roads to Water* by Jonathon Packer and Associates, it stated that, potentially, some 3.5 per cent of UK road freight traffic could be diverted on to the waterways. To encourage this modal switch to short-sea and coastal shipping, and to encourage greater use of the inland waterways, where this is a practical and economic solution, the freight grants regime (see Chapter 10) was extended to cover these modes.

2.1.3.1 Sustainable distribution

The White Paper, *New Deal for Transport*, spawned a number of so-called daughter documents; an important one in the context of freight transport being *Sustainable Distribution: A Strategy* published in March 1999 (DETR – now the DfT). Among many key provisions devised to show the Government fulfilling its commitment to the White Paper, *New Deal*, by setting out an integrated strategy for sustainable distribution of goods, are those concerned with promoting intermodal integration. Particularly, the document notes the vital need for building sustainable distribution systems that achieve better utilization of our railways, ports, and shipping. It reiterated its intention to:

- set up the SRA (which came into being from 1 February 2001 under provisions included in the Transport Act 2000);
- promote improvement of rail-freight services;

- promote greater use of rail freight through incentives such as the freight grants scheme;
- extend the coverage of the freight facilities grant scheme to the coastal and short-sea shipping sector;
- encourage waterborne traffic where practical and economic;
- work with British Waterways and others to identify realistic market opportunities for inland waterways (i.e. against a significantly declining market between 1986 and 1997);
- establish a Shipping Working Group (in late 1997) to identify measures to help reverse the decline in UK shipping and foster greater use of short-sea and coastal shipping around the UK, and beyond to the rest of Europe.

It will be appreciated from the content of the two important policy documents outlined above that the Government has on its mind the essential need to improve policies to ensure the success of our freight transport industry and to deliver a transport system that is sustainable and can support economic development. It particularly supports combined road–rail transport in the belief that in the right circumstances this form of intermodalism offers real economic and environmental benefits, and is a key means by which the volume of freight travelling by road can be reduced.

With specific regard to combined transport, the Government's sustainable distribution strategy document sets out a policy framework for major freight interchanges, recognizing that the performance of these facilities in our distribution networks is vital to promoting greater use of intermodal freight. The policy framework does the following:

- Promotes the contribution of the UK's major freight interchanges to national and regional competitiveness by pursuing policies of fair competition in the UK and throughout Europe; by giving due weight to the need for efficient trans-shipment between the different transport modes; and by providing efficient access to and from major interchanges.
- Aims to improve the operational and environmental performance of existing interchange facilities by promoting greater use of less damaging modes for onward distribution.
- Encourages full and efficient utilization of existing interchange facilities in preference to expansion in cases where suitable spare capacity exists or can be created.

2.1.4 Transport 2010

The Transport Ten-Year Plan (dubbed Transport 2010) published in 2000 sets out the Government's long-term strategy for delivering a quicker, safer, more punctual, and environmentally friendly transport system. It was said at the time to be an ambitious but realistic view of what could be achieved over the following 10 years. The key to this plan was the proposed 'massive new investment' in the transport system of no less than £180 billion to be delivered through public and private partnership ventures. Basically the idea was, and remains still, to:

- modernize the system;
- achieve integration between modes;
- cut congestion on the roads (set to grow by 15 per cent over the following 10 years, i.e. 2000–2010);
- reduce pollution;
- boost the choice between modes (this in particular being aimed at the concept of motorists choosing to leave their cars at home and use public transport, which, of course, has a direct relevance to the congestion affecting road haulage operators).

In the rail context, the aim is to achieve an 80 per cent growth in rail-freight traffic; modernize the East and West Coast main rail lines; complete the Channel Tunnel Rail Link (CTRL); and improve rail safety.

In a road haulage context, the aim is to widen 360 miles of trunk road; bring in a number of key road improvement schemes; wipeout the backlog of road maintenance projects; introduce new technology for traffic management and provide real-time information for drivers; and speed up the introduction of cleaner fuels and vehicles.

2.1.5 Other policy measures

A number of other policies documents, among a great many studies and reports relating to waterborne and rail freight, are worthy of mention here.

2.1.5.1 Freight on Water – A New Perspective

The Government-sponsored report, *Freight on Water – A New Perspective* (DEFRA, 2002), sets out its policies for the future of the inland waterways in England and Wales including their use for freight transport. This study was developed from the 'Better Deal' White Paper that proposed the notion that the Government would encourage greater use of the inland waterways for freight transport where it was practical and economic to do so.

2.1.5.2 UK Marine Motorways Study

In 2002/2003 the DfT along with the Engineering and Physical Sciences Research Council (EPSRC) funded the *UK Marine Motorways Study*, under the LINK Future Integrated Transport Programme, which was carried out by the Tri Maritime Research Group, Transport Research Institute, Napier University, and the Logistics Research Centre at Heriot-Watt University. The aim of the study was to establish the operational and commercial viability of fast freight ferry services on UK coastal routes as an alternative to long-distance road transport. However, the conclusion of the study appeared to be that on a purely cost basis alone such an operation could not compete directly with road haulage.

2.1.5.3 SRA strategies

Since so much emphasis has been placed on the need to achieve a modal switch of freight traffic from road to rail, the ability of the rail network to handle increased volumes of such traffic comes into question. This is especially so when one considers the fact that, as is commonly known, any significant increase in freight on rail would jeopardize passenger rail services. In fact, it is said that doubling of the amount of freight carried by rail would virtually bring passenger services to a standstill on many parts of the network. But more significantly, it would only reduce the amount of goods carried by road by a mere half of 1 per cent.

Nevertheless, the efforts of the rail business to improve performance and increase its freight carrying capacity are important if we are to see any significant switch of freight from road to rail. So we need to consider the policies of the SRA, which was originally envisioned in the White Paper, '*Better Deal*', and set up in shadow form prior to being substantively established in February 2001 under provisions contained in the Transport Act 2000. It issued a *Freight Strategy* document in 2001 followed by its first Strategic Plan in January 2002, *The Way Forward for Britain's Railway*, which in turn was followed in 2003 by a document called *Everyone's Railway – The Wider Case for Rail* and another Strategic Plan, *Platform for Progress*. The *Freight Strategy* document of May 2001 described the benefits of freight on rail, stressing in particular its contribution towards reducing both road congestion and carbon dioxide (CO_2) emissions, and highlighting its 10-year freight target of 80 per cent growth from 1998/1999 levels.

As might be expected, since the passenger business is rail's bread and butter while freight appears to be little more than a hindrance, rail freighting features much less in the 2002 and 2003 Strategic Plans or the 'Everyone's Railway' document than keen proponents of a modal switch of heavy traffic from road to rail might have hoped. The first Strategic Plan notes that in 2002 rail had a 7.5 per cent share of the total UK freight market against the Ten-Year Plan target of a 10 per cent market share by 2010. When this report was published the SRA was proposing to spend £300 million under its Freight Small Schemes Fund over 10 years to secure small- and medium-sized freight infrastructure schemes which would provide valuable improvements to the network, quickly and within a fixed annual budget. The 'Everyone's Railway' document of 2003 was intended to reaffirm the SRA's commitment to developing the rail network by showing how it had laid the foundations for the future. In particular, it showed how it had aligned the rail network to the growing logistics industry whereby, in addition to its traditional bulk freight market, it was now moving high value freight requiring speed and reliability. It identified too its current agenda on freight as being to tackle the problem of controlling costs while improving performance and reliability.

The second Strategic Plan (2003), *Platform for Progress*, set out the rail industry's next steps towards the original goals established on formation of the SRA. Again, this document is substantially oriented towards the performance and goals of the rail passenger business, being woefully short on good news for the freight industry and for intermodalists, apart from noting that freight tonne-kilometres had increased by 7.1 per cent over the previous 12 months, and promising a review in 2003 of the strategy for Channel Tunnel freight services in the medium and longer term. However, there was just one snippet of good news in so far as intermodal operations are concerned; namely, the development of an enhanced gauge route to accommodate 9 feet 6 inches high ISO shipping containers en-route to and from the port of Felixstowe. Also encouraging, was mention of proposed reductions in track access charges for freight operations worth about £170 million in 2003–2004 and the fact that the 80 per cent goal for growth of freight traffic would be largely achieved from general unit-load traffic, including premium logistics – thus meaning, substantially from intermodal traffic.

2.2 Intermodal policy in the EU

As the UK Government's White Paper, *New Deal for Transport* (outlined above), clearly indicates, it cannot succeed in delivering an integrated transport policy in isolation from Europe and it acknowledges that the EU has an important role to play in setting the framework of policy and law on intermodalism. This the EU has done, at least since 1970, as the accompanying chronology at the end of the chapter clearly shows. However, a number of key policy documents have been published and legislative measures taken in more recent times; most notably publication in 2001 of its important White Paper, *European Transport Policy for 2010: Time to Decide*, which is described in greater detail later in the chapter, and on the historic date of 15 March 2003, when open access was declared for international rail-freight services across the EU as envisioned in the so-called Railway Directive of 1991 (i.e. Directive 91/440/EEC), see below. This step effectively completed the integrated market for rail-freight services that was described in Brussels as being one of the cornerstones of the White Paper on European transport policy for 2010, *Time to Decide*.

Two other key moves towards further integration of European transport occurred in 2003 when the EC became actively involved in issues regarding navigation on the rivers Rhine and Danube (both vital for trans-Europe inland waterway freight traffic) and concluded accession to the Convention concerning International Carriage by Rail (COTIF), thereby being in a position to facilitate international rail transport by participating in the development of uniform legal rules for the whole of the continent of Europe – yet another measure that will effectively facilitate cross-border rail freighting and consequently intermodalism.

2.2.1 *The Railway Directive*

The so-called 'Railway Directive' of 1991 (Directive 91/440/EEC) was intended to instigate a status of independent operator on national rail systems throughout the Union so that they behave in a more commercial manner adapting to market needs, all in the interests of improved efficiency – in other words promoting the concept of privatization. Its specific aim was to open up the railways for use by independent operators and facilitate access to rail networks throughout the EU for those organizations engaged in the international combined transport of goods – which has now happened across Europe since 15 March 2003.

In practical terms, this significant step means allowing a freight consignor in the UK, for example, to hire a freight train, load it with intermodal transport units, then run it through the Channel Tunnel on the through-rail service and into the European rail network, direct to its destination – and all under his direct control. Hitherto this was not possible in international rail freighting, with responsibility largely, and control totally, being passed into the hands of the former British Rail, the UK state-owned rail operator, as soon as the loading units were on board the train – a scenario frightening enough to dissuade many consignors from shipping their goods by this means.

The Directive required the management of rail transport services and of the railway infrastructure to be separated (with separate accounting systems) and this we have seen in the UK with the establishment of the track operator (formerly Railtrack and now Network Rail) and a whole clutch of independent regional passenger train operating companies and freight train operators (primarily EWS, formed from three former rail-freight companies, and Freightliner, now in the private sector, plus also now, Direct Rail Service (DRS) and Great Britain (GB) Railfreight). The point is that with this new structure there is greater facility and incentive for private operators (e.g. road hauliers) to incorporate rail trunking within their operating plans. This in turn encourages the use of intermodal transport units, whether complete semi-trailers, swap body or container traffic, and consequently contributes further to the overall ideology of switching long-haul freight traffic from road to rail.

2.2.1.1 The railway packages

Development of EU railway policy has focused on what are termed the three 'railway packages'. The first of these was the launch on 15 March 2003 of the Trans-European Rail Freight Network (TERFN), based on the provisions of the Railway Directive 91/440/EEC of 1991, permitting open access for international rail services across the EU. This was followed, on 23 January 2002, when the EC adopted the so-called 'second package' of measures to revitalize the railways by building an integrated European railway area, in particular by opening up more quickly the international rail-freight market, by proposing a new directive on railway safety and by establishing a European Railway Agency. The third railway package, which was adopted in March 2004, proposes opening up of international passenger rail services within the Community from 2010 and the introduction of a certification system for locomotive drivers.

2.2.2 The 'General Plan' for combined transport

However, important though the documents described above are to the development of intermodalism in Europe, we need to go back further in time to really understand the background to the policy developments for combined transport particularly. In fact, although pro-combined (i.e. road–rail) transport legislative measures date back to at least 1970, it is only in more recent years that the first positive steps were taken, when the Council of the European Community commissioned a report on combined transport developments. This was in 1990 when the Council gave a mandate to the EC (Council resolution of 30 October 1990) to present a 'General Plan' for combined transport that was to include a definition of the network and the conditions for its smooth operation. The Commission in turn appointed a High Level Working Group to assist in this task comprising representatives of Member States, the railway industry, and professional organizations (e.g. road–rail companies, road hauliers, and own-account carriers). The Council resolution highlighted a number of points that were to receive special attention as follows:

- Implementation of a policy for intermodal transport ensuring free choice of transport methods for users.
- Economic and commercial aspects of intermodal transport.

- Definition of a rational plan of routes and terminals, including clearance and railway track gauges.
- Harmonization of technical and operational aspects including terminal and rolling stock management.
- Organization and management of the administrative, financial, quality of service, and responsibility aspects according to decision-making level (i.e. European Community, Member State, or combined transport operator).
- The coherence of legislative, statutory, and administrative regulation of intermodal transport at both Community and national level.
- Access to the combined transport system, particularly for the peripheral regions of the Community.
- The contribution that combined transport can bring in reducing the problems linked to road transit through third countries (e.g. Switzerland, and the former Yugoslavia).

2.2.3 Working Group Report

The Working Group's intermediate report of 23 April 1991 addressed these main issues and explained the main points highlighted during its discussions. It also gave the following outlines of the solutions that could be adopted.

2.2.3.1 Freedom of choice

While the Group recognized and agreed the principle of free choice for the transport user, it commented that this should not be distorted by variable allocation of the infrastructure costs depending on the transport mode, and that external costs related to the different modes are not always given the same consideration and may therefore cause distortion in the competitivity of modes. It proposed progressive reintroduction of the real cost of transport – albeit a long-term project that would need to be fair for all modes.

2.2.3.2 Economical and commercial aspects

Three thresholds of profitability were proposed for study: that of the operators, of the investment in infrastructure, and a broad economic profitability integrating part of the external costs, all of which should lead to a profitability model of a network. The Group further proposed that analyses should be carried out to determine the calculation methods by which external costs could be integrated and a decision made as to which of such costs should be taken into account.

2.2.3.3 Scheme for routes and terminals

The establishment of a minimum common profile to be achieved by the year 2000 either by gauge development or by setting up specific railway rolling stock was suggested. Due to the diversity of multimodal platforms (i.e. terminals), the importance of establishing a coherent network and the need for their optimal location, the Group proposed further studies on this subject, first by obtaining relevant data on existing and proposed terminals, second by bringing together, in the form of a seminar, owners and operators of terminals with the objective of promoting the implementation of a network, and third by establishing a management and location model for a goods handling station to see how this might be reproduced for all terminals.

2.2.3.4 Technical and operational aspects

The (High Level) Group recognized the many problems of technical compatibility that exist in multimodal transport and the high costs of investment in equipment which inhibits the participation of small and medium companies in combined transport. It stated the need for study of the specific standards for rolling stock used exclusively in combined transport and, more particularly, for road equipment, and it commented on new technical developments in bi-modal transport, of which there are currently about 10 versions operating in the Community, all of which are 'perfectly incompatible between them'. The Group proposed a demonstration of these new technologies on a selected route, so a study could be made of the commercial viability of the equipment tested. Manufacturers and operators would be asked to help by developing compatible equipment.

2.2.3.5 Organization and management

The need for quality of service was seen as paramount, but the Group recognized that it is based on a large number of elements which have neither the same status nor structure, objectives nor operating modes between the large monopolistic structures (such as the railways) and what it calls the 'atomized market' (such as small road hauliers). It proposed the launching of an experimental combined transport service on one, two, or a maximum of three routes with a high traffic density potential taking advantage of the best possible conditions (not current normal conditions), so an audit could be carried out to determine to what extent and according to which modalities this experiment could, in future, be reproduced on other routes. This would involve partnership agreements (taking account of competition rules) between all the parties involved, and sponsorship from the Commission.

2.2.3.6 Statutory, legal, and administrative aspects

Member States were requested to provide information on the present status of their domestic legislation regarding combined transport. This information was to be analysed to obtain a picture of the regulatory situation for combined transport within the Community, and to identify any discrepancies still existing between Member States.

2.2.3.7 Connection with the peripheral regions

The Group proposed that routes connecting the peripheral regions of the Community should be incorporated in the European network plan despite the fact that traffic along them may not be sufficient to make them profitable. It also proposed that the pilot scheme referred to above should be extended to some of these peripheral regions. Finally, the Group added two riders to its report:

1. That the term 'combined transport' did not hold the same meaning for all participants, and that the concept of 'container' or 'loading unit' is sometimes misleading.
2. That it was now necessary to extend the scope of work to the routes connecting with Eastern European countries by taking account of present and future traffic towards these states.

2.2.4 The Master Plan

In June 1992 the Commission of the European Communities adopted a blueprint 'Master Plan' for developing a European combined transport network based on the previously mentioned recommendations of the Working Group. In its preamble the Commission stated that although combined road–rail transport had expanded swiftly in the previous 5 years, it still accounted for less than 4 per cent of international goods traffic in the Community. This was despite the fact that over longer distances (not less than 600 kilometres) combined transport exhibits fully the modal advantages of both road and rail systems; namely, being flexible, of high quality in terms of reliability, environmentally friendly, and energy saving.

The Plan went on to say that in order to cope with the challenges presented by the expected substantial rise in demand between then (1992) and the end of the century (35–40 per cent at least), and in the interests of compliance with the Community's undertakings on vehicle exhaust (especially CO_2) emissions, the development of combined transport is vital. It pointed out also that as intermodal systems can lower costs over long distances, it has an important role to play in lessening the handicaps affecting the peripheral regions, and for the Community's central regions, it should provide an effective means of tackling congestion. Being aware of the major distortions in the transport market (principally because road transport, allegedly, does not reflect its current track costs), the Commission proposed action to back an intermodal system with a package of measures for both the short and medium term, representing the minimum necessary to ensure the system's future success in a free market.

2.2.4.1 *Establishment of networks*

Most importantly, among these intended measures, the Commission proposed the adoption of a rail network covering the routes at present heavily used by long-distance road traffic to be established by the year 2005, with the first 5 years being devoted to actual establishment of the network and the second stage being concerned with bringing rail track clearance and loading gauges up to the minimum standard needed to take containers and swap bodies standardized at the European level. The Master Plan detailed those rail routes that were to receive priority consideration for the provision of adequate rail track clearance and loading gauge.

2.2.5 The Common Transport Policy

In December 1992, the EC further signified its concern for the development of combined transport systems when it issued a communication, (*The Future Development of the Common Transport Policy*) – COM 92/494 (final) – which, it said, would gain a new impulsion from the implementation of the Treaty on EU agreed at Maastricht (the Maastricht Treaty of 1993). In particular, it would include measures for improving trans-European networks, a new basis for development of the transport infrastructure, and the challenge of integrating environmental objectives, not least of which was the expansion of combined transport operations. The Commission's communication went on to deal with a number of points relevant to this book and which are of interest for the insight they gave to the then current concerns in Europe's corridors of power. These points are elaborated here. In general, it will be apparent that rail is the underutilized sector and that the road mode is the 'bad guy' producing all the problems, and this is the imbalance that the Commission is trying to address.

2.2.6 Modal disequilibria

The Commission's disquiet was stated as follows:

> The expansion of the transport sector (within the economy) has, however, brought problems in its wake: growth has produced or exacerbated a number of imbalances and inefficiencies in the transport system taken as a whole and also within transport modes. Despite the considerable success of transport in meeting consumer demand, inadequate integration of modes is denying users [of] some practical alternatives to current services. Inadequate capacity in some modes is producing congestion and environmental damage, while, in others, underused capacities exist. Inadequate or incomplete networks cause bottlenecks in some areas; in others, they prevent fuller integration of peripheral regions into the single market.

2.2.7 Road

The Commission went on to say in its communication:

> One of the important reasons why imbalances and inefficiencies have arisen is because transport users have not been adequately confronted with the full costs of their activities

and because the construction of transport infrastructure has been lagging behind what was needed.

2.2.8 The challenge

And it finally outlined the challenge thus:

> The challenge for the Community's transport system is how to provide, in the most efficient manner, the services that are necessary for the continued success of the single market, while continuing to reduce the inefficiencies and imbalances of the system and safeguarding against the harmful effects that increased transport activity generates. It is possible to meet this challenge, while respecting the basic tenets of the free market, by the introduction of economically efficient transport policies.

The document suggests that one of the ways in which the shortcomings in infrastructure, for example, could be overcome was to look for ways to increase resources available for investment, particularly by the private sector. We have seen this happen in the UK with the Channel Tunnel, a massive £10 billion private enterprise venture, and with privatization of the railways. It also suggests the necessity of tackling the fundamental question of the true costs of transport. However, this specific issue is outside the scope of this book. What is within its scope is the further suggestion that it is:

> essential to examine how under-used capacity in the transport system, particularly in the modes other than road, could be brought into service, while respecting the free choice of the user.
>
> Policy measures and advanced technologies can seek both to increase the attractiveness of the individual modes in question and promote the combined use of different modes. These intermodal operations should provide the best combination of different systems to meet the needs of operators and users.

The document says that besides facing up to the infrastructure costs problem:

> … complementary measures to improve the quality of different modes and intermodal services … will need to be considered.

It identifies some of these measures: first that the development of more attractive intermodal services depends to a considerable degree on the efficiency of the management and supervision that is applied. This is particularly true of international routes where existing services have often not met performance requirements. In this context, the so-called 'Railways Directive' (Directive 91/440/EEC) permits new railway operators to enter the combined transport market, thus stimulating a higher quality service from all concerned. And second, that intermodal services may require co-operation between different transport enterprises (particularly in the light of the Directive mentioned above) and agreements made by such enterprises will have to comply with EU competition law. In connection with this, the Commission has to consider modification of its competition regulation (Regulation 68/1017/EEC).

The third set of measures to be examined is that of technical harmonization, particularly to guarantee the interconnection of intermodal systems and the inter-operability of mobile equipment. Additionally, compatible information systems must be developed. As the report says, the field is vast and complex. The Commission sees the EU's role in this as being to direct technical harmonization through Directives, with technical specifications being achieved through reference to European standards developed by recognized standardization bodies.

The fourth step that is necessary is to overcome the problems associated with the different liability regimes that apply to intermodal operators, not only between Member States but also between the different modal stages of an intermodal journey. And finally, the fifth step concerns the development of trans-European networks for different modes and their progressive integration.

2.2.9 The European Directive

Hard on the heels of the Commission's December 1992 communication on the Common Transport Policy, which identified policy thinking, came the European Council Directive 92/106/EEC of 7 December 1992 – a document which set out to establish common rules for the combined transport of goods. This particular Directive was, in fact, a 'recast', in the interests of clarity, of the previous Directive 75/130/EEC of 17 February 1975 and its subsequent amendments. The preliminaries to the Directive set out the basic precepts, namely that with an increase in traffic due to the formation of the internal market (i.e. from 1 January 1993), the Community (now the EU) must do what is necessary to ensure optimum management of its transport resources in the interests of all citizens – which means using combined transport. It says that the increasing problems of road congestion, the environment and road safety, and the public interest too, call for the further development of combined transport as an alternative to road transport.

The purpose of the Directive was stated to be the encouragement of combined transport by freeing it from all quantitative restrictions and by elimination of various administrative constraints that still exist in the field of road transport. These liberalization measures were intended to relate to road journeys of limited distance in order to result in a real reduction in road congestion. Among the liberalization measures indicated were so-called 'stimulation measures', the reduction of taxation on the use or possession of commercial vehicles to the extent that they are carried by rail, and exemption of the initial and final road haulage journey legs from compulsory tariff regulations. Additionally, it said, the access of own-account operations to combined transport must be facilitated. To ensure that the Directive is working, the Commission of the EU is required to submit a report on its application every 2 years.

The Directive contains 11 articles as follows:

Article 1 states that the Directive applies to combined transport operations without prejudice to EU Regulation 881/92/EEC of 26 March 1992, which relates to access to the market in the carriage of goods by road within the Union to or from the territory of a Member State or passing across the territory of one or more Member States (in other words the establishment of the community authorization system, see Chapter 15).

It goes on to define 'combined transport' as meaning: the transport of goods between Member States where the lorry, trailer, or semi-trailer (with or without the tractive unit), swap body or container of 20 feet or more in length uses road transport on the initial or final leg of the journey, and on the other leg uses rail, or inland waterway or maritime services where this section exceeds 100 kilometres as the crow flies. The initial and final legs of the journey are defined as being from the point where the goods are loaded to the nearest suitable rail loading station, and from the nearest suitable rail unloading station to the point where the goods are unloaded, or in the case of transport by sea, within a radius not exceeding 150 kilometres as the crow flies from the inland waterway port or sea port of loading or unloading.

Article 2 requires all Member States to liberalize combined transport operations from all quota systems and systems of authorization by 1 July 1993.

Article 3 requires the use of a combined transport document that fulfils the requirements of Council Regulation 11/60 of 27 June 1960 (Article 6) concerning the abolition of discrimination in transport rates and conditions (i.e. implementing Article 79 (3) of the Treaty of Rome). This document must specify the rail loading and unloading stations relative to the rail leg of the journey. These details must be recorded before the transport operation is carried out and must be confirmed by means of a stamp affixed by the rail authorities when the rail part of the journey has been completed. This system is now in use in the UK as one of the conditions under which 44-tonne lorries may operate on combined transport journeys.

Article 4 specifies that all hauliers established in a Member State who meet the conditions of access to the occupation and access to the market for the transport of goods between Member States (i.e. in the UK, those who hold international operator's licences), shall have the right to carry out, in the context of a combined transport operation between Member States, initial and/or final road haulage legs which form an integral part of the combined transport operation and which may or may not include the crossing of a frontier.

Article 5 requires the Commission of the EU to draw up a report to the Council every 2 years (in the first instance by 1 July 1995) on:

1. the economic development of combined transport;
2. the application of Community law in this area;
3. the definition, where necessary, of further measures to promote combined transport operations.

When drawing up the report the Commission must call on Member States to collect the information needed for this purpose and analyse the information and statistics, in particular, relating to:

1. the transport links used in combined transport operations;
2. the number of vehicles (a road train counting as a single vehicle), swap bodies, and containers transported over the various transport links;
3. transported tonnages;
4. services carried out, in terms of tonne-kilometres.

The report was also required to propose solutions, where appropriate, for the subsequent improvement of such information and the situation in the combined transport sector.

Article 6 requires Member States to take such measures as are necessary to ensure that relevant taxes which are applicable to road vehicles (e.g. vehicle excise duty in the UK) when routed in combined transport are reduced or reimbursed either by a standard amount, or in proportion to the journeys that such vehicles undertake by rail, within limits and in accordance with conditions and rules they fix after consultation with the Commission. These reductions or reimbursements shall be granted by the state in which the vehicles are registered, on the basis of the rail journeys (carried out) within that state, or on the basis of rail journeys that take place partially or wholly outside the Member State in which the vehicles are registered. Without prejudice to the provisions resulting from possible reorganization of national taxation systems for commercial vehicles at Community level, vehicles used exclusively for road haulage in feeder or final delivery carriage by combined transport may be exempted, if taxed separately, from the relevant vehicle taxes listed below.

Article 7 specifies that where a trailer or semi-trailer belonging to an own-account operator is hauled on a final leg sector by a tractive unit belonging to a road haulier, the document specified in Article 3 is not required; however, another document must be provided giving evidence of the journey covered or to be covered by rail.

Article 8 specifies that initial or final road haulage legs of combined transport journeys shall be exempted from compulsory tariff regulations.

Article 9 relates to own-account carriage of goods on road haulage journeys and specifies that notwithstanding the definitions contained in other provisions relating to own-account carriage (i.e. EC Directive 2205/62 and EC Directive 86/647), where the initial leg of a combined transport journey is carried out by the own-account use of vehicles of the dispatching company (whether owned or hired by it), the final leg may be carried out by the receiving company using its own tractive unit (whether owned or being purchased on deferred terms) or one hired without a driver. Similarly, this article also allows a situation where the dispatching company uses its own tractive unit to draw a trailer or semi-trailer owned or hired by the receiving company.

Article 10 sets the date, namely 1 July 1993, by which Member States had to bring into force the laws, regulations, and administrative provisions necessary to comply with the Directive. Reference to the Directive had to be incorporated in any such provisions made to comply with it, and the main provisions of the relevant national law had to be communicated to the Commission.

Article 11 repeals the earlier combined road–rail Council Directive 75/130/EEC.

2.2.10 The combined network

Further progress was made by the European Council with its Decision of 29 October 1993 on the creation of a trans-European combined transport network (Council Decision 93/628/EEC). For all the reasons why more freight should be consigned by combined road–rail transport (particularly the environmental issue) the Council has adopted a union-wide network of routes which, it says, should have been brought into service by 2005, and which would allow the passage of the standard loading units authorized within the union. It acknowledged that creation of the network required a series of works to be undertaken to ensure its rapid introduction and full market operation and that some of these works were particularly urgent and would require launching as soon as possible in their first phase. The document identified a number of projects which,

> wherever possible and taking account in particular of the financial constraints on Member States should be completed, or failing that, nearing completion within 6 years and 12 years, respectively (i.e. from October 1993).

2.2.10.1 For 6-year completion

The following railway axes were to be adapted so as to provide sufficient clearance and the necessary conditions for the transport of containers and swap bodies compatible with Council Directive 85/3/EEC of 19 December 1984:

> on the weights and dimensions and certain other technical characteristics of certain road vehicles.

- Taulov–Northern Jutland
- Hamburg–Padborg–Taulov–Copenhagen
- Hamburg–Berlin
- Hanover–Berlin
- Nuremburg–Berlin
- Berlin–Dresden
- Frankfurt–Würzburg
- Betuwe line (Rotterdam–Ruhr) and connections in the Netherlands for Hengelo and Venlo
- Rotterdam–Antwerp/Zeebrugge–Brussels–Luxembourg–Bettembourg
- Antwerp–Aachen
- Rotterdam–Antwerp–Brussels–Aulnoye
- Aachen–Liège–Erquelinnes
- Lisbon–Madrid–Barcelona
- Lisbon–Burgos
- Port Bou–Barcelona–Valencia–Murcia

- Madrid–Almeria/Algeciras
- Le Havre–Paris
- Dijon–Modane
- Paris–Strasbourg
- Kehl–Dijon
- Nancy–Avignon
- Marseille–Genoa
- Avignon–Narbonne
- Paris–Dijon
- Paris–Hendaye
- Aulnoye–Metz
- Tarvisio–Udine–Bologna
- Brenner Axis–Bologna
- Udine–Trieste
- Iselle–Turin/Milan–Bologna
- Modane–Turin–Milan
- Chiasso–Milan
- Verona–Trieste
- La Spezia–Fidenza
- Leghorn–Florence
- Patras–Athens
- Athens–Larissa (Volos)–Thessaloniki–northern border (former Yugoslavia and Bulgaria)

These axes are shown on the accompanying map (Figure 2.1) of the European corridors of combined transport.

In order to complete the network referred to in the Council's decision, further works, which are still to be determined, need to be undertaken on the following rail axes:

- Madrid–Albacete–Valencia
- Madrid–Irún–France
- Bologna–Bari/Brindisi–Greece
- Igoumenitsa–Patras
- Patras–Athens
- Athens–Larissa (Volos)–Thessaloniki–northern border (former Yugoslavia, Bulgaria, and Albania)
- Igoumenitsa–Volos
- Igoumenitsa–Thessaloniki
- Thessaloniki–Alexandroupolis–Ormenio (Greece/Turkey–Greece/Bulgaria borders)
- Bologna–Naples
- Naples–Reggio de Calabria–Messina–Palermo/Catania
- Genoa–Leghorn–Rome
- Civitavecchia–Olbia–Sassari–Cagliari
- Antwerp–Ruhr
- Hengelo–Osnabrück
- Venlo–Cologne
- Berlin–Frankfurt/Oder–border/Poland
- Berlin–Stralsund
- Dresden–border/Czech Republic
- Dresden–Görlitz–border/Poland

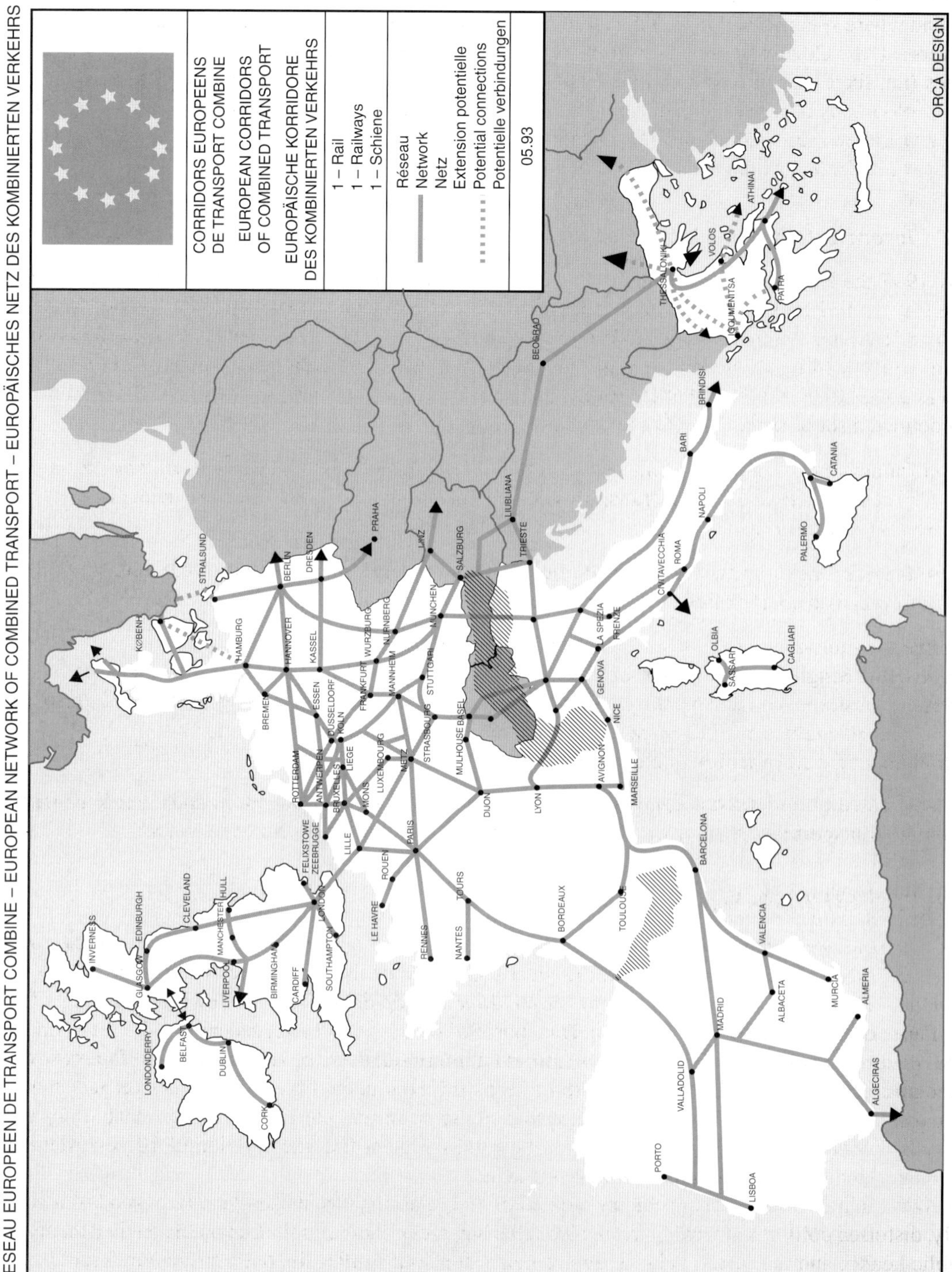

Fig. 2.1 European network of combined transport. (*Source*: European Commission, via Internet.)

The list in Chapter 11 indicates those axes identified for pilot action and where detailed examinations were necessary. It also identifies connections outside the EU and where further pilot actions would be useful and where special wagons could be utilized. Besides the actual routes or connections identified for improvement, the Council was anxious that attention should be given to the provision of trans-shipment facilities (i.e. fixed and movable equipment) and for the provision of appropriate rolling stock for the rapid development of combined transport links where called for by the nature of the infrastructure. In other words, concentration on the rail infrastructure alone is not enough; there must be adequate provision of suitable terminals and sufficient numbers of the right types of rail.

2.2.11 Intermodality and Intermodal Freight Transport in the EU

In May 1997 the EC issued a Communication entitled *Intermodality and Intermodal Freight Transport in the EU* (COM 97/243). This document recognized that a 'business as usual' approach would not solve the future problems associated with transport. It acknowledged that the efficient and balanced use of existing capacities in the transport system was a key challenge and defined key actions necessary for what was hoped to be the development of a step change from the then current approach. This communication defined a set of objectives for the development of:

> a framework for an optimal integration of all different modes so as to enable an efficient and cost-effective use of the transport system through seamless, customer-oriented, door-to-door services …

It set out four key strategies to stimulate the development of intermodal transport in the overall context of the Common Transport Policy. These were steps to:

1. refine the trans-European networks and nodes;
2. realise the Single Transport Market by harmonizing the regulations and competition rules;
3. remove obstacles to intermodality and the associated friction costs (or costs that make intermodal transport uncompetitive in comparison to uni-modal, i.e. single-mode transport);
4. implement the Information Society in transport.

It also advocated an intermodal transport system which encourages co-operation and complementarity between the transport modes, and which favours competition between transport operators.

2.2.12 The EU White Paper: 'Time to Decide'

The EC has been very active in developing transport, and specifically intermodal transport, policies over recent years as we have seen above. But the most significant of these measures was publication of its *White Paper, European Transport Policy for 2010: Time to Decide*, in 2001 (following on from its first White Paper on the future of the Common Transport Policy issued in 1992, see page 24). The 'Time to Decide' document defined a number of very important intermodal-related issues. First and foremost was the recognition that as the demand for transport keeps increasing, the Community's answer just cannot be to build new infrastructure (i.e. especially roads) and open up new markets. The transport system needs to be optimized to meet the demands of both enlargement (i.e. the addition of 10 new Member States from May 2004), and sustainable development.

The White Paper reviews the problems created by the relatively unchecked growth of road transport, namely, distorted competition, traffic congestion, the harmful effects of pollution on the environment and on public health, and the heavy toll of road accidents. It also identifies the need for integration of transport in a climate of sustainable development as envisioned in the Common Transport Policy (originally set out in 1957 in the Treaty of Rome). After setting out the case for shifting the balance between modes of transport, the White Paper goes on to examine the way in which the various modes may be linked. It especially identifies the need for integration between sea, inland waterways, and rail where it notes there is a big missing link. These modes in particular are seen as offering considerable potential because they are both under-utilized and provide the opportunity to reduce road congestion and air pollution. As the document states, the EU has 35 000 kilometres of coastline and hundreds of sea and river ports (i.e. which could be more effectively utilized).

Overall, the White Paper proposed, as we have already seen, some 60 specific measures to be taken at Community level under the transport policy, including action programmes extending up to the year 2010. The key issues in regard to intermodal transport (bearing in mind that the White Paper covers all aspects and forms of transport) are as follows:

- revitalizing the railways,
- improving the quality of the road transport sector,
- promoting transport by sea and inland waterway,
- turning intermodality into reality,
- building the Trans-European Transport Network,
- improving road safety,
- adopting a policy on the effective charging for transport.

Since publication of the White Paper, a number of other key policy and legislative documents have emerged.

2.2.12.1 Intermodal loading units (ILUs)

The EC published a proposal for a Directive on *ILU* in April 2003 (COM 2003 155 final) as part of its programme for the *Promotion of Short-Sea Shipping*. The purpose of the proposal is to overcome the diversity of ILUs and their handling and securing devices, a factor that hampers the efficiency of trans-shipment operations. The intention is to increase efficiency, to ensure safety, and to minimize risks to persons and property by introducing more uniform designs and requiring that such equipment is subject to specified and uniform maintenance obligations and periodic inspections (see also Chapter 13).

2.2.12.2 The Marco Polo programme

Another measure of considerable significance is the 'Marco Polo' funding initiative (officially designated as a Community Support Programme) designed to promote intermodal transport, especially short-sea shipping and inland waterway modes, and the motorways of the sea concept by providing financial assistance (i.e. grants) to improve the environmental performance of the freight transport system. This programme for 2003–2010, established via Regulation 1382/2003/EC of 22 July 2003, replaces the earlier Pilot Actions for Combined Transport (PACT) funding programme that came to an end in December 2001 (see Chapter 10). By mid-2004 further emphasis was being put on the quest to switch more freight traffic from road to other modes, particularly shipping. According to the EC, taking account of the 10 new Member States admitted to the EU from May 2004 would ensure that more funds would be available for investment in fast shipping links and traffic avoidance technology. Against a 2003–2006 budget of 75 million, the EC is proposing a 2007–2013 budget of 750 million, which besides being a significant increase, also takes the whole programme 3 years beyond its original completion date.

2.2.13 The EU Case for Intermodalism

Thus we can see from these various documents that within Europe, both operationally, where the greater potential for long-distance freighting makes it a more economically viable proposition, and in EC policy making circles in Brussels, where there is concern to promote growth in rail traffic which has a less harmful effect on the environment than road freighting, intermodal transport, and especially combined road–rail transport, has received more attention and been viewed with more enthusiasm than has hitherto been the case in the UK. For many years, for example, the French (SNCF) and German (DB) railways – and other European rail networks, notably those of Switzerland and Austria – have operated effective combined transport services, carrying large numbers of unaccompanied lorry semi-trailers piggyback-style, and substantial volumes of container and swap body traffic on long-haul rail journeys. Similarly, for a long time there has been recognition within the political corridors of the EC that efficient systems of combined transport have the potential to resolve some of the major problems brought about by an ever-growing demand for road transport.

In particular, greater use of intermodal transport is seen as a means of reducing harmful atmospheric pollution, high levels of energy consumption, intolerable levels of traffic congestion, too many road accident deaths and casualties, and excessive use of land space taken up with road building all emanating from a road-biased freight system. The EU view is clear: the balance must be changed from a freight system where road accounts for 80 per cent or more of all movements to one where a much higher percentage than at present is long hauled by rail or on the waterways. An efficient Euro-wide intermodal transport system is seen as the obvious solution.

2.3 Chronology of reports and legislation

1970 Regulation 1107/70/EEC authorizes state aid to modernize combined transport infrastructure and fixed and movable trans-shipment facilities. (Later amended by Regulation 1658/82/EEC and by Regulation 1100/89/EEC which extend this to operating costs through non-EU countries.)

1975 Directive 75/130/EEC exempted road journeys to and from combined transport terminals from quota and permit requirements.

1982 Directive 82/603/EEC allowed governments to make tax concessions to vehicles used in combined transport operations.

1986 Directive 86/544/EEC extended this tax liberalization to the own-account sector.
Directive 86/647/EEC relates to the initial legs of combined transport journeys being carried out by own-account carriage.

1990 (October) Council of European Community mandated the Commission to present a 'General Plan' for a combined transport network and its operation.

1991 Directive 91/440/EEC on the development of the Community's railways (i.e. the 'Railway Directive').
(April) High Level Working Group established by EC Commission following above mandate presents its report.
(April) the International Road Transport Union (IRU) and the International Union of Railways (UIC) present a report, *The Conditions for the Development of Combined Transport in Europe*, and announced their intention to operate as partners in the development of combined transport in collaboration with Intercontainer and the International Union of Road–Rail Transport (UIRR).

1992 EC communication, *The Future Development of the Common Transport Policy*.
(June) EU Commission adopts Master Plan based on High Level Working Group's April 1991 report.
Directive 92/106/EEC (i.e. the European Directive) allows initial and final legs of combined transport operation to be opened to hauliers from any EU country – previously this was the illegal practice of 'cabotage'. Also, such operations are exempted from any compulsory tariff system.
Directive 881/92/EEC 'on access to the Market' (i.e. Community Authorization).

1993 Commission Decision 93/45/CEE on the granting of financial support for pilot schemes to promote combined transport (see Chapter 11).
EU Council Decision 93/628/EEC creating a trans-European combined transport network (see also EC publication *Trans-European Networks – Towards a Master Plan for the Road Network and Road Traffic*). Brussels and Luxembourg 1993 (ISBN 92-826-4881-8).

1994 EC report on PACT (see Chapter 11).
The Royal Commission on Environmental Pollution report on *Transport and the Environment*, which set out objectives for a sustainable transport policy for the UK.

1995 (May) IRU–UIRR Joint Declaration on the Promotion of Combined Transport.
New department created for transport research and development policy within EC Directorate-General VII (Transport).
Cross-Directorate Task Force for intermodal transport established within EC.
UK DoE/DTp *PPG 13* which set out the Government's far-reaching aims for land use and transport planning in England to be achieved by the use of alternative means of transport.

1997 The EC launches its Communication document, *Intermodality and Intermodal Freight Transport in the EU*. This was a vital step forward in the development of intermodalism in Europe.

1998 UK Government publishes the White Paper, *A New Deal for Transport: Better for Everyone*, which forecast a radical change in transport and set out ideals for achieving a more integrated transport system.

1999 UK Government publishes *Sustainable Distribution: A Strategy* reiterating its commitment to the White Paper, New Deal, and defining the need for sustainable distribution systems that better utilize the rail network, ports and shipping.

2000 The UK Government launches its Transport Ten Year Plan 2000, *Transport 2010*. This is described as a long-term strategy for delivering a better and more environmentally friendly transport system and it promised an investment of £180 billion in the system over 10 years.

2001 Publication of the SRA's *Freight Strategy* document sets out for the first time its freight strategy based on its Strategic Agenda.
Publication of the EC's important White Paper, *European Transport Policy for 2010: Time to Decide*. This presented a key stepping-stone in the quest to switch more traffic from road to rail and on to the inland waterways and short-sea shipping.

2002 Launch of *Freight on Water* – A new Perspective report by DEFRA on policies for the future of inland waterways in England and Wales – including their use for freight transport.
Publication of the SRA's first Strategic Plan, *The Way Forward for Britain's Railway*.
Publication of: *UK Marine Motorways Study* into fast-freight ferry services.
EC adopted (on 23 January) 'second package' of measures to revitalize the railways.

2003 Publication of the SRA's Corporate Plan: *Everyone's Railway – The Wider Case for Rail* reaffirming the SRA's commitment to developing the UK rail network.
EC Regulation 1382/2003/EC establishing the Marco Polo programme which provides financial assistance for improving the environmental performance of the freight transport system.
Publication of SRA's second Strategic Plan, *Platform for Progress* setting out the next steps towards achieving its stated goals.
Historic launch on 15 March 2003 of the TERFN permitting open access for international rail services across the EU.

2004 Third 'railway package' presented by the EC (3 March 2004).
Publication of the UK Government's White Paper, *The Future of Rail*, in July, which sets out a blueprint for improving rail performance and gives Network Rail a stronger role. It gave notice of

the winding up (within 12–18 months) of the Strategic Rail Authority and the transfer of its functions and financial obligations to the DfT. While being an important document in its own right, it contained no specific measures relating to intermodal transport.

Publication of the UK Government's White Paper, *The Future of Transport*, (also in July), which promised nothing in regard to intermodalism apart from its intention to 'encourage freight traffic to be shifted from road to rail or water where this makes sense, and where appropriate we will offer financial support'.

'Intermodal Transport and Logistics – An overview of current activities' published in September 2004 by the EC's Directorate for Energy and Transport, Unit G3 – Motorways of the Sea and Intermodality.

3

Intermodal Developments in the UK

It is widely accepted that Europe has traditionally been ahead of the UK in the development and application of intermodal transport, being well supplied with strong national rail systems, an extensive inland waterways network and a burgeoning road haulage industry; whereas in the UK, until quite recently according to the Department for Transport (DfT) in its publication, *Transport Statistics for Great Britain – 2004*, our railways and inland waterways have been in decline leaving a freight market heavily dominated by road haulage – in 2002 accounting for no less than 62 per cent of all goods moved (i.e. the weight lifted multiplied by the distance carried) and 82 per cent of all goods carried.

However, since the early 1990s the situation in the UK has changed somewhat with the Government, in particular, taking up the reins, so to speak, and increasingly campaigning on environmental grounds, for a switch of a freight from road to more sustainable modes of transport such as rail, inland waterway, and short-sea shipping. Since that time it has been possible to see the development of UK intermodal freighting as being attributed to a number of key factors which can be identified as having provided the essential impetus. For example, there has been the build-up of UK–European trade following the opening of the Single European Market in 1993. This was followed by the opening, in 1994, of the Channel Tunnel which, by directly linking the UK and European road and rail networks, has provided scope for economic long distance freighting direct from inland UK locations through to international destinations across Europe and vice versa. Then came the opening-up of the rail network, first to access by private operators and subsequently by full-scale rail privatization in 1996 – a government policy decision that was not specifically intended to promote intermodalism, but one that was certainly conducive to the general aim of switching freight traffic from road to rail.

Other key factors that have contributed to a growing interest in the benefits of intermodalism include the availability of grant aid from the UK Government to help fund intermodal infrastructure developments and to encourage the switch of freight traffic off our congested road system and onto the railway or the inland waterway system. In 1999, the Strategic Rail Authority (SRA) launched a national competition to foster the development of innovative, commercial, intermodal logistics solutions. This is a project that met with considerable initial success in terms of highlighting various different intermodal possibilities, but one could say that the jury is still out on the long-term feasibility of the winning schemes. We also saw the introduction of legislation permitting the use of 44-tonne gross weight heavy lorries, which are more operationally suitable and economically viable for carrying heavily laden ISO shipping containers and intermodal swap body units. There are of course many other contributory factors to the growth of intermodalism in the UK and it is important to note that while some particular developments have made the freight news, other intermodal schemes have been operating with significant success, albeit with little publicity. In reviewing developments in the UK it is useful to examine in more detail the factors mentioned above and some of these other successful intermodal schemes.

3.1 Euro-trade and the Channel Tunnel

Trade is obviously the catalyst to a burgeoning intermodal scenario, and there is no doubt that trade in the UK is booming. According to the DfT, our rising national wealth is being reflected in a growing demand for goods and services with an associated increase in freight and commercial traffic. It says that over the past 20 years the amount of goods moved has grown by around 70 per cent, and despite the number of licensed goods vehicle falling by one-fifth, there has been a more than a 40 per cent rise in goods vehicle traffic according to the DfT's report *Managing Our Roads* (July 2003). As trade growth means increased transport activity and increased transport activity means greater potential for goods shippers to opt for the operational and environmental benefits that intermodalism undoubtedly delivers, then intermodalism has not been short of opportunity. Again, according to Department of Trade and Industry (DTI) figures published in the UK Balance of Payments publication (i.e. the so-called *The Pink Book 2004*, from The Stationery Office (TSO)), the value of Britain's trade in goods with Europe (i.e. with the 14 other pre-2004 EU Member States) has increased by over 50 per cent since the opening of the Single European Market in 1993. This expansion in trade has had an obvious impact by way of a substantial growth in the volume of freight being shipped between the UK and Europe by all transport modes and much of this growth, in fact, is accounted for by intermodal freighting, particularly combined road–rail transport. There is no doubt that the Channel Tunnel has been the one of the encouraging factors in these increasing traffic flows. Since becoming fully operational in 1994, the Tunnel has opened the gateway for a flow of freight on rail between the UK and Europe amounting, in 2003, as we have seen in Chapter 1, to nearly 1.3 million trucks carried on its shuttle service and more than 1.7 million tonnes of freight on rail through the Tunnel – the latter figure being partially beset by the still recovering scenario from severe interruption caused by the stowaway saga of 2001 and 2002.

Growth in intermodal freighting through the Channel Tunnel has been accounted for by the two main forms of road–rail transport; namely, heavy lorry traffic via Eurotunnel's 'rolling motorway' freight shuttle service between terminals at Folkestone in the UK and Coquelles, near Calais in France, and intermodal swap body and container traffic carried on through-Tunnel rail-freight services operated by English, Welsh, and Scottish Railway (EWS) jointly with the French State rail operator Societe Nationale des Chemins de Fer (SNCF) Fret, between inland freight terminals in the UK and in Europe.

Following the inauguration of through-Tunnel services in 1994, a number of specialist combined transport operators (CTOs) were founded specifically to operate combined transport services. These included Combined Transport Limited (CTL), subsequently replaced by Norfolk Line; Allied Continental Intermodal (ACI), now part of Intercontainer-Interfrigo (ICF); Unilog NV, which operates services from the UK to its Belgian hub at Muizen; Spanish operator Transfesa, which provides logistics services for the Ford Motor Company; and Environmental Freight Services, set up in 1998 to serve the Italian market. Similarly, a number of major road haulage firms, such as Davies Turner, Hay Pollock (now part of ABX Logistics), the Potter Group, and Harris Distribution (part of the Transport Development Group (TDG)), to name just a few of the key players, have established international rail-linked combined transport services using the Channel Tunnel as a primary route.

3.2 The rail scene

3.2.1 Open access and privatization

In August 1991, the UK Secretary of State for Transport announced plans to break the state monopoly of UK domestic rail-freight services that had existed as British Rail since the Transport Act of 1947, by opening up the network for use by private freight operators. This was in response to the European Railway Directive (440/91/EEC) (see Chapter 2). The intention, so said the Secretary of State's announcement, was to do away with red-tape bureaucracy and to invest in a programme to encourage the development of more road–rail combination schemes, principally through hauliers establishing their own freight terminal facilities. However, on a negative note, it has to be recognized that changing from road to road–rail operations is not a simple matter of choice; there are problems in the provision of suitable infrastructures such as terminals, transfer facilities and equipment, and both road and rail vehicles, all of which requires substantial capital investment. Besides this, the former British Rail had neither the versatility to handle all the different types of road vehicle trailer and swap body in common use, nor a rail infrastructure which allowed sufficient gauge clearance to carry road trailers on piggyback-style rail operations. Similarly, most British road hauliers did not, and most still do not, have the specially built trailers needed for piggyback loading onto rail rolling stock, or suitable swap bodies built to international standards – normal domestic-use demountable bodies are not designed to withstand the stresses imposed by repeated transfer between road and rail transport systems. The Government has said that it expects to see more effort being put into overcoming these limitations and progressive development of intermodal transport for both its economic benefits to users and the environmental benefit to the community as a whole.

The rail privatization plan involved breaking down the British Rail network into a number of separate entities. First and foremost, the rail infrastructure, comprising largely the track, the signalling and the stations, is now managed by Network Rail, which makes track access charges, under contract, to the train operating companies to run their services. These train operators comprise the passenger service network, which was divided into regional companies and sold off, and British Rail's freight operations, including its trainload businesses, which were sold to the US private rail operator, Wisconsin Central Transportation, and which have subsequently been formed into a new firm, EWS Limited; while Freightliner, the domestic container business, was sold to its management and has retained its Freightliner name. The new railway freight system works by private companies either owning or hiring fleets of rail wagons (frequently whole block trainloads) to suit their needs which they then contract with the train (locomotive) operator who undertakes the rail haulage operation.

According to the pro-rail body, the Rail-Freight Group (RFG), 1995 proved to be a turning point in the fortunes of UK rail freighting, and it was projecting expansion well into the twenty-first century. In a letter to *International Freighting Weekly* (*IFW*) in January 1995, RFG highlighted the contrast between British Rail's declining freight services under public ownership and the initiatives being taken at that time for privatization, which showed how major freight companies were keenly examining the rail option and running trial operations, often prompted by customers who wished to use, or to be seen to be using, rail as a major constituent of their freighting operations for environmental reasons.

What is important, at least so far as this book is concerned, is that the quite dramatic changes described above have subsequently allowed private companies such as road hauliers and others, either individually or in partnership, to own or hire rail rolling stock and to operate freight services across the rail network, negotiating and paying access charges as appropriate and where necessary obtaining Track Access Grants (TAG) (see page 42) to enable them to do so.

3.2.2 The freight train operators

Rail privatization, as was clearly intended by the then UK Conservative Government, has resulted in the formation of a number of key rail-freight traction companies – the so-called 'freight train operators' (FTOs). Principal among these are EWS and Freightliner, both outlined in more detail below, but there are others; namely, Direct Rail Services (DRS) which provides rail haulage, primarily for carrying nuclear materials on behalf of British Nuclear Fuels Limited (BNFL), but also for the private freight sector between the English Midlands and Scotland in association with road haulage and logistics provider WH Malcolm (more of which later); Great Britain (GB) Railfreight, which, it says, aims to provide high-quality flexible bespoke rail-freight logistics solutions; and Advenza Freight, the latest firm to join the ranks of rail-freight operators, which ran its first 'FreightBus' services for palletized goods on the rail network between Willesden (London) and Mossend (Scotland), in October 2004.

3.2.2.1 English, Welsh, and Scottish Railway

Formed on rail privatization in 1996 following its acquisition of the four previous British Rail-freight divisions; namely, Rail Express Systems, Loadhaul, Main Line Freight, and Transrail Freight, EWS became the largest of the UK rail FTO. Now running some 8000 services weekly with nearly 500 locomotives and over 14 000 freight wagons listed among its rolling stock assets, the company claims to haul over 100 million tonnes of freight every year.

3.2.2.2 Freightliner

Freightliner is the largest intermodal rail operator in the UK, primarily concerned with the movement of deep-sea containers between container ports and inland terminals. It moves around 650 000 containers per year with its fleet of over 100 locomotives and serves 19 terminals. In 2004, a year that was showing booming growth in its intermodal business, Freightliner achieved a record when it moved 2635 containers in a single day. Through its Heavy Haul division, Freightliner is increasingly operating domestic services carrying chemicals, finished automotive products, and coal and rail infrastructure materials.

3.2.3 Piggyback developments

Another major intermodal initiative in the UK was undertaken by the aptly named Piggyback Consortium. This grouping of interests from the public and private sectors (including such bodies as local authorities, transport trade associations, port authorities, Railtrack and Eurotunnel, and a number of private companies) was formed in 1993 to study the potential for piggyback services between the UK and the Continent, and to determine whether a commercially viable network could be established between key markets in GB and the Continent after meeting the cost of engineering works. It made a detailed study, which was part funded by the European Commission Transport Directorate (DGVII) under the Pilot Actions on Combined Transport (PACT) programme (see Chapter 11), and a report was published in April 1994.

It should be explained here that piggyback operation is just one form of intermodal combined road–rail transport. In this case it is the carriage of 4-metre high, road-going articulated semi-trailers on special rail wagons, principally designed to provide a low travel height to ensure loading-gauge clearance through existing rail bridges and tunnels. The Consortium saw this as an important segment of the freight market, which had received scant attention in the UK despite the fact that, at that time, three-quarters of unit-load traffic between GB and the Continent was carried in semi-trailers via the roll-on/roll-off (RO/RO) cross-Channel ferries. Initial analysis of the potential UK market for intermodal services carried out for the Consortium suggested that as many as 400 000 unit movements per year could be captured by rail by the year 2003 – that would have meant between 20 and 40 full trains per day.

However, there were shortcomings to this as the Consortium was well aware, not least because of the physical structure of Britain's rail network. Its restricted loading gauge (lower and narrower bridges and tunnels, and station platforms being closer to the track) meant that the operation of piggyback in the UK, as practised in Europe, would not be feasible without a very substantial investment in rail network restructuring to increase gauge clearance by at least 160 millimetres to allow the passage of 4-metre-high semi-trailers on low-built (i.e. piggyback) rail wagons. For example, a projected investment of around £70 million would have been needed at that time to provide adequate gauge clearance on the Ireland – Scotland – Continent via the Channel Tunnel route. It is worth noting that this multi-million pound estimate escalated

Fig. 3.1 Typical piggyback tank trailer being unloaded by overhead gantry crane. (*Source*: Interferry.)

alarmingly as the years passed, in fact until the Piggyback Consortium finally gave up on the project and alternative solutions were beginning to be developed (see Chapter 7).

Before the loading-gauge saga was fully played out, another obstacle reared its head. This was the fact that the UK's standard road-going semi-trailer fleet was not, and indeed is still not, generally suitable for two reasons: first, most semi-trailers are not constructed to withstand the stresses of lifting under load for loading/unloading from piggyback rail wagons; and second, they do not have the built-in lifting pockets necessary to engage the crane's lifting arms. Most standard UK semi-trailers are also built to at least 4.2-metre-high (rather than to the European 4-metre limitation on maximum height), which adds further complication to the piggyback scenario. Yet another shortcoming at that time (i.e. the early 1990s) was that UK law did not permit vehicles used in either piggyback or rolling motorway services through the Channel Tunnel to operate at the 44-tonne gross weight necessary to provide efficient and economical intermodal operations. They were restricted to the then current standard UK maximum gross weight limit for articulated vehicles of only 38 tonnes. (Figure 3.1 shows a typical piggyback semi-trailer showing how the grapple arms of the overhead crane match with the sockets in the trailer sides and giving an indication of the stresses imposed when lifting loaded units.)

In response to the Consortium's April 1994 report, in which it called for the development of prototype piggyback wagons, an October 1995 update brochure indicated that four separate manufacturing groups were working on such wagons. These were anticipated to be in 'revenue' service in 1996 carrying low-height semi-trailers within GB and through the Channel Tunnel, and full-height semi-trailers onto the Continent where the higher loading gauge required is already available.

By 2004 the original enthusiasm for piggybacking semi-trailers on rail had largely dissipated for reasons of gauge limitation as described above and also for economic reasons. There is no doubt that despite the superficially heartening impression given by seeing a trainload of semi-trailers that, 'here's another load of heavy vehicles removed from our motorways', it is a fact that lifting, loading, and carrying semi-trailers on rail is an extremely high-cost operation on many scores. First, the fact that the wheels and suspension forming the semi-trailer running gear adds considerable weight, which limits payload capacity within road vehicle maximum legal weights, and also produces stability problems. Second, the additional capital costs involved in building both the special low-height rail wagons and the extra strong road semi-trailers needed for piggyback operations. Third, the capital cost of the special lifting equipment needed, and its operating costs. And fourth, and not least, the demurrage costs of expensive semi-trailers that are out of circulation while on long-haul train journeys.

Some successful piggyback operations, nevertheless, do still exist. This is particularly so, it seems, where a physical barrier exists such as the Channel Tunnel, where the Eurotunnel freight shuttles provide an economical rolling motorway-type operation for complete road vehicles with drivers, and similarly on trans-Alpine routes such as those through Austria and Switzerland where local environmental legislation sets severe limitations on the passage of heavy road vehicles on transit journeys. There are other instances, too, where piggyback can be successful; for example, in the Blue Circle/Lafarge cement operation described below, which is very specific in that positive outside factors overrule the negativity aspects of basically 'carrying wheels'. Apart from these relatively isolated examples, the current trend is very much towards the use of ISO containers – commonly referred to as 'boxes' – and intermodal swap bodies with predominance being given to the former because of their stackability. In fact, in most world markets outside of Europe, the standard ISO shipping container reigns supreme, principally because of relative cheapness, its inherent strength and durability to withstand many handlings, and particularly because of its capabilities in regard to top lifting and stackability, neither of which apply to the conventional swap body.

3.2.4 The central railway project

An ambitious £10 billion project to develop a dedicated 'lorries on trains' freight railway, linking the Northwest of England with mainland Europe via the Channel Tunnel, which was destined to remove significant numbers of heavy goods vehicles from UK motorways, appears to have finally hit the buffers in 2004. This was when the UK Government decided against supporting the necessary Parliamentary Bill because of potential risk to the public purse should private financing arrangements fall short. A substantial case had been put forward by Central Trains plc, the Franco-British Consortium leading the project, for a 420-mile long railway running between Liverpool and Lille in Northern France, and mainly using underused and disused railways along existing transport corridors. The scheme was claimed to have been based on a proven, commercially viable, US model and designed to carry some 3 million heavy lorry trailers a year. Certainly, from the environmentalists' point of view, this was a very laudable proposal, but without the surety of vast amounts of private finance and the consequent likelihood that Government aid would be required to help defray the massive costs projected, it was, disappointingly, an almost inevitable non-starter.

3.2.5 Gauge enhancement

An essential aspect of the development of intermodal freight services, and particularly piggyback systems, on Britain's rail network – as we have seen above – has been the limitation imposed on such operations by the restricted rail gauge clearance on the network. While this issue is discussed in greater detail in Chapter 7, it is useful here, in this review of intermodal developments, to mention the fact that over recent years gauge restrictions on, for example, the West Coast Main Line (WCML) have inhibited attempts at running north-to-south piggyback trains and vice versa, and restricted the routes on which the latest generation of 9-foot 6-inch-high, so-called 'tall', maritime containers can be carried.

At the time of completing this book, significant steps are being taken to develop clear routes for such traffic, with one in three containers currently passing through UK ports being 9-foot 6-inch-tall and the prospect, according to the SRA, that by 2010 this number will rise to one in two containers. In fact, as recently as November 2004, the completion of major engineering work on a rail tunnel at Ipswich has allowed the first services for 9-foot 6-inch containers carried on standard rail wagons to operate from the Port of Felixstowe to the Midlands, Northwest, and Scotland. At the same time the key rail corridor between Nuneaton and Birmingham has also been opened up for this traffic. Earlier in 2004, a vital 'tall container' link between the port of Tilbury (London) and the WCML had also been opened up meaning that virtually half of all the containers that currently land at Tilbury can be transported by rail to the Midlands and the north of England.

In November 2004, the SRA published a consultation document on gauging policy in which it was seeking public views (by 18 March 2005) on what its gauge policy should be for the future, with the objective of setting a target gauge configuration for the whole network. While the consultation covered many relevant issues, freight traffic obviously featured largely, in particular because of the increasing trend towards taller and wider containers as mentioned above.

3.3 Tall and long boxes: the new container revolution

In terms of traditional ISO container use, the trend is away from the standard 8-feet-high box, of which there are many thousands in use worldwide, towards the use of 9-foot 6-inch-high containers as mentioned above. This predominantly maritime-led revolution for so-called 'high cube' containers has caused problems in such areas as rail loading-gauge clearances in the UK; but as we have seen above, this shortcoming is progressively being overcome. However, along with this, we must also consider the use of 45-feet-long containers, which also have had a somewhat adverse initial impact on freight systems, most notably when carried on road vehicles. The problem here was that under current European overall vehicle length legislation, the carriage of conventional, square-ended 45-feet-long containers is illegal in the UK because either the maximum 2.04-metre dimension between the semi-trailer king pin and the front of the trailer, or the 12-metre maximum overall length dimension between the king pin and the rear of the semi-trailer, would be exceeded. The problem has largely been overcome by a new design of tapered corner casting, developed by Geest North Sea Line B.V. – by which the container is secured to the semi-trailer – which allows the box to meet the crucial 2.04-metre swing clearance requirement specified in EU Directive 96/53 (this is dealt with in greater detail in Chapter 13). Now, the latest generation of 45-feet-long by 9-foot 6-inch-tall containers are also built to the 2.5-metre maximum external width dimension of heavy road vehicles, instead of the hitherto standard container width of 8 feet, thus facilitating the loading of 33 (800 millimetres × 1200 millimetres) Euro-pallets to provide a similar loading capacity to a standard articulated dry-box semi-trailer.

3.4 Government grant aid

Despite the EU policy of not supporting lame dog industries, or individual commercial firms come to that, it does, conversely, have various Community Support Programmes to aid developments that help to reduce road congestion, improve the environmental performance of freight transport within the Community, and enhance intermodality, thereby contributing to an efficient and sustainable transport system. In a UK context, as opposed to EU schemes, grant aid has been available since 1974 to help fund projects that are designed to facilitate a switch of freight from road to rail. In particular, the grants are provided to assist with the purchase of equipment or to fund the development of infrastructures that would not be necessary for road-only transport operations. It is important that the grants, since they involve public money, are seen to meet certain key criteria as follows:

- The project must produce an environmental benefit – principally taking lorries off the road.
- The project must meet only a 'financial need' – defined as being the difference in total cost between the rail and the road operations.

From an EU perspective, grant aid must be restricted to the amount that is defined as being 'necessary', that it does not lead to unacceptable distortion of competition and complies with Community rules.

Four different types of grant are currently available from the UK Government, subject of course to satisfying relatively onerous criteria, as follows:

- *Rail-Freight Facilities Grant (FFG)* provided to part-finance the capital costs of providing rail-freight facilities.
- *Waterborne FFG (WFFG)* made available to part-finance the capital cost of providing waterborne freight facilities at ports and wharves.
- *TAG* which helps offset the track access charges levied on rail-freight operators for use of the rail in return for generating environmental benefits by switching freight traffic from road to rail.
- *Company Neutral Revenue Support (CNRS)* which is a relatively new scheme designed (launched in February 2004) to further growth in the movement of deep-sea, short-sea, and domestic rail freighting of intermodal containers. It will replace around 80–85 per cent of the TAG.

These schemes, which are described in greater detail in Chapter 10, have resulted in varying amounts of aid being awarded to qualifying projects; for example, at their peak in 2001, £32.7 million was awarded for rail projects and £15 million for inland waterway, and coastal and short-sea shipping projects. There was something of a hiatus in the grants scheme in 2003 when the SRA suspended the FFG and the TAG because budget constraints meant that it no longer had sufficient discretionary funding to support new schemes. However, the new CNRS scheme was launched for 2004/2005 with a budget of £22 million and is intended to continue, with roughly similar annual budgets, for 3 further years until 31 March 2007.

3.5 Commercial developments

Over recent years many commercial firms have adopted intermodal systems for various economic, operational, and environmental reasons. While it is still early days to assess whether these current operations, some still in their trial stages, will eventually manifest themselves as being permanently sustainable, at least the present scenario is showing an encouraging trend, which hopefully, will be followed in due course by other schemes from other freight shippers.

3.5.1 WH Malcolm

Formerly a well-known Scottish road haulage company WH Malcolm, the Malcolm Group has increasingly developed combined road–rail intermodal services and boosted its profitability at the same time.

In 2002, it opened a new £2.75-million state-of-the-art rail-freight terminal complex at Grangemouth, Scotland, from where it is providing a high-speed overnight transport service between Scotland and England, and claiming to be removing more than 10 000 lorry journeys a year off Britain's congested motorways. The company now operates up to 12 trains weekly, six in each direction, between Grangemouth and its other road–rail complex at Crick, near Rugby (part of the Daventry International Rail-Freight Terminal). The north–south link is part of a major strategy decision by the Malcolm Group to move more freight from road to rail, minimizing the effects of fuel costs as well as helping to cut down levels of pollution and reduce congestion on roads and motorways.

3.5.2 The Potter Group

Another key name in UK intermodalism is the Potter Group, a leading road and rail-freight logistics service provider, and independent rail-freight distribution centre operator with terminals at Ely in the Southeast, Selby in the North, and Knowsley in the Northwest of England. They provide important road and rail transfer points for conventional wagons, swap bodies, freight containers, and ISO tanks, in either full trainloads or as single wagonloads, and are connected to the international rail network via the Channel Tunnel. These rail-freight distribution centres are all equipped with shunting locomotives, intermodal handling equipment, bulk discharge facilities, and covered areas for unloading high-capacity freight wagons.

3.5.3 The Innovation in Freight: SRA competition

One of the most interesting developments in intermodalism within the UK has been the *Innovation in Freight* competition sponsored in 1999 by the SRA. With a £6 million budget, the idea was to bring together companies and in consortia to develop 'Innovative Solutions for Rail-Based Logistics'. The remit went beyond funding studies and research papers; in fact, it encouraged the physical development of ideas and concepts to commercial trial stage. Three projects were declared winners of the competition; namely, the Lafarge (formerly Blue Circle Cement) intermodal project; the Exel Freight Multiple Unit (FMU) project; and the Minimodal project. It is useful to look more closely at these projects, which are described in outline below, because they clearly illustrate very different approaches to the intermodal conundrum. As we can see here, the Lafarge project justifies the carriage of wheeled semi-trailers despite conventional wisdom suggesting that such an approach is not normally considered viable, for reasons described above. But in this particular case other economic factors came into play; namely, the opportunity to dispense with fixed storage silos.

3.5.3.1 Lafarge intermodal project

The Blue Circle Consortium project involving Blue Circle Cement (now part of the Lafarge Group), Babcock Engineering, and tanker builder, Feldbinder, was to develop a new type of rail wagon and road tanker trailer to allow piggyback movement of product from Blue Circle Cement works to customers in the UK allowing Blue Circle to reduce reliance on remote silos and strip out cost from their distribution operation as a result. £2.9 million was awarded to the consortium in 2000 as one of the winners of the first SRA Innovative solutions competition and a 2-year commercial trial commenced during 2003.

This project involved the development of a 'piggyback' concept for intermodal distribution of cement by rail and road. In addition to a special rail wagon, the Babcock Mega3, two trailer types were developed with the capability to be used on either road or conveyed by rail. The two types are a tank trailer and a bag trailer. The design also had to be able to carry the maximum payloads required for the customer and crucially to fit within the smallest common UK rail gauge profile to allow distribution by

Fig. 3.2 Lafarge bulk cement tanker semi-trailer being offloaded from rail wagon. (*Source*: Author.)

rail to any part of the network. Despite challenging established design principles and technical constraints, and securing the necessary railway approvals for the project, driving the need for Blue Circle was to achieve operational savings by making direct deliveries to customers from a central cement plant and dispensing with the need for storage silos in remote locations. It had been identified that rising land values, especially in the Southeast of England, were increasingly impacting on the costs of distribution but while this new intermodal solution was likely to prove more expensive in pure transport terms to some areas, it was expected to reap overall financial and operational supply chain efficiencies for the Group. Figures 3.2 and 3.3 illustrate the concept of this unique operation.

3.5.3.2 FMU project

The Exel Logistics, Amec Rail and Isotrak project was to develop a rail-freight product more appropriate for the logistics industry, using track and trace technology, short engineering trains (FMUs) with acceleration and speed characteristics similar to passenger services and for less reliance to have to be placed on achieving full train payloads. £2.1 million was awarded to the consortium as one of the winners of the first SRA Innovative solutions competition in 2000.

The aim of this project was to demonstrate how fast, compact, intermodal trains could be operated on the UK rail network. Using a smaller flexible train based on road technologies allied to modern track and trace programmes, and linked to the Internet, it aimed to offer levels of service and flexibility in line with road-based solutions. The trial included the use of Isotrak command and control technology that allows cargo to be traced in transit using global positioning satellites (GPS). As a result, total transparency is achieved as to where a client's goods are located at any given time. For the logistics industry this has become a 'standard' and its application in real time in a live rail transport environment, independent of the network, is likely to prove to be a real step forward for the rail-freight industry.

Fig. 3.3 Lafarge/Blue Circle intermodal train loaded with both bagged-cement curtain-sided semi-trailers and tank trailers. (*Source*: Author.)

Unlike the other two competition winners, which involved the physical development of new equipment, the FMU project took existing rail equipment and focused on adapting, applying, and trialling services with a number of customers. The trials, conducted in the spring of 2002, were run for Marks and Spencer (M&S), Bulmer (the cider makers), and Goodyear Tyres, amongst others, to prove the integrity of the equipment used and the performance levels achievable using the equipment.

3.5.3.3 The Minimodal project

This project involving Minimodal, TDG, DRS, The Engineering Link, and Boalloy was to develop a new type of intermodal container, which could easily be moved on conventional passenger platforms between road and rail using conventional forktrucks making rail more attractive to the consumer goods market. £1 million was awarded to the consortium as one of the winners of the first SRA Innovative solutions competition in 2000. Key to the project was the development of a new type of mini container designed to meet the needs of the parcel and retail distribution sector making deliveries in an urban environment.

The Minimodal container with its common platform which maximizes floor area, gives four way access for forktrucks and the ability to load onto both road and rail vehicles, can be configured for many applications from conventional ambient box, to tank container or refrigerated unit. The final design, after much consultation with potential customers in order to optimize the size, provides a 2.55-metre-square base, which enables five units to be conveyed on a standard 13.6-metre road trailer and six on a standard rail wagon. The design in ambient form with two roller doors allows through loading on a conventional loading dock, and enables individual units to be sectioned off for specific store deliveries or transshipment onto shorter-length urban distribution vehicles using conventional forktrucks, without directly re-handling the product. The Minimodal box is approved for 75 miles per hour operation on the rail network, a speed common to the majority of intermodal services.

These three projects, as has been made clear above, are all trial projects. The question that remains is whether they will still be viable over the longer term; or indeed, could they be sustainable on a permanent basis?

3.5.4 The supermarket projects

While 'the High Street', and the retail sector, has traditionally been cited by the roads lobby as an example of where and how the railways cannot perform with the speed, efficiently and flexibility of road haulage – basically because there is simply no rail connection with the High Street – now, in these enlightened days of intermodalism, this fallacy is being proven to be well and truly wrong. Stores hardly become more 'High Street' than Asda, Argos, Superdrug, M&S and Safeway, and yet these leading logistics-oriented firms are proving that intermodal operations can add significant benefits to their operations, not only on environmental grounds, but also on grounds of speed, reliability, and regularity factors, which help to keep them competitive and their operations cost effective. It has to be mentioned also, in late 2004, that with the prospect of the EU's proposed Road Transport Directive (RTD) due to severely hit at lorry operations from March 2005, these firms, and many others of a similar nature need to be looking very closely to see how they can overcome the problems of reduced driver availability, and the national lorry driver shortage that goes hand in hand along with it – now proving to be fact rather than mere conjecture. Switching long-haul trunking operations to rail instead of road is proving to be a worthwhile solution to the problem.

The *Freight on Rail* Internet web site (www.freightonrail.org.uk) cites a number of case studies showing how these supermarket firms are successfully integrating rail operations into their supply chains. These are reproduced here in outline, and with acknowledgement.

Asda, Britain's second largest food retailer, was planning to cut 4 million lorry miles by 2004 as it sought to expand its use of rail freight. Since October 2003 it has been importing goods through the port of Felixstowe, with GB Railfreight providing a 5-day service to and from Daventry for Asda locations in the Midlands. Over a year this contract will remove up to 2000 containers from our congested road network. A partnership with WH Malcolm set up in 2002 has enabled Asda to take advantage of rail's fast and punctual services to Grangemouth and Coatbridge in Scotland. Rail operations for Asda alone on this route to Scotland have already removed over 6000 lorries from the congested M6 and M74 motorways.

Superdrug, the leading health and beauty retailer, is using EWS Intermodal Express rail services for the movement of goods imported through the port of Felixstowe to the EWS-run terminal at Wakefield. This service offers the company fast, reliable, and regularly timetabled services enabling them to connect rail with their own distribution network and improve its availability.

Argos, the UK's largest non-food retail chain, is using EWS services to move containers of mixed cargoes from Southampton to the EWS intermodal terminal at Willesden, Northwest London. The containers then go forward (sometimes by rail via other railheads) to Argos' high street and out of town discount stores.

M&S has been using rail freight for imports from France and Spain since early 2001 as well as daily deliveries to its distribution centre in Scotland. According to M&S, rail freight is part of a mix of supply chain solutions that assist in keeping the company competitive. Security of supply is vital to the company with no one route or mode being totally secure so having the rail alternative is important in providing diversity in the supply chain that would not otherwise exist. M&S says it has found rail to be 15 per cent more efficient (per pallet) to unload than a road vehicle and that it gives better utilization of warehouse resources in the UK as the rail wagons do not clutter the exporting truck loading docks.

Safeway, at the leading edge of many transport initiatives, was the first to integrate rail into a daily part of its supply chain in 1998. It currently uses an EWS daily service to take goods from depot almost

Fig. 3.4 Rugby Cement/ISO-veyor intermodal tank container on road vehicle delivering in Heathrow Terminal 5 site. (*Source*: Rugby Cement.)

directly to its stores across Northern Scotland from Mossend Glasgow to Inverness and Georgemas Junction in the far North. One of the big advantages that Safeway claims for rail operations is that of speed compared to a road operation, as the trains run at 75 miles per hour and that brings a significant time saving. The temperature-controlled intermodal swap bodies used to carry the Safeway traffic look almost identical to conventional trailers but are fitted with strengthened lifting points.

3.5.5 The Heathrow Terminal 5 project

Intermodalism is playing a key role in the massive Heathrow Terminal 5, £4.2 billion, construction project. Commenced in 2002, after years devoted to the planning application and a 3-year, 10-month planning enquiry, the project is due for completion in 2011. RMC Rugby Cement, as one of the main materials suppliers to the project, is contracted to supply 130 000 tonnes of pulverized-fuel ash (p-fa) – a processed residue from power stations used as a cement replacement – over a 5-year period. This material is shipped from West Burton Power Station in Nottinghamshire, England to Heathrow by rail using a new and highly innovative 'environmentally friendly' bulk transport system developed in conjunction with Bulk Technologies. The system comprises a cylinder-shaped ISO-Veyor container with a 34-cubic-metre capacity, constructed within the frame dimensions of a standard 30-foot commercial container. As a result, the unit can be easily transferred between road-going flatbed trucks, rail-based rolling stock, and a variety of ships. The containers are filled at source, remaining sealed until the point of delivery, removing the need for intermediate handling or storage of the contents. At the Heathrow site the ISO-Veyor units are transferred from rail onto Rugby Cement vehicles for the short haul to the on-site batching plant. Since the containers are also stackable, this provides an additional storage option. Figures 3.4 and 3.5 show the ISO-Veyor container tanks on a road vehicle and as part of trainload.

Fig. 3.5 A trainload of Rugby Cement/ISO-veyor tank containers en-route to Heathrow Terminal 5 project. (*Source*: Rugby Cement.)

3.6 Combined transport vehicles

It would be remiss to ignore the potential effect that increased lorry weights under UK legislation has had on intermodal freighting. There is no doubt that efficient intermodal transport is dependent on the availability of suitable road vehicles to carry heavily laden intermodal swap bodies and ISO shipping containers. In particular, the determining factor is the maximum gross weights at which road vehicles are permitted to operate by law. Prior to 1994 the maximum permitted weight for heavy goods vehicles in the UK was 38 tonnes (subject to specified technical standards being met). However, in order to boost both rail and intermodal freight, from March 1994, legislation was introduced which raised the maximum permitted weight of heavy goods vehicles to 44 tonnes subject to such vehicles being engaged solely in combined road–rail operations.

To qualify for combined transport use at up to 44 tonnes maximum gross weight, goods vehicles were required to conform to the specific requirements of The Road Vehicles (Construction and Use) (Amendment No. 2) Regulations 1994. This meant that only articulated vehicles and drawbar–trailer combinations meeting the technical specification set out in the regulations qualified for such use. Bimodal articulated vehicles (i.e. those able to operate as rail wagons when mounted on rail bogies) were also included in this provision. The regulations required these combined transport vehicles, whether articulated or drawbar combinations, to have at least six axles of which three had to be on the tractive unit or drawing vehicle. Driving axles needed to be equipped with twin-tyred wheels. Furthermore, either every driving axle had to be fitted with 'road-friendly' suspension or alternatively no axle was permitted to have a weight exceeding 8.5 tonnes. A road-friendly suspension system, according to the regulations, was either one which relies on air suspension (i.e. with at least 75 per cent of the spring effect caused by an air spring), or one which although not being air suspension, could be regarded as being equivalent to air suspension for the purposes of EU law. The minimum distance (i.e. axle spacing) between the rearmost axle of the tractive unit and the rearmost axle of the semi-trailer of an articulated combination was based on a sliding scale relating to the permissible maximum weight, ranging from the then current 6.3-metre 38-tonne gross weight operation to 8.0 metres for 44-tonne gross weight operation.

3.6.1 The combined transport operation

Besides meeting the technical specification requirements for vehicles as outlined above, the actual combined transport operation was also required to comply with the provisions set out in the regulations. In the first place, the vehicle, if it was a drawbar combination, needed to be carrying on both the drawing vehicle and the trailer a 'relevant receptacle', and if it was an articulated vehicle it had to be conveying a 'loading unit'. These terms were defined as follows: A relevant receptacle is a receptacle (i.e. container or swap body) at least 6.1-metre-long and constructed for repeated use for the carriage of goods on, and for transfer between, road vehicles and railway vehicles. A loading unit was defined as either a relevant receptacle as described above or a bi-modal vehicle, which is a semi-trailer that can be adapted for use as a rail vehicle (i.e. designed to mount directly onto a rail bogie) (see below).

To qualify for the weight concession a combined transport vehicle, as defined above, needed to be on a journey either to a railhead where the receptacle or bi-modal semi-trailer was to be transported by rail under a contract made prior to the commencement of the road journey, or from a railhead to which the receptacle or bi-modal semi-trailer had been transported by rail. A railhead in this context was defined as a place with a facility for the trans-shipment of bi-modal semi-trailers and/or receptacles between road and rail vehicles. While on a combined transport journey the vehicle driver was required to carry in the vehicle cab a document specifying the railhead to which he was destined, the date the contract for the rail movement was made, and the names of the parties to the contract. Conversely, if returning from a railhead, the document had to specify its location, and the date and time that the receptacles (or bi-modal semi-trailer) were collected from the railhead. There was no limit on the distance that may be travelled by a combined transport vehicle either to or from a railhead.

The regulations made two important points clear:

1. No goods were to be added to or removed from loading units during the road journey (i.e. between its beginning and end) prior to or following the rail sector of the operation.
2. These provisions (i.e. permitting 44-tonne road vehicle operation) did not extend to operation where the semi-trailer itself was to be, or had been, carried piggyback-style by rail, or where the whole vehicle was to be, or had been, carried on a rail vehicle (i.e. a rolling motorway type of operation). This clearly excluded any possibility of 44-tonne road vehicles using Eurotunnel's shuttle service through the Channel Tunnel.

From 1 January 1999, under the provisions of EU Directive 96/53, the facility to run at 44-tonne gross weight was extended to five-axle articulated vehicles in which the tractive unit had three axles and the semi-trailer just two axles. However, it is important to note that the Directive only covered such vehicles carrying 40-feet ISO shipping containers to or from a railhead as part of through-international journeys by rail and that appropriate documentation as described above had to be carried on the vehicle.

3.6.2 The 44-tonne vehicles for general haulage

The use of 44-tonne gross weight goods vehicles for general haulage operations has been permitted in the UK since 1 February 2001 under The Road Vehicles (Authorised Weights) (Amendment No. 1) Regulations 2000. Such vehicles are required to meet the following technical standards:

- Drive axle weights must not exceed 10 500 kilograms.
- Drive axles to be fitted with road-friendly suspension, or if not must not exceed 8500 kilograms.
- The trailer to be equipped with road-friendly suspension.
- Both tractive unit and semi-trailer must each have three axles.

- The vehicle engine to comply with at least the 'Euro II' emission standard or the vehicle to be fitted with a gas-powered engine.
- The distance between the coupling centre on the tractive unit and the centre of the rearmost axle on the semi-trailer must be at least 8.0 metres.

This welcome legislative change further encourages intermodal operations since it is no longer necessary to confine 44-tonne road intermodal haulage operations merely to rail-head journeys or to be able to prove the existence of pre-arranged movement contracts and carry relevant documentation.

3.7 Working time and fuel prices

Two further important factors are expected to have an increasingly encouraging impact on the growth of intermodalism in the UK. This prediction is confidently based on the fact that these two measures will both have an undoubted adverse impact on road haulage operations. The first such issue is legislative and concerns the likely effects of the EU's Working Time Directive (WTD) on the road haulage industry. Under this Directive, effective from 4 April 2005, road haulage drivers will fall within scope of the 48-hour maximum working week imposed by the RTD (Directive 2002/15/EC), basically an offshoot of the main WTD (Directive 93/1104/EC), and bringing within scope of the rules those vehicle drivers not previously covered by the Horizontal Amending Directive (Directive 2000/34/EC). Clearly these far-reaching rules will affect road transport operations in many ways, but in broad terms the principal effect will be to reduce the availability of suitably licensed heavy goods vehicle drivers (an estimated shortfall of some 80 000 drivers has been predicted by the transport trade associations and the trades unions) and to restrict the flexibility of road haulage operations, thus forcing more road hauliers to think increasingly in terms of finding alternative methods for delivering goods to their customers.

The second factor concerns the extortionately high cost of diesel road fuel in the UK, particularly when compared to the prices reigning in Europe, due to the fact that some 70 per cent of the price per litre goes in Government excise duty. High road fuel prices are a common factor among the various reasons cited by converted intermodalists when defining the justification for switching freight loads from road haulage onto rail or waterway. And with no foreseeable likelihood that there will be any change in official policy on this issue, one can only expect that it will remain a driving force towards increased intermodalism in the future.

4

Intermodal Transport in Europe

Mainland Europe could be called the heartland of freight intermodalism; principally because of the extensive use that is made of combined road–rail transport and inland and short-sea shipping (SSS) across the whole of the Continent, but particularly within the 25 Member States of the European Union (EU). With extensive inter-continental rail systems and a large network of commercially navigable inland waterways, Europe has the essential infrastructure necessary to facilitate the switch of freight from its heavily congested road systems on to these more environmentally friendly transport modes.

4.1 Euro-enlargement

By far the biggest development Europe-wise in recent times has been the enlargement of the EU which took effect from 1 May 2004 when the then 15 Member States increased to 25 (i.e. from EU15 to EU25 in Brussels' terminology) with the inclusion of Cyprus, the Czech Republic, Estonia, Hungary, Latvia, Lithuania, Malta, Poland, Slovakia, and Slovenia. These additional states, apart from Cyprus and Malta, neither of which have a terrestrial connection with mainland Europe, are mainly former communist countries from Eastern Europe, which with their combined population of 105-million people, are expected to significantly bolster intra-European trade and importantly boost the two-way flows of cross-border transport: a growth figure of 38–50 per cent has been suggested by Groupe Transport Combiné (GTC), a body within the International Union of Railways (UIC).

This additional traffic is likely to be predominantly road-borne in the early stages since these new Member States are unlikely as yet to have had the opportunity or encouragement (or indeed the perceived need) to develop intermodal systems on the same scale of sophistication as the rest of the EU. But with so many national boundaries opened and their customs controls now abolished the scope is enormous for road haulage and logistics firms in the East to develop links with their Western European counterparts. And similarly we may well see large numbers of EU firms seeking to establish routes, terminal connections, and logistics operations in the East, thus creating a reverse west-to-east flow of freight traffic – possibly on a scale to equal the anticipated east-to-west flows. Initially, these are likely to be predominantly road haulage operations, but with continued pressure from Brussels to reduce road traffic congestion and air pollution, and with financial aid available under the Marco Polo Programme (see Chapter 10) to help switch freight away from road-borne systems, a strong intermodal network of combined road–rail and inland waterway routes may be expected to develop quickly; in fact, this is one of the pre-conditions of EU membership. But there is still a lot of work to do; many road haulage fleets in the East need to be modernized and their operation aligned with EU legislation, particularly on such issues as mechanical condition and safety standards for heavy vehicles, and working times and tachograph controls for drivers; and rail freight systems need to become fully integrated in terms of locomotive power and braking systems, signalling equipment compatibility, crew training and safe operating standards, to identify just a few of the key issues which currently contribute to the lack of interoperability between existing rail networks.

It is anticipated that the addition of 10 new Member States to the EU is likely to boost freight carrying by some 358-million tonne-kilometres, most of it, as we have come to expect, being accounted for by road haulage which carries three-quarters of the total. Overall, freight carrying in the EU25 is expected to increase by about 2 per cent over the period 2002–2030 according to the Energy and Transport Directorate-General's 'Trends to 2030' publication (available on the Internet at: www.europa.eu.int/comm/dgs/energy_transport/figures/trends_2030/index_en.htm)

4.2 IRU/UIC position statement on combined transport

As we have seen in the preceding chapters much of the governmental and official drive towards intermodalism has emanated from within the corridors of Brussels, but not all of it.

While the powerhouse of Europe was determining policies for intermodal transport, commercial interests in the form of the International Road Transport Union (IRU), representing road haulage firms, and the UIC, representing national railway systems, declared a partnership in April 1991 to foster the development of combined road–rail transport in Europe in collaboration with international rail operator Intercontainer (since 1993 merged with Interfrigo to form Intercontainer-Interfrigo (ICF)) and the International Union of Road-Rail Transport (UIRR). In a statement that these organizations jointly presented to the European Transport Commissioner on 25 April 1991 – just 2 days after publication of the report by the so-called EC (European Commission) High Level Working Group on combined transport (see p. 21) – they expressed their wish to make a contribution to the process of the Commission's thinking and that of the Working Group. In this seven-page document entitled *The Conditions for the Development of Combined Transport in Europe*, the following key issues were raised and specific points made.

4.2.1 Needs of consignors

First, and of fundamental importance, was the conclusion that developments in combined transport must take account of current and foreseeable needs of freight consignors for both intra- and intermodal interchangeability by means of standardized, and if possible, modular transport units (i.e. swap bodies and containers) so as to reduce part loads and empty journeys. This was thought to be particularly crucial since three-quarters of road–rail traffic in most countries comprises swap body movements.

4.2.2 Freedom of choice

Freedom of choice for users of transport modes was seen by the IRU and UIC as being a key factor in any future development of the system. They said that the commercial vehicle operator must be able to choose between the various combined transport operating companies without any constraint from whatever source. Furthermore, they emphasised the need for the haulier to retain the right and freedom to maintain control of the transport operation from start to finish even if he entrusts his vehicle or its load to a combined transport operator. Clearly, this is a message to the effect that the intervention of national governments or even the bureaucratic channels in Brussels is neither required nor desirable beyond establishing the infrastructure.

4.2.3 No discrimination

A point was made in the report about the special position of the railway networks in the combined transport system and the need for them to respect a position of neutrality in that they will not (and in this the railway companies apparently agreed) discriminate in favour of certain types of transport by offering commercial advantages (i.e. by way of discounted freight rates) which would be discriminatory. The same degree of neutrality is required of haulage companies where the railway companies have marketed and secured traffic. The railways were also called upon to respect the anonymity of loads entrusted to them by road transport operators.

4.2.4 Equality of treatment

The document goes further in demanding equality of treatment. For example, the railway networks should accept that where road hauliers and forwarding agents have sufficient traffic to purchase traction for block trains without going through a combined transport operator, they should be allowed to be a direct client of the railway network in the same way as a combined transport operator. Moreover, as far as identical or equivalent railway services are concerned, the railway companies are committed to identical or equivalent tariffs for such traction.

4.2.5 Harmonized tariffs

The IRU and UIC, in fact, advocate a harmonization of tariff structures at both national and international levels – to be agreed between all the parties, the railway networks, the combined transport companies, road hauliers, and freight consignors – in the interests of providing long-lasting and stable relationships between the parties to protect their respective investments.

4.2.6 Harmonized conditions of liability

A major objective to be achieved according to the IRU/UIC is the harmonization at European level of the general conditions of liability in combined transport. To attain this objective the combined transport partners should, they say, draw up a standard combined road–rail transport contract, establishing properly balanced contractual relationships between operators and users, the form of which could be based on the United Nations layout. They further suggest that rules on liability should be adopted for combined road–rail transport companies based on the contractual responsibility of road transport undertakings (i.e. CMR – Convention Marchandises Routiers, the Convention on the Contract for the International Carriage of Goods by Road), and the CIM rules of COTIF (i.e. the Convention on the International Carriage of Goods by Rail) – discussed in more detail in Chapter 14.

4.2.7 Concentration on the swap body

In outlining its views on the conditions for the development of combined transport the IRU/UIC said that a first precondition was the avoidance of excessive dispersal of efforts to encourage techniques whose requirements and performance are very different. Instead, they said, stress should be laid on the use of the swap body, which, like containers, constitutes the technique preferred by both loaders and transport operators. They suggested that the development of such techniques as the moving highway (i.e. the rolling motorway where whole vehicles drive on to special rail wagons), on account of its (lack of) productivity, falls more within political logic than either commercial or technical logic aimed at responding to the needs of the transport market. With this thought in mind the IRU/UIC suggested that technical progress should be used in order to take full advantage of new opportunities such as those which permit the horizontal transfer of containers and swap bodies from road to rail by means of integrated devices (i.e. eliminating the need for gantry cranes and other complex, and capital intensive items of lifting equipment).

4.2.8 *Effective measures*

Among the measures that the IRU/UIC called for, if combined road–rail transport was to be fully effective, were the following:

- Increased international co-operation between railway networks, aimed at devising an overall strategy in both the technical and commercial fields.
- Creation of a pan-European combined transport network by adapting terminal capacities and the dimensions of railway lines to transport units.
- Stabilized tariffs.
- Harmonization of the conditions of liability.
- The optimal and harmonized adaptation of transport techniques to the characteristics of freight units and transloading techniques.
- Development of timetables adapted to users' needs.
- Improved journey time reliability.

4.2.9 *Less discrimination*

At the same time the document called upon European governments to cease regulatory discrimination hampering the development of combined transport. In many countries, international road–rail transport is not considered as international transport from start to finish, notwithstanding the fact that intermodal transport units are loaded under customs control. This anomaly results in the following problems:

- Requirements to apply tariff rules specific to each country, where they exist, for initial and terminal journeys, since these are considered as domestic journeys, which results in discrimination distorting competition.
- Requirements in certain countries for carriers who are responsible for the tractive element on approach journeys to hold domestic licences, giving rise to the same consequences as those mentioned above.
- Non-harmonized conditions of contract (liability systems), of fiscality (taxation and detaxation) and regulations (maximum weights and dimensions permitted for vehicles) in Europe. These conditions must be amended and harmonized to provide equality of treatment for all combined transport users.
- By divergent regulations on the maximum weights and dimensions for road vehicle combinations used for initial and terminal journeys.

4.2.10 *Optimization of weights*

The IRU/UIC opinion was that it is imperative to optimize the weights and dimensions of road vehicles, taking particular account of the maximum weights and dimensions permitted for standardized transport units (i.e. containers and swap bodies) by raising the maximum permissible weight of road vehicle combinations used for what are described as initial and terminal combined transport journeys to 44 tonnes in all European countries (a measure which was sanctioned in the UK from 1 January 1999 for vehicles solely engaged in combined transport operations – as mentioned above – and from 1 February 2001 for general haulage vehicles).

The reader should note that many of the provisions identified in the IRU/UIC document described above have in fact been adopted and implemented by various means to the overall benefit of intra-European intermodalism.

4.3 Research and further action

Such was the continued concern in Brussels to promote intermodal, and particularly combined road–rail transport that in 1995 a new research department was set up and an Intermodal Task Force was created.

First, regarding research; a new department called Development of Transport, Research and Development Policy was created within the Directorate-General VII (Transport – now the Directorate-General of Energy and Transport) of the EC on the initiative of the EU Commissioner for Transport and Trans-European Networks. Creation of this department was in line with the EU's research and development framework programmes comprising various specific programmes in several fields, including transport.

Currently, at the time of writing this book in 2004, we are within the so-called 'Sixth Framework Programme' which, among many planned research activities due to be carried out during the period 2002–2006, contains two specific intermodal tasks under the respective headings of: 'Intermodal Freight Transport Management' and 'Intermodal Freight Transport Systems, Technologies and Strategies'. In a call for proposals published in June 2004 and at a TREN-3 Info Day held in Brussels on 22 September 2004, the Directorate-General for Energy and Transport set out the principal objective to be achieved in regard to management systems which is to improve the management of intermodal, process-oriented logistics chains through integrated and interoperable information and management services. The first task identified was to improve the quality of alternative modes particularly in regard to their efficiency, reliability, and responsiveness, and for the 25 Member States of the EU (generally referred to as EU25) to have a common approach to such issues as organization, planning, monitoring, and communication within the alternative modes to road transport. The Directorate-General set out various challenges to be tackled of which managing information flow and integration of transport with other logistics tasks are a priority, as is the need to focus on the process of moving goods through supply chains rather than just on the mode or transport vehicle – meaning, it suggests, greater customer focus. It is also seeking to attach greater importance to such issues as security, energy consumption, and emissions aspects.

The Directorate-General's aims in regard to technologies and strategies are stated to be;

> to co-ordinate and disseminate European research in order to create synergies and to encourage a more wide spread uptake of research results and to stimulate further improvement and implementation.

In particular, it identifies the key topics to be included within scope of this task. These are to include: terminal management systems; trans-shipment technologies; information systems; vehicle technologies; loading units; the creation of inventories; performance indicators, benchmarks and quality standards; the identification of innovative concepts, best practices and business models; and finally comparison of the economical viability of solutions. According to information published on the EC's Europa Internet website (www.europa.eu.int/comm/transport), successful proposals for undertaking these tasks are due to start work in 2005.

4.4 The EU's Intermodality Task Force

Now to the Task Force concept mentioned above. In the mid-1990s a number of EU Commissioners combined their efforts and resources to create Task Forces, one of which was the Intermodality Task Force, set up in September 1995 to work towards improvement of intermodal transport operations. With Task Forces having previously been set up with, for example, such titles as Cars of Tomorrow, Trains and Railway Systems of the Future, and the Plane of the Future, the EC wanted to ensure that available resources were compatible with the multimodal transport networks and that transport systems being developed were in line with the objectives of the lasting mobility policy. Overall, the aim of this Task Force was to promote relations between the organisations involved in intermodal transport.

In an up-date of the Task Force's activities published on the Internet on the Community Research and Development Information Service (CORDIS – www.cordis.lu/transport) website, it was stated that from the time it was created in 1995, an effective and continuous co-operation developed between all Commission services concerned and that a common understanding had been fostered of the problems and challenges involved in developing intermodal transport in Europe. A so-called 'Diagnosis Report' produced in March 1996 by the Task Force highlighted the modal imbalance of the EU's transport system and identified the obstacles that were preventing the development of user-oriented door-to-door intermodal transport services. The report confirmed the need to focus future Research and Technological Development (RTD) activities on six priority themes of major interest for the improvement of intermodal transport in the EU, namely:

- Transfer point efficiency;
- Intermodal network efficiency;
- Information and communication technologies;
- Improvement of transport means and transport equipment;
- Market conditions;
- Training and market-oriented strategies.

The Task Force also made an inventory of all past and on-going intermodal-related RTD projects at the EU level as well as some projects at Member State level and in a report on this issue it concluded that a gap existed between research and the implementation of its results and that innovations only find their way into the market at a slow pace. Following consultations by some 200 technical experts in six working panels, the Task Force drafted its final report in September 1996 in which it defined a number of priority RTD actions, not all concerned with freight intermodalism since passenger transport also featured in the Task Force's deliberations.

4.5 Rail interoperability

One of the major operational developments in intermodal Europe has been the important progress in open-access rail freight operations. An efficient Euro-wide rail freight network is obviously the key to increased intermodalism and with the developments that have taken place over recent years we are beginning to see, in the early 2000s, the emergence of a truly 'European Railway'. The first steps were taken in 1991 with the adoption of EC Directive 91/440/EEC (the so-called 'Railway Directive') which introduced, by way of a first package of measures, a degree of liberalism into railway operations and prompted the railways to develop a more competitive stance. In effect, the Directive made railway companies more independent of the nation States that owned them; in other words, enabling them to begin freeing themselves from the shackles of state ownership, especially in their accounting and budgeting systems. It also guaranteed rights of access to rail transport operators in other Member States to international combined transport services.

A second major package of railway revitalization measures was launched in January 2002 with the objectives of developing a common approach to rail safety; bolstering the principles of technical interoperability between separate National rail systems; setting up a new steering body for European railways – the European Railway Agency; speeding up the opening of the entire rail freight market – intended by 2006; and by acceding to the Convention on the Intergovernmental Organisation for International Carriage by Rail (COTIF), the body which makes regulations on such matters as the carriage of dangerous goods by rail. The EC adopted a third regulatory railway package on 3 March 2004, but this does not concern us in this book because it deals with passenger rail transport issues.

However, a momentous date in European rail history occurred on 15 March 2003 when, as a result of the first of the so-called 'Railway Package' of 1991 (as mentioned above) intended to open up access to the international rail freight market, the first new pan-European commercial rail freight operations were launched. Basically, this meant that from this key date national, state-owned, railway companies were to be in open competition with private, commercial, rail operators a number of which were, in fact, already operating on trial bases. While it has to be acknowledged that not all of these new 'open-access' operations were, or indeed still are, purely intermodal services, many of them will carry combined road–rail traffic, and therefore, with the opening up of the European rail freight market in this way so too will cross-border and Euro-wide intermodal traffic develop and expand. The reader will find that Chapter 7 of this book examines the subject of rail freight developments and combined road–rail transportation in greater detail.

4.6 Inland waterways

Europe is renowned for its inland waterway network of great rivers and man-made canals which, for centuries, have carried significant volumes of trade goods across the Continent as well as providing extensive leisure facilities and a great deal of hydroelectric power. Nowadays, with greater emphasis than ever before on the concept of removing freight traffic from an increasingly congested road system and switching it to more environmentally acceptable transport modes, such as inland waterway barges, the waterways of Europe are becoming major intermodal arteries. According to Europa web site data (www.europa.eu.int), more than 35 000 kilometres of waterways connect hundreds of cities and industrial regions across Europe. Eighteen out of the 25 Member States have inland waterways of their own, and 10 of these have an interconnected waterway network with other Member States. Currently Europa shows the modal share of river transport to account for 7 per cent of the total inland transportation in the EU, which in 2003, amounted to some 125-billion tonne-kilometres of freight. Waterborne transport plays a vital role in transport throughout Europe, but particularly so in the Northwest where, within the hinterland of the largest seaports of the EU, the modal share of inland waterway transport can reach as high as 43 per cent.

In development terms, vessels of newer design, greater sophistication and expanded carrying capacity play their part with many of these being specially built to accommodate intermodal freight units ranging from euro-pallets to maximum capacity International Standard Organisation (ISO) containers and swap bodies. Routes too have been developing, none more so than the Rhine-Main-Danube Canal, linking these three great European rivers, which was finally completed in 1992 to form a 3500-kilometre (2200-mile) long inland waterway link along which freight barges, and leisure traffic, can travel from the North Sea right through to the Black Sea passing through no less than 15 mainland European countries en-route. Even within in its first year of opening it was estimated that approximately 4.7-million tonnes of goods traffic was moved along the canal in a virtually silent, environmentally friendly manner; and on a long-term basis it is predicted that traffic to and fro between the Black Sea and Western Europe could experience an enormous boom.

4.7 Current activities: an overview

In September 2004, the EC through its Directorate-General Unit G3 (Motorways of the Sea and Intermodality) published a 'Power Point' presentation entitled, 'An Overview of Current Activities', in which it summarized intermodal developments; first by defining why freight transport is so important: precisely because inefficient freight transport leads to higher costs which reduces the international competitiveness of Europe's goods; and because it is a major part of the economy and a key service to the rest of European industry. The presentation highlighted the growth in freight volumes between 1990 and 2002, which showed that these increased faster than overall economic activity and that over the period 1970–1999 the balance of modal carrying changed significantly; road increased from 31 per cent to 45 per cent, SSS increased from 35 to 40 per cent, while rail diminished from 21 to just 8 per cent, inland waterways from 8 to 4 per cent and pipelines from 5 to 3 per cent. And, the EC predicts, heavy goods traffic will increase by more than 60 per cent by 2013, most of it being carried by road which will, it says, 'remain the backbone of inland surface transport.'

4.7.1 Motorways of the Sea

Gloomy though these predictions about increases in road freighting will be to the pro-environmental bodies, at least the EC's message on the principles of the 'Motorways of the Sea' should bring some relief. As it says in the Power Point presentation mentioned earlier, the unique geography of the EU with its widespread accessibility by sea brings both significant opportunities as well as significant challenges, particularly to reduce congestion on roads and improve access to peripheral and island regions by greater use of the SSS mode. The Motorways of the Sea project is in fact one of the Trans-European-Transport Networks (TEN-T) funded projects (see Chapters 8 and 10 for more information) and for these purposes four key shipping development regions have been identified as follows:

1. Baltic Sea, linking the Baltic States (Estonia, Latvia, and Lithuania) with Member States in Central and Western Europe including the route through the North Sea and the Baltic Sea canal.
2. Western Europe, leading from Portugal and Spain via the Atlantic Arc to the North Sea and the Irish Sea.
3. South east Europe, connecting the Adriatic Sea to the Ionian Sea and the Eastern Mediterranean, including Cyprus.
4. South west Europe, where the Western Mediterranean connects Spain, France, Italy (including Malta) and links with the South east route and the Black Sea.

Completion of all of these development routes is scheduled for 2010 with the funding being available from 2005. In the meantime, work is going ahead on developing integrated links between these regions and existing inland corridors.

4.7.2 Intermodal loading units

Part of the EC's overview concerned the problems associated with the different types of intermodal loading units (ILUs) in use; principally, swap bodies, ISO containers and semi-trailers. This multiplicity of equipment results in complicated handling procedures in terminals, a lack of common maintenance or inspection procedures and no common identification system; plus other fundamental problems such as swap bodies being non-stackable and unable to resist horizontal forces in SSS. The ideal solution according to the EC is the 'one box for all modes' concept which, it suggests, would save 2 per cent in logistics costs through better equipment utilization and integration, and up to 10 per cent reduction in transport costs, depending on the commodities carried and the corridors concerned.

While the EC recognizes that existing ILUs need to be used until the end of their working life – and in the meantime need to be inspected once within 5 years and then at intervals of every 30 months – with new ILUs certain common features need to be adopted; not least compatibility with road, rail, inland waterway, and SSS modes, stackability to at least four units high in sea conditions, top lifting capability, and internal dimensions optimized for euro-pallet loading. The optimum solution is seen to be the so-called 'Euro-Intermodal Loading Unit' (EILU), in other words, the 45-foot (13.716-metre) ISO container with internal dimensions of 2.4 metres × 13.2 metres that will allow for loading 33 euro-pallets (i.e. europalettes in French). These proposals for standardization of ILUs were first set out in a Commission COM document (155/2003) in April 2003 and are due to enter into force, after EU Parliamentary process, by the end of 2006 (see also Chapter 13).

4.8 Financial support

The EC is most anxious to support modal switch of freight, via grant aid funding of qualifying projects, principally from road to other more sustainable and environmentally friendly modes such as rail and waterway. (This topic was briefly mentioned in Chapter 3 and is covered in greater detail in Chapter 10.)

In fact, it has been doing so for some years past; initially under the Pilot Actions for Combined Transport (PACT) programme which, in the years 1996–2001, it supported with an injection of some €35 million (i.e. 35 million euros) for innovative mode-shift actions only; and subsequently under the Marco Polo programmes. The first of these, which runs from 2003 to 2006, is budgeted to aid modal shift projects, catalyst actions and common learning programmes to the tune of some €102 million. This is due to be followed by Marco Polo II, initially proposed in July 2004 and destined to run from 2007 to 2013 with a projected budget of some €740 million.

The PACT programme of 1996–2001 was deemed to be a considerable success, especially in terms of the reduction in CO_2 emissions, and hopes are high in Brussels for further modal shift from road freight to alternative modes with proportionate reductions in CO_2 emissions. But, of course, account needs to be taken of the fact that the EU now comprises 10 extra Member States from which calls for funding will inevitably arise, which may not have the same degree of success in this direction (i.e. emissions reduction). For instance, under the rules for Marco Polo II, subsidies for so-called 'Actions' will primarily be forthcoming for robust projects which merely shift freight off the road. Supported 'Catalyst Actions' will be those that help to overcome structural market barriers and which are of a highly innovative nature causing a real break through; while funding for 'Common Learning Actions' is intended to support co-operation and sharing of know-how, and the setting up of mutual training programmes designed to cope with increasing complexities in transport and logistics. None of these programmes, it will be noted, having a direct aim to further reduce CO_2 emissions, for example.

Finally, in its Overview the EC outlined progress with its 'Freight Transport Management Systems (FTMS)' concept. These systems focus largely on such issues as the different standards, investments, languages, and cultures encountered in European transport and the challenges of integration between transport and other logistics tasks and between alternative transport modes, all with the primary objective of increasing quality and efficiency.

4.9 Operational developments

While intermodal policy-making activity has been buzzing away within the corridors of Brussels and elsewhere among the thinkers and deciders, as we have seen above, on an operational level Europe has been benefiting equally from practical developments in much the same way as the UK, as described in Chapter 3. These developments include such technical measures as the introduction of 44-tonne lorries capable of carrying maximum weight swap bodies and ISO containers, and new-design articulated semi-trailers which can carry, within legal limits, the 45-foot long shipping containers now predominantly favoured in maritime circles.

4.10 The way forward

It will be clearly seen from this chapter that, besides any commercial or market pressures which may be driving forward the intermodal transport revolution, considerable will power is being applied from within the corridors of Brussels and elsewhere to ensure development of the right networks and infrastructures for such systems to operate effectively on a pan-European basis. National governments for their part too are willing progress in this direction. To give just two examples: firstly, the UK government finally conceded (as the Chapter 6 shows) that heavier lorries are a necessity to enable combined transport systems to develop and expand on an efficient basis, so it enacted appropriate legislation to allow 44-tonne vehicle operation. And secondly, in France, a government-commissioned report into combined road–rail transport recommended the creation of a national agency to co-ordinate development of intermodal traffic in that country; the establishment of a network of combi-hubs; and, presumably with Brussels' approval, the injection of a degree of state aid.

While combined transport accounts for only about 4 per cent of freight traffic movements now and may only grow, according to various estimates, to about 10 per cent by the year 2015, it is still seen as sufficiently significant in terms of its ability to staunch the growth in road freighting and the environmental problems it engenders for both the central powerhouse in Brussels and individual national governments to consider implementing unpopular legislation and investing substantial aid to further its development.

Legislation in other forms is also likely to have a consequential impact on the future growth of intermodalism; namely, the introduction of the EC's Working Time Directive (Directive 93/1104/EC) and particularly its off-shoot, the Road Transport Directive (Directive 2002/15/EC), and the reinstatement of the German LKW-Maut motorway tolling system from 2005. Both of these 'direct-hit' measures on heavy vehicle operations will inevitably drive shippers to seek alternatives modes by which to move their freight traffic; the first because it will reduce available working hours for heavy vehicle drivers and ultimately lead to severe shortages of skilled driving staff and reduce vehicle availability, and the second because as other EU Member States follow Germany's lead, as they inevitably will do so (the UK, for example, is planning to introduce its own motorway tolling system – the highly controversial lorry road user charge (LRUC) – by 2007–2008), the increased costs and potential delays at tolling stations will cause goods shippers to think that there must be better ways of consigning freight than by road transport.

5

Intermodalism in North America and World Markets

Intermodalism is virtually a worldwide concept; it is widely practised in North America, in the Baltics, parts of the Middle East, Asia, and Australasia – in fact in most world trading markets – very much as we have come to know it in Europe. But the basic ingredients and the degree of sophistication in the equipment used and the operating procedures adopted tend to differ according to region. Basically, Europe is alone in being strong on swap body and piggyback operation as well as containerization while the rest of the world – apart from North America, which also strongly favours piggybacking of road trailers on rail wagons – focuses predominantly on the shipping container; in other words containerization. In fact, the most common scenario in most developed territories is the high level of dependence on this ubiquitous item of transportation hardware. From its origins in the USA in the 1950s, the container can now be seen almost everywhere, in many different guises. Statistics from UNCTAD (the United Nations Conference on Trade and Development, www.unctad.org) indicate that in 2002 containers totalling about 240 million teu (i.e. a 20-foot equivalent unit) were transhipped worldwide, with about 127 million teu being handled by just 20 of the world's largest container ports, six of which are in the Pacific–Asia region underlining the importance of this area as the manufacturing centre of the global economy.

However, containers have many uses other than merely fostering global trade; besides being an intermodal loading unit (ILU) in international trans-ocean shipping, for which it was originally conceived, discarded units have been adapted to a multitude of other uses; for example, they are ideal for making secure garden and site storage sheds, many provide warm and dry animal shelters on farms and small-holdings; and in third-world countries some people even live in them.

5.1 North America

As is customary in many spheres of technological development here in Europe and the UK we are inclined to look at progress being made in similar fields across the Atlantic in North America, even if the intention is not necessarily to follow their lead. This applies as much in transport operations as in other fields of industrial activity, and the case of intermodalism is no exception. The container revolution is just one such example. It is reputed to have started in the USA in 1956 when a converted tanker ship sailed from Newark, New Jersey to Houston, Texas loaded with 58 trailer vans on its specially adapted decks. This new concept was reputedly the brainchild of US trucker Malcolm McLean, founder of Sea-Land Services, one of the world's largest container shipping lines. Sea-Land's first trans-Atlantic container ship docked in Rotterdam in 1966 starting a trend in deep-sea services that now cover the globe. But, by this time, steel containers built to internationally agreed standard (ISO) dimensions were coming into common usage in the UK and Europe.

However, this is only one part of the US intermodal story. The switching of containers between rail and waterway has a history dating back to at least the middle of the nineteenth century and by the 1920s intermodal container services in the US were regularly operating by road. The first recorded piggyback trailer operation commenced in 1926. Meanwhile, here in the UK wooden containers, as mentioned in Chapter 1 of this book, were being interchanged between horse-drawn road vehicles and rail wagons before the start of the twentieth century.

This shows signs of parallel development of intermodalism on a common theme albeit, obviously, on a larger scale in the USA. But where the North American experience clearly benefits is from its huge landmass, which offers great potential for long-haul intra-continental operations. The economics of this scenario make good sense. Rail freighting a whole train-load of containers hundreds, if not thousands, of miles across the continent *en bloc* compared with road-hauling them all the way as individual unit loads, each one requiring a separate truck and driver, is obviously much more efficient and cost effective, even taking account of the additional costs and delays inherent in transferring unit loads from road to rail and then back again at the end of the rail journey. However, while this potential was being progressively exploited in the US by the development of long-distance piggyback services with road-going semi-trailers being carried on flat railcars, there were problems between the 1920s and the 1950s arising from the fierce competition between US truckers and the rail companies for long-haul freight. This inhibited co-operation between the two modes and tended to restrict the growth of combined road–rail services. So too did regulatory controls which complicated the calculation of freight rates in general and hindered the application of through-rates for combined road–rail services.

But by the mid-1950s, with a softening of the Interstate Commerce Commission's (ICC's) regulatory controls (in response to the requirements of the US Transportation Act 1940 for the development of a national transportation system), the establishment of structured tariff systems, and clearly delineated roles for all the individual parties to combined transport operations, namely freight shippers, the railway companies, private and contract truck operators, and freight forwarders, this form of transportation really began to expand. Further transportation deregulation in the USA in the late 1970s (as has happened in Europe from 1993) has since resulted in the opening of new markets and new transportation opportunities, particularly for the intermodal sector.

In his very detailed book *Intermodal Freight Transport* (3rd Edition 1995, published jointly by the Intermodal Association of North America (IANA) and the ENO Transportation Foundation Inc., Lansdowne, Virginia, USA), American expert Gerhardt Muller outlines the history and development of this mode of transportation in the USA. He names three particular companies that were largely responsible for developing piggyback road-rail services in the 1950s and 1960s. These were Railway Express Agency (now defunct), Flexi-Van and Trailer Train (now TTX Company). The latter, in 1994, owned 36 000 intermodal railcars capable of carrying some 125 000 semi-trailers or containers, and expected that number to grow to over 150 000 by 1998. It is anticipated that, as a whole, industry growth in the USA will double in volume every 10 years. Muller's book clearly shows the very wide acceptance of intermodal operations and the advanced state of sophistication with which this industry operates in his country, even to the operation of double-stack railcars on which containers can be carried two high, well within the recommended 23-foot national clearance (gauge) standard. However, while much of the equipment and many of the intermodal operations he describes for his own country are paralleled in the UK and Europe, one of the significant differences is in the favour given to loading units.

While, on this side of the Atlantic, we have moved towards widespread acceptance of standard dimension swap bodies, their use being very actively promoted here, in the USA they are rare, a situation that Muller says is not likely to change in the next decade. Instead the American industry favours the use of semi-trailers of varying lengths for piggyback operations (being carried either on flat railcars – there being no height restrictions on many routes – or on specially-built spine cars), bimodal semi-trailers such as the RoadRailer (for a detailed description of this system see Chapter 13), a semi-trailer which converts to ride on rail bogies, and conventional ISO-type containers – the latter progressively moving ahead so far as intermodal rail carryings are concerned.

5.1.1 *Trucks and trailers*

Another anomaly between the USA and European experience is the wide variety of road vehicle configurations, axle numbers, overall dimensions, and gross weights between individual American States. However, while, for example, Member States of the European Union (EU) tend to have varying domestic maximum vehicle weight limits (e.g. contrast the UK's normal 40 tonnes, or 44 tonnes under certain conditions, on five or six axles and the Netherlands's 50-tonne limit for the same type of vehicle), most feature the EU's recommended standard limit of 40 tonnes for five- and six-axle articulated vehicles and road trains, and a permitted 44-tonne maximum weight limit for combined transport vehicles. Overall length and width dimensions are standardized under EU rules (i.e. 16.5- and 18.75-metre maximum lengths for articulated vehicles and road train combinations respectively and a standard 2.55-metre maximum width). Most European countries (but not the UK) are standardized on a 4-metre height limit, which conveniently suits road-rail operations within European rail loading gauge limitations.

The American trucking industry has a basic vehicle fleet which mirrors that of the UK and Europe, with conventional rigid and articulated vehicles at varying lengths and maximum weights. But additionally, it utilizes vehicle configurations that are not permitted here, for example a 'double' or 'double bottom' combination comprising an articulated tractive unit pulling two semi-trailers, the second one being mounted on a single-axle steerable bogie (a 'towing dolly' in UK parlance). This provides two 28-foot load-platform lengths. In some US States, an extended version of this type of combination allows the carriage of 45- or 48-foot-long containers on the front semi-trailer and a 20-footer on the second trailer while an even longer 'Turnpike Double' carries two 45- or 48-foot containers. Some states permit a triple combination (again based on an articulated tractive unit and linked semi-trailers) that can carry three 20-foot containers.

5.1.2 *Rail vehicles*

Two notable developments in combined road–rail technology which have crossed the Atlantic and are now being adapted to UK/European-style operations are the bimodal trailer (e.g. the Wabash 'RoadRailer') and a more recent development on similar lines, Canadian National's 3R International system, and the 'spine' type rail wagon. The former is extensively used in the USA today. The Chesapeake & Ohio railway company originally developed it in the late 1950s by fitting a pair of steel rail wheels to a standard road semi-trailer. The design has gone through a number of major changes since then, the most significant being that rail wheels are no longer carried on the semi-trailer. The latest version has the semi-trailer resting on free standing rail bogies, one at each end, with the road-wheel bogie being pneumatically raised for rail track operation. For road haulage at either end of the trunk haul the road-wheel bogie is lowered, the semi-trailer is hitched to a tractive unit and is driven away, a transfer that takes literally minutes only and requires no additional equipment or facility. Presently, in the US, up to 75 of these RoadRailer semi-trailer units can be linked to form a train; that number may be increased in due course to up to 125 units.

The Thrall Car Manufacturing Company's so-called spine wagon (spine car in US terminology) is based on the concept of linking front and rear rail bogies by a rigid steel box beam (the spine) with outrigger platforms to support the semi-trailer bogie of two- or three-axles. These wagons are claimed to offer reduced weight, length, and cost over existing wagon types. A number of these special wagons may be linked together to form a four-segment train utilizing just five bogies (one at each end and three intermediates) to provide full articulation. Such a train can carry four semi-trailers, one on each section, or a combination of two semi-trailers and two 40/45-foot ISO containers.

Muller states that these cars are available in large numbers in the USA where they are used in five sectors to carry five semi-trailers or containers of all sizes. Prototype versions of this concept been built and tested in the UK and Europe by a Thrall subsidiary under the trademark name of EuroSpine. The advantage in UK operation is that the low loaded height of these wagons and their narrow profile (since their main structural member is in the centre, passing under the semi-trailer axles, compared with standard piggyback wagons in which the semi-trailer bogie sits inside the main wagon frame) allows the carriage of standard 4-metre high semi-trailers within Network Rail's limited bridge, tunnel, and platform loading gauge.

5.1.3 The political scene

On the political front, as in the UK and within the EU, the US Federal Government is anxious to promote the concept of an efficient, energy saving, and environmentally sound transportation policy. In fact, in the 1980s, its Department of Transport identified six key needs to be addressed as follows:

- To maintain and expand the nation's transportation system.
- To support a sound financial base for all transportation modes.
- To keep the transportation industry strong and competitive.
- To ensure that the transportation system supports public safety and national security.
- To protect the environment and the quality of life.
- To advance US transportation technology and expertise for the twenty-first century.

In 1991, the US Congress passed the Intermodal Surface Transportation Efficiency Act (ISTEA) in which the current transportation policy of the USA is clearly stated:

> It is the Policy of the USA to develop a National Intermodal Transportation System that is economically efficient, environmentally sound, provides the foundation for the nation to compete in the global economy and move people and goods in an energy efficient manner. The National Intermodal Transportation System shall consist of all forms of transportation in a unified, interconnected manner including the transportation systems of the future, to reduce energy consumption and air pollution while promoting economic development and supporting the nation's pre-eminent position in international commerce …

A clearer message to promote the concept that transport modes must work together and integrate their services and technologies in the interests of efficiency, energy conservation and the environmental good could not be given – whether in the USA or in the UK and Europe.

Muller says that the enactment of the ISTEA Act mentioned above and its implementation was the beginning of a revolution in transportation in the USA. In particular, it has resulted in the development of the US highway system comprising, today, some 3.9 million miles of road of which 155 000 miles are major roads and 44 000 miles are Interstate highways which link the USA from coast to coast. Planning guidelines have been clearly set so that transportation decisions cannot be made in isolation of national, state and regional concerns. Management systems monitor the effectiveness of the nation's transportation system, especially in regard to maintaining the efficiency of its investment in infrastructure, and in identifying deficiencies in the transport network and rectifying them to gain more capacity and provide greater efficiency for the mobility of passengers and goods.

Among all these ISTEA provisions, the issue of air quality has received special attention, with the US 1990 Clean Air Act requiring the transportation community to provide improved mobility while still adhering to national clean air goals, and, in areas where established air-quality standards have not been achieved, calling for reduction in the growth of vehicle miles travelled and congestion levels. Funding to the extent of some $155 billion was authorized (starting from 1992) to support the ISTEA provisions.

We can perhaps thank the North American experience for at least showing us part of the way in combined road–rail operations and bringing us some of its technology. Clearly, in a country where land space for terminal development is not at the premium it is in the UK and most of Europe, where greater demands for long-haul freighting have provided the incentives for seeking intermodal initiatives, where there are not the restrictions arising from an old-established limited gauge rail infrastructure, and where, most of all, the railways operate in a free enterprise commercial environment, it is to be expected that the USA should be ahead in its exploitation of the whole range of intermodal possibilities, and combined road–rail transportation in particular. But Europe has been catching up fast; combined road–rail freight volumes have increased, the technology to suit our particular operational needs is developing, and both shipper and public opinion is turning towards the view that intermodalism is *the* transport solution for the future.

However, to show that not everything moves in a west-to-east direction across the Atlantic in terms of technological know-how, North America had its own tunnel 'engineering feat' of the 1990s. The 1-mile long St Clair Tunnel under the St Clair River, linking Ontario in Canada and Michigan in the US, was constructed using, apparently, similar design techniques to those employed in building the Channel Tunnel between England and France. This new tunnel was intended to streamline the annual flow of some 275 000 freight cars and provide a high capacity corridor for double-stack container traffic between the two markets.

Yet another area of intermodalism where we are showing America the way forward is in short-sea shipping (SSS). In a speech to a Maritime Administration Conference in New York in October 2004, where he was promoting a SSS initiative, Charles G. Raymond, CEO, Chairman and President of Horizon Lines, America's largest domestic ocean carrier recommended, firstly, the need for shippers to develop a trilogy of water, rail, and highway systems that complement rather than compete with each other. Mr Raymond then went on to say that SSS entails using existing vessels to move freight between coastal ports and also between coastal ports and inland ports as a means of reducing congestion on America's highways and rail systems. The SSS plan, he said, envisions waterways as a complementary mode to traditional modes of inland freight movement. He also suggested that in the longer term, SSS would require new, technologically advanced vessels and infrastructures to meet capacity demands, to help revitalize the American maritime and shipbuilding industries. He added that the environmental benefits included reducing emissions and energy use. In a reference to SSS in Europe he noted that this transportation mode had been used to mitigate significant surface transportation problems for more than 10 years, and that in excess of 44 per cent of present freight movements in the EU are waterborne and that EU policy makers have placed short-sea and coastal shipping, in close co-ordination with rail and highway freight improvements, at the top of their transportation agendas. In Europe, Mr Raymond said, what is called 'the motorways of the sea' is already a vital part of the transportation system, clearly a growth industry and the only transportation mode that has kept pace with the growth of road transportation.

5.1.4 Expansion of trade

As an indicator of the ever-growing expansion of North American container trade, the international freight press, which reports regularly about North American intermodalism and containerization, has frequently relied on one key word – congestion. It appears that US ports are congested and US road networks are also congested. For example, writer, James Graham in the 20 October 2003 issue of *International Freighting Weekly* (*IFW*) suggested that:

> it is unlikely that sufficient highway capacity can be built to handle future volume growth over the next two decades.

He believes intermodalism to be the answer; 'it is more cost effective and environmentally desirable than maintaining the country's present over reliance on the highways', he says.

One year later, in October 2004, *IFW's* Leigh Stoner was reporting that peak season traffic from Asia was being diverted from congested US west coast ports such as Los Angeles-Long Beach (the main US intermodal gateway) via the Panama Canal to east coast ports such as New York and New Jersey, and further south, the ports of Charleston (Georgia) and Portsmouth (Virginia).

Finally, if proof was needed of North America's growing trend towards intermodalism, a look at the statistics published on the Internet by IANA at www.intermodal.org show that for 2003 (the most recent figures available) a total of 11.9 million intermodal units (i.e. containers and semi-trailers) were transported by rail, representing an increase of about 1 million units (6.4 per cent) over the corresponding 2002 figure. International containers totalled 6 470 080 movements (54.3 per cent of the total) while semi-trailers totalled 2 400 558 movements, representing 20.1 per cent of total movements – a 2.4 per cent increase over 2002.

5.2 Canada

Canada, like the USA, is heavily dependent on containerization and piggyback intermodalism. By way of example, an interesting report in the December 2003/January 2004 issue of *CILT World* (the World-wide Newsletter of the Chartered Institute of Logistics and Transport) showed how a Canada–mid-west USA cross-border rail link is carrying large volumes of lorry trailers piggyback-style – much on the lines once envisioned here in the UK. Canadian Pacific Railway, following 3 years of trials, launched its 'Expressway' scheme in 2002; it runs regular services between Montreal, Toronto, Windsor (Ontario), and Detroit in the USA. The hardware statistics are impressive; trains are up to 1.8-kilometre long and are hauled by three or four diesel locomotives at 100 kilometres per hour. At the heart of this operation is the inland Port of Montreal, 1000 miles along the St Lawrence River and serving as a gateway for trade to and from mid-west USA. It handles more than 1 million containers annually. Much of the popularity of the trailer-only piggyback route is attributable to the shortage of truck drivers in both Canada and the USA, a problem increasingly being faced in the UK and Europe – only two shunter drivers are needed, one at each end. The service helps to overcome another equipment problem; namely, that it allows for continued utilization of Canada's 3 million plus unstrengthened road semi-trailers still in circulation since there is no lifting requirement in the loading or unloading operation; its all done by conventional shunting with conventional dockside tractors.

5.3 The Baltic States

It is easy to think of Europe and the EU purely in terms of the central and southern parts of the continent, completely forgetting that since 1 May 2004 EU territory extends virtually as far north of Brussels as it does southwards, and encompasses the key Baltic States of Estonia, Latvia, and Lithuania, as well as Poland. The importance of this region lies in its role as a transit corridor for shipments to and from Finland and Russia, as well as being a consumer market in its own right. Substantial numbers of containers are shipped through the ports of these States and reports are, in fact, of increasing numbers, especially through the Estonian port of Tallin, which virtually doubled its container throughput between 2000 and 2003 with nearly 100 000 teu being handled. These regions are reputed to be among the fastest growing in the world with a forecast growth of some 6 per cent in gross domestic product (GDP) compared with that for the rest of Western Europe of only some 2 per cent. The point here is that growth means demand, and consumer demand means transport demand and this means potential growth in intermodalism. A report in the weekly freight trade journal IFW in December 2004 shows, for example, that freight traffic accounts for about 90 per cent of Lithuania's national railway income, much of which is transit cargo for the Russian Federation and Belarus – amounting to some 48.4 million tonnes in 2003 and likely to have been well exceeded in 2004. In the case of Latvia, EU grants amounting to € 150 million have helped the country invest in its rail infrastructure between 1998–2004. The railway reports growing demand for container trains for neighbouring markets such as Belarus and Kazakhstan, which use Latvia as a gateway.

While Latvia transportation is very much rail oriented, Poland on the other hand, which is considered to be a rapidly growing market since acceding to membership of the EU from 1 May 2004, conversely, is heavily dependent on road haulage with little likelihood of any redressing of the balance with rail. This is principally because; firstly, two major new road projects are currently under construction to improve the east-to-west link across the country (funded by the EU) and to provide a new north-south highway from the port of Gdansk down to the Czech Republic; and secondly, because modal choice for transit freight is largely determined by shippers located in eastern European States who are seen to have a vested interest in trucking freight by road rather than shipping it by rail.

5.4 Asia

In shipping terms, Asia is home to six of the world's busiest container ports. In fact, they vie with each other to see which is the busiest (*Note*: World rankings are quoted in terms of the throughput of 20 teu containers (i.e. 20-foot equivalent units) rather than by the physical size of the infrastructure). Hong Kong currently holds the number one ranking with a throughput of more than 18 million teu – vastly outranking the throughput performances of both Felixstowe and Rotterdam, for example, but Singapore (around 17 m teu), currently in second place, Shanghai (11 m teu) in third place and the Korean port of Busan (9 m teu) in fourth are catching up fast – Felixstowe scrapes in at 20th place in the rankings and Rotterdam is 7th.

To give some indication of the sheer size of these Asian ports, and the volumes of international shipping sailing back and forth across the world's oceans, Singapore claims to have about 1000 ships in port at any one time and to be the focal point for some 400 shipping lines linking the port to more than 700 other ports in 130 countries worldwide. As I have said before, not all of this trade is container traffic, but much of it is, and 'boxed freight' – to use a colloquialism, is the growing trend for reasons of security, speed, and efficiency of handling and economy of both dockside and ship-hold space. And, Hong Kong, for example, predicts future container through-put growth on a scale of 30 million teu by 2010 (from the 2000 base figure of 18 million teu) and 40 million teu by 2020. To accommodate this virtual explosion in trade, the port has built a new container terminal (CT 9), which was due to become fully operational by 2004 and it has been reviewing its strategy for coping with the situation when that reaches its capacity.

Shanghai port, too, is growing at an unprecedented rate to match China's 'scorching economic expansion'. The port's ocean trade is being moved away from the shallow Yangtze River to a deep-water island site located 31 kilometres off the mainland coast and is being transformed, with a $20-billion project, into what may become the world's largest container port by 2020 (note the competing claims with Hong Kong!). New double-deck expressways have been built to connect the islands of the Qiqu archipelago – and a 32-kilometre-long causeway is due to connect the island terminal to the mainland where a new multimodal transport hub is under construction as well as a new town and a new 42-kilometre-long highway to speed the flow of freight into the hinterland.

Such is the growth of Far Eastern shipping that a report in IFW on 12 July 2004 highlighted the worries of shippers who are concerned that demand on Asia-Europe routes is outstripping available capacity and that shipyards cannot produce new vessels fast enough. In the meantime, vessels are being switched from other routes to cope with this trade boom, which is reputed to be returning 100 per cent utilization figures on westbound sailings to Europe. Fortunately, the east-to-west trade imbalance is reported to be narrowing as Asia, and especially China, increases the volume of its imports from European markets.

5.5 The Middle East

Middle Eastern ports such as, for example, those run by the Dubai and the Sharjah Ports Authorities are seeing increased container throughputs. Dubai now in fact, claims to be ranked the world's 11th busiest container port operator, regularly serving more than 100 shipping lines. In terms of container numbers, even smaller ports like Sharjah reported a throughput of some 2 million teu in 2003/2004 while the Dubai ports of Jebel Ali and Port Rashid were reported to have handled around 6.4 million teu in 2004. While these numbers are not large by the standards of the world's major container ports, they are significant because they show year-on-year increases reflecting the continuing boom in Middle East trade and the development of the Gulf region as, reputedly, the world's largest warehouse. Jebel Ali, the busiest port in the Gulf region, has a so-called 'Freezone' that allows shippers to store goods 'tax free' under bonded conditions until they are required for onward distribution – many well-known European and worldwide companies make use of this facility. The port has become, in effect, a major trans-shipment hub for east–west container traffic. The Dubai Port Authority is in the process of developing its operations even further in 2004/2005 with a four-phase expansion project to construct new wharfs and container yards and to deepen the present docks to handle the latest mega-container carriers in the 9000/10 000 teu range – the world's largest. By 2020, the port expects to handle 21.8 million containers annually.

5.6 Australia

Like many other countries outside of Europe and the US, intermodalism in Australia predominantly means containers on trains for the long haul and on trucks to and from collection and delivery locations and major seaports. It is here that the advantage of intermodalism lies more in such widely cited benefits as door-to-door cargo movements on a single bill of lading, and a single transportation company providing everything from consolidation services and liability coverage to inland distribution, than in the environmental pluses of reduced air pollution and traffic congestion that we are seeking in Europe.

As with many other markets, Australia depends heavily on long-distance road transport for its internal freight movements, especially for traffics other than bulk commodities such as coal, ore, other minerals, and grain which are predominantly rail borne. But there are exceptions; on the 3400-kilometre route east-to-west from Melbourne to Perth, for example, rail carries some 70 per cent of all freight. Similarly, between Brisbane on the east coast and Cairns in the north as much as 40 per cent of freight is carried on the narrow gauge rail line. Otherwise rail freighting accounts for only about 15–25 per cent of general traffics – a figure that is reputedly falling even further as road transport efficiency increases while rail-freight developments fall behind. A notable problem in this respect, as pointed out by consultant John Hearsch in his Paper: *Australia's Rail Transport Policy – Towards a Sustainable Future*, presented to UIR seminar, New Delhi, India, in 1998, is that some of the origin and destination points for long-distance general freight cannot be competitively serviced by rail because the rail network does not directly parallel major highways that link these points by the shortest practicable route. Hearsch cites by way of example the Melbourne–Brisbane corridor on which rail has a less than 20 per cent share of the freight traffic due to the fact that the current 1930-kilometre circuitous coastal rail route via Sydney is almost 250 kilometres longer and much more congested (i.e. with passenger traffic) than the competing inland road corridor. 'Consequently', he says, 'effective door-to-door transit times on the corridor are almost 24 hours longer by rail than by direct road'.

As we have seen above, containerization is the predominant intermodal method used in Australia and most general traffics are conveyed by this means, with container trains operating between dedicated terminals located, or being developed, in most major conurbations. Normally, container movements are arranged by rail-based forwarding agents who deal directly with the rail company, or by market-oriented rail operators themselves who act as the primary contractor and hire-in road haulage for the local collection and delivery operations (referred to as 'drayage' – a term commonly used in Australia, but much less so it seems, if at all, in the UK or Europe).

A draft governmental 15-year transport plan for South Australia, *South Australia's Approach to Intermodalism*, presented by Ian Lovell, Freight Transport Co-ordinator, Department of Transport and Urban Planning, Government of South Australia, at the Ausintermodal Conference, 23–24 September 2003, Sydney, Australia, determined among its various targets, that freight transport was to be made more efficient, that 75 per cent of interstate freight should move by rail and sea, and that intermodal interchanges had to be made efficient and effective. In fact, a Strategic Intermodal Terminal Plan has been drawn up as a key initiative to co-ordinate and support the establishment of intermodal terminals which is seen as being a private sector responsibility while the Government's role is to facilitate the establishment of terminals and to ensure that they are located at strategic sites that fit in with the State's economic development, transport and land use plans. Attention is being particularly focused on developing innovative systems to reduce the time and present cumbersome procedures for transferring cargo (basically, containerized freight at present, but potentially road-railer and swap body systems in the future) between road and rail modes. There is concern too that regulatory barriers should not inhibit intermodal efficiency. In fact, modal and intermodal standards and regulations for the weights and dimensions of containers and other ILUs have been identified as one means of improving intermodal efficiency. And, in a glance towards the west, the South Australia Government noted that the EU has been pressing ahead with its ILU Directive (see p. 32) – and it is worried too that once fully implemented the standards defined by this Directive may compromise its own trading efficiency.

6

The Road Haulage Role in Intermodalism

Road haulage is a vital constituent in virtually all intermodal movements. In fact, very little cargo moves intermodally without, at sometime being carried on the back of a lorry, either before being transferred onto a rail wagon, a waterway barge or a coastal ship, or after shipment by one of these alternative modes; the so-called 'initial' and 'final' legs of an intermodal journey. It is by far the most predominant modal choice for freight movements, accounting, across the whole of Europe, for some 75 per cent of the total freight moved – on a weight basis alone it accounts for about 82 per cent of freight lifted; the difference in the figures is that when travel distance is brought into the equation, freight on rail, while less in total weight, tends to travel greater distances, apart from coal and other quarried and bulk materials which invariably travel a very short distance.

Road haulage is also a complex business, being burdened with a large number of restrictive governmental directives and regulations. Nevertheless, in terms of its role in intermodalism it offers, on the plus side, a degree of convenience and flexibility unmatched by any other mode. The principal tools of its trade, (heavy lorries), can go virtually anywhere, by comparison with other modes, to load or deliver goods. Lorries can be de spatched at very short notice to destinations either near or far, being reasonably assured that these will inevitably be served by a road, even if not by rail or waterway. Also, the operator or the customer, or both, can give precise instruction about their delivery requirements directly to the lorry driver in the reasonable belief that he will carry them out to the best of his ability. In many instances, this is seen as a considerable plus point in favour of lorry transport.

But there is also a downside. Generally, for a start, people do not like heavy lorries – 'juggernauts', they are often disparagingly called – and bemoan their presence in congested town centres, on narrow urban roads and on motorways. This is despite a recent study by Professor Alan McKinnon of the Heriot-Watt University, Edinburgh, *Bigger is better when it comes to road freight* reported in *The Scotsman*, 10 January 2005, which shows that the introduction of the heaviest, 44-tonne, lorries in 2001 has actually helped in the quest to reduce traffic congestion and air pollution. According to the report, this legislative measure has resulted in 34 per cent fewer lorry journeys and 36 per cent fewer carbon dioxide emissions than had been expected (i.e. 134 million vehicle-kilometres against 100 million projected kilometre savings and 136 000 tonnes of carbon dioxide emissions saved against 100 000 tonnes projected). The study also predicted that such benefits might further increase by some 27 per cent by 2006/2007. The influential Freight Transport Association agreed wholeheartedly with McKinnon's conclusions. Spokesman Geoff Dossetter was reported as saying; 'we are moving more goods in fewer vehicles … but no-one has noticed.'

People (i.e. the general public) also question why more goods (all goods even) do not go by rail and, in the extreme, why roads cannot be turned into railways, little realizing that were this to occur the local supermarket would not be able to stock their favourite brands of breakfast cereal, no more than the local garden centre or do it yourself (DIY) store would have on their shelves the goods customers are looking for on a Saturday morning shopping trip. As we have already seen (p. 59), even the European Commission (EC), which is a very keen proponent of intermodalism (i.e. getting freight off the roads, and on to rail and inland waterways) admits that road transport will remain the backbone of inland surface transport. However, a growing number of trucks on our roads, along with increasing private car and light van use, is putting a great strain on the road networks in the UK and Europe, causing daily traffic jams affecting an estimated 10 per cent, or 7500 kilometres, of the European Union's (EU's) motorway network. Road congestion also leads to vehicle accidents which across Europe are said to result in 41 000 deaths annually, a figure that the EC is desperate to reduce by half by the year 2010 according to its 2001 White Paper on transport; *European Transport Policy for 2010: Time to Decide*. It should be noted that not all of these road deaths are attributable to heavy lorries, although precise Euro-wide statistical data on the breakdown of lorry involvement in road accidents is not currently available.

The full legal burden on road transport operations is far too extensive to explain in depth here; neither it is appropriate because road haulage is not the *be all* and *end all* of intermodalism. Any reader wishing to gain a full understanding of transport law is recommended to consult the author's work on the subject, *The Transport Manager's and Operator's Handbook* published annually by Kogan Page Limited, London. However, it is useful to consider here just a few of the vital issues that feature prominently on the road haulage fleet manager's agenda, particularly when he is contemplating the intermodal conundrum – for example, issues such as maximum vehicle dimensions and weights; operator licensing, community authorizations, and professional competence; exhaust emissions, energy consumption, and noise limits; limits on driver working times; safety law in regard to carrying containers and working in docks; and proposed measures for tolling lorry use of motorways; namely, the dreaded Lorry Road-User Charge (LRUC).

6.1 Lorry sizes and weights for intermodal operations

For a start, we should consider the physical constraints on road haulage vehicles of the type that are normally encountered in intermodal operations; namely their maximum size and gross weight as defined in EC Directive 96/53/EEC and for Great Britain (GB) are defined in both *The Road Vehicles (Construction and Use) Regulations 1986* and *The Road Vehicles (Authorized Weight) Regulations 1998*. Most road vehicles used in intermodal operations are either articulated combinations comprising a tractive (i.e. motive) unit and a semi-trailer, or are road trains comprising a load-carrying rigid vehicle towing a drawbar trailer (alternatively referred to as a drawbar combination, or colloquially in old English haulage parlance as a 'wagon and drag'). There are important operational differences between these two vehicle types, but the most significant in terms of their role in intermodalism lies in the length of the loading platform, a vital dimension in relation to carrying containers and swap bodies. Maximum-length articulated vehicles at 16.5 metres usually have a load platform length of 12.2 metres on which a standard 40-foot International Standards Organization (ISO) container or a 13.6-metre swap body can be loaded, or a 45-foot container which has Geest-designed corner castings to allow for the necessary swing clearance defined by regulations (see also p. 164). Alternatively, such vehicles can carry two 20-foot ISO containers. The drawbar vehicle combination, on the other hand, can manage only two 20-foot ISO containers or two 7.15 metres long swap bodies within its 18.75 metres maximum overall length despite having a combined maximum platform length potential between the drawing vehicle and the trailer of 15.65 metres. A common width limit of 2.55 metres applies to all heavy vehicles except those with refrigerated bodies designed for carrying controlled-temperature products which may be 2.6 metres wide. Europe has a 4-metre maximum height limit for all vehicles, but the UK has no such limit.

The 44-tonne maximum weight limit for heavy road vehicles has already been mentioned a number of times and it would generally be a vehicle of this weight capability that would be used in an intermodal operations. However, in circumstances where the load density permitted (i.e. such as with high-volume, low-weight loads like products made from plastics and polystyrene, for example) vehicles of lower maximum permitted weights (e.g. 38- or 40-tonne gross weight, or even less) may be suitable without infringing the law on maximum weight limits.

6.2 Operator licensing, community authorizations, and professional competence

Road haulage operations in the UK are controlled by a system of 'quality' licensing which requires hauliers to prove their ability to maintain and operate their vehicles safely and within the law, and be able to show that they are of 'good repute' meaning that they must not have previous convictions for road transport related offences. Broadly similar provisions apply in Europe, albeit the words are different. The system in the GB is called 'Operator' ('O') licensing. There are two main categories of licence applying to professional road hauliers who carry goods for hire and reward; namely, standard national licences for GB domestic operations only and standard national and international licences for haulage firms operating internationally (i.e. beyond the country in which their business is registered). Firms that carry goods only in connection with their own trade or business, such as manufacturers or suppliers, who run their own vehicle fleets require only a restricted licence. International 'O' licence holders in GB are automatically supplied with a document known as a 'Community Authorization', which conforms to EU legal requirements and puts them on the same basis as road hauliers in other EU Member States. Certified copies of this document must be carried on relevant vehicles when they are travelling on international journeys.

Another crucial aspect of operator licensing is that 'O' licence holders in the UK and Community Authorization holders in Europe must be 'professionally competent'. This qualification applies personally to the licence holder if the business is registered in their name and the licence or authority is held personally (i.e. if it is not a corporate entity) or, where the licence is held by a corporate entity, the company must employ a person who is professionally competent. This provision, implemented under EC legislation (Directive 98/76/EC), requires the individual (the scheme only applies to individuals) to qualify either by examination or by past experience to prove their knowledge of road transport law and safe operating procedures.

6.3 Exhaust emissions, noise limits, and energy consumption

Heavy lorries (trucks in common terminology) are notoriously persecuted by public opinion because of their reputation for emitting noxious exhaust fumes, making noise, and creating undue vibration, especially in confined urban streets, and for their energy consumption. But it is important to recognize that in all of these aspects today's lorries are a vast improvement on those that populated our roads only a few years ago. Driven by legislative pressure from government and the constant griping of conservationist groups the latest products from heavy truck manufacturers are surprisingly quiet, fume and vibration free; and they are very energy efficient. Driving cabs in top-of-the-range models are akin to saloon cars in terms of their equipment, appointments (e.g. air conditioning, compact disc (CD) players, bunk beds, etc.) and comfort, even their in-cab quietness (a recent press article said drivers of some of the latest top-of-the-range heavy vehicles are complaining that they are so quiet inside they cannot judge their gear changing effectively) making the driver's job a great deal easier and much less stressful than in the old days.

Relevant legislation on vehicle exhaust emissions, noise and even vibration is extensive and extremely complex, but the reader may find it helpful to understand the following outlines. Firstly, so far as emissions are concerned, both UK and EU law is relevant; this is to be found in the UK's *Road Vehicles (Construction and Use) Regulations 1986*, as amended, and in EU directives 91/542/EEC (of 1991) and 99/66/EEC (of 1999) which respectively specify what are commonly referred to as Euro I and II, and Euro III, IV and V standards. Euro IV is due to come into force in 2005 to be followed by Euro V which will apply from 2008. The net effect of these ever tightening standards is to achieve overall reductions in the four main poisonous constituents of diesel vehicle exhausts; namely, nitrogen oxides (NO_x), particulates (i.e. particulate matter, PM), hydrocarbons (HC), and carbon monoxide (CO). In the case of nitrogen oxides and particulates, truck manufacturer Volvo reports that emissions of these substances from new trucks over the past 20 years have dropped by 75 per cent and 85 per cent, respectively. For instance, the Euro IV standard which comes into effect in 2005 is intended to reduce NO_x emissions from the present level of 5 grams/kilowatt hour to 3.5 grams/kilowatt hour and by 2008 the Euro V standard will have reduced this further to just 2.0 grams/kilowatt hour. Clearly these stringent controls on emission standards are to the public benefit, but while we are actually seeking to reduce road transport use by switching more freight to alternative modes, where road haulage remains the preferred or the essential mode at least we (the public) can rest assured that everything possible is being done, especially by the heavy truck makers, to clean up the air we breathe.

Vehicle noise (i.e. external noise as opposed to in-cab noise mentioned above), too, is subject to strict legislative control. However, here we are dealing with a much more subjective topic than exhaust emissions, for example, which are specifically controlled by legislation. Noise in this context is not just the vehicle's engine exhaust note, which invariably with heavy trucks is only a minor constituent of the whole noise package, provided the silencer is properly maintained, but it includes the noise made by the tyres on the road surface (quite a considerable element) and the vehicle bodywork and equipment which again can be considerable in the whole scheme of things, especially if the driver does not properly secure container doors, securing chains or even the load itself. There is controlling legislation under the vehicle construction and use provisions mentioned above, which makes it an offence to use a vehicle on a road that causes excessive noise which could have been reasonably avoided by the driver. Noise standards set out in EU Directive 92/97/EEC (of 1992) apply to vehicles made on or after 1 October 1995 and first used on or after 1 October 1996. For post-October 1995 goods vehicles in the category with which we are concerned in this book the maximum noise level in decibels (written as dB(A)) is limited to 78 dB(A) for vehicles with an engine power output between 75–150 kilo Watts, and 80 dB(A) for vehicles with engines exceeding 150 kilo Watts.

Measurement of energy consumption for heavy lorries is quite simply a matter of diesel fuel consumption in miles per gallon (mpg) in imperial measure or, in metric measure, in litres per 100 kilometres. Generally, as may be expected, the heavier the vehicle the greater the fuel consumption and similarly, the higher the engine power the greater the fuel consumption; thus a 44-tonne articulated vehicle with an engine producing, typically, some 500 horse power (bhp) would, on average, be expected to return a fuel consumption figure of around 8.0–8.5 miles per gallon which in metric measure equates to approximately 35–33 litres/100 kilometres.

6.4 Limits on driver working times

One of the major constraints on road haulage operations is the legal restriction on lorry drivers' working times. There are, in fact, two quite separate pieces of legislation for the operator to contend with; first, the new set of rules introduced by the EU's Road Transport Directive (Directive 2002/15/EC) commonly referred to as the RTD, which became effective on 4 April 2005; and second, the long-standing EU drivers' hours law which has been in existence since 1986 under the provisions of EC Regulation 3820/85/EEC and which remains in force despite overlapping the RTD. Both of these sets of rules have to be observed in full making the driver's life a nightmare when it comes to working out what hours he has left to drive and work, when he must take a statutory break and when his rest periods are due. It is, of course, right and proper that heavy vehicle drivers should be restricted in the hours they are allowed to work for reasons of safety, both of the drivers themselves and of other road users. A heavy lorry is a potentially lethal weapon when in the charge of a tired driver or, indeed, one who is over the limit on drink or drugs.

The RTD, which may be described an off shoot from the main EC Working Time Directive (Directive 93/104/EC) which applies to most workers in Europe, deals specifically with working time provisions for those persons performing 'mobile transport activities' (mainly LGV drivers) – within scope of the EU drivers' hours rules (mentioned above), but with the exception of self-employed mobile workers (such as owner–driver road hauliers) who remain exempt from the Directive until 23 March 2009. The most important, and one of the most heatedly debated provisions of these working time directives is the restriction to a weekly maximum of 48 hours' work, with a maximum of 60-hours work allowable in any 1 week. These weekly totals may be averaged over a 4-month (i.e. 17 weeks) reference period, which can be extended by agreement between employer and employee to 6 months. The total working time is calculated excluding breaks during the working day and waiting time (i.e. 'periods of availability'), but unlike the main working time directive, the RTD does not offer an 'opt out' from the maximum working time. There are special provisions in the Directive relating to break periods, periods of availability, otherwise known as waiting time, night work, worker's holiday entitlements (i.e. at least 4 weeks paid holiday), the need for employers to keep detailed records and the rights of night workers to be given health assessments by a qualified health professional, and even a medical examination if necessary.

Self-employed workers are exempted from inclusion in the RTD until 23 March 2009, as already mentioned. However, to qualify as self-employed these drivers must be genuinely self-employed in accordance with both RTD and Inland Revenue rules on self-employment meaning:

- having as their main occupation the transport of goods by road for hire or reward under an 'O' licence;
- being entitled to work for themselves;
- not being tied to an employer under an employment contract or other hierarchical working relationship;
- having the freedom to organize their own work;
- their income must derive from the profits of their business;
- they must have the freedom, either alone or in co-operation with others, to work for more than one customer.

The limitation on LGV drivers' working times and the self-employed provisions included in the Directive could have a significant long-term bearing on intermodal transport operations. Firstly, because the reduced working time measure is virtually crucifying the road haulage industry which for years has depended on the willingness of truck drivers to work and drive for the maximum that the EU law allows (60–65 hours weekly depending on circumstances). This means a 20–26 per cent reduction on average in driver's available working time and, by simple calculation, a similar reduction in the numbers of LGV drivers likely to be available in the future. The transport trade press, the trade associations, and the trades unions have bandied around various numbers for the likely LGV driver shortages, but a figure of 80 000 seems to be a reasonable and commonly accepted 'guestimate' for the shortfall. The second significantly potential effect of the EU directives is that changing work practices such as reduced driver working hours, and the obvious increase in costs to the transport operator, could very likely lead to firms hiving off their vehicles and employed driving staff to a freelance, self-employment, status so they can continue to operate for the firm but on the exempt owner-driver's available 60–65 hours weekly basis rather than the restricted 48-hour employed driver's basis.

In intermodal terms, these significant changes could prove beneficial because firms that operate their own delivery vehicle fleets will, in the future, undoubtedly seek to save the costs, inconvenience and lost time of vehicles being away on long-haul journeys when they could so easily switch this traffic to alternative modes. Indeed, in practical terms, with fewer drivers available, firms may just not be able to man vehicles for two, three, or more day journeys away from base; better to keep their own drivers for the local collections and deliveries, and seek intermodal possibilities for the long haul.

6.5 Safety law for carrying containers and working in docks

The carriage of ISO containers and swap bodies can be accomplished very safely if legal provisions and correct procedures are followed. Otherwise, as past horror stories have shown, great danger can arise, especially if such units are not properly secured on vehicles. Equally, lorry drivers are at great risk if they do not follow the rules laid down explicitly to protect their safety. Generally, ISO containers should not be carried on vehicles unless specifically designed securing twist locks are fitted on the vehicle and/or trailer and are secured to the container by inter-locking into the corner castings. For maximum safety at least four twist locks should be used (one at each corner) and these should be fully engaged. Regrettably, one often sees vehicles carrying containers where the twist locks are not properly engaged, the driver merely relying on the weight of the container to hold it in place over the twist locks; a dangerous practice adopted solely on the grounds of laziness and disregard for safe practices.

Container owners and those who lease or are in control of freight containers must, in fact, ensure that they comply with the *International Convention for Safe Containers* agreed in Geneva in 1972. *The Freight Containers (Safety Convention) Regulations 1984*, which is the legislation that relates to this issue, applies to containers; 'designed to facilitate the transport of goods by one or more modes of transport without intermediate reloading, designed to be secured or readily handled or both, having corner fittings for these purposes and which have top corner fittings and a bottom area of at least 7 square metres or, if they do not have top corner fittings, a bottom area of at least 14 square metres'.

Containers in the UK that fall within scope of these legal provisions must have a valid approval issued by the Health and Safety Executive (HSE) or a body appointed by the HSE (or under the authority of a foreign government which has acceded to the Safety Convention mentioned above) for the purpose of confirming that they meet specified standards of design and construction and should be fitted with a safety approval plate to this effect. If they are marked with their gross weight such marking must be consistent with the maximum operating gross weight shown on the safety approval plate. The law also requires that containers must be maintained in an efficient state, in efficient working order and in good repair. Details of the arrangements for the approval of containers in GB are set out in a document *Arrangements in GB for the Approval of Containers*, available from the HSE. The safety approval plate (issued by the HSE) as described in the regulations must be permanently fitted to the container where it is clearly visible and not capable of being easily damaged and it must show: the date of manufacture of the container; its identification number; its maximum gross weight in kilogrammes (kg) and pounds (lbs); its allowable stacking weight; and its Racking Test Load Value.

Also originally published in 1972, like the container safety regulations, and now available in its third edition, the UK Department for Transport (D*f*T) Code of Practice (CoP), the *Safety of Loads on Vehicles*, recommends that ISO containers should only be carried on vehicles fitted with twist locks and these must be maintained in a serviceable condition. The CoP warns that:

> Unlike normal box type loads that spread their weight over a large area, containers are designed to stand on the twist lock sockets or feet that protrude down at each corner. With heavy containers this produces high point loading that could over-stress a normal platform floor. Other platform vehicles may have raised or wide section side ledges, which would prevent the container from resting on the platform floor. The resultant interface between the side ledges and the container feet would offer little frictional resistance making it virtually impossible to secure the container on to the vehicle safely and the practice should be avoided.

Where containers are carried on a vehicle not fitted with twist locks, the Code advises that a safe and secure retention system must be used such as that set out in the Code. (As a general word of warning; where twist locks are not available chains should be used to secure a container through the corner casting apertures, but never ropes.)

In the case of swap bodies, which the Code also classes as a type of container, it suggests that these may be fitted with special attachment brackets or lashing rings which should be secured as shown in the Code. In particular it warns against allowing containers (i.e. swap bodies) to project beyond the rear or sides of the vehicle load platform because permanent distortion of the container may take place if part of its base is left unsupported. It says that lashings or other securing devices should only be attached to those points on the container intended for the purpose or for lifting or for mechanical handling when laden, such as lashing rings or special brackets.

6.6 Safety in docks

There are strict rules concerning the activities of lorry drivers when they are collecting or delivering containers to or from the docks. The *Docks Regulations 1988* made under the authority of the *Health and Safety at Work etc Act 1974*, specify that when goods vehicle drivers work in or visit docks premises including roll-on/roll-off ferry ports they must be provided with high-visibility clothing to be worn when they leave the vehicle cab. The clothing may take the form of fluorescent jackets, waistcoats, belts or sashes and must be worn at all times when out of the cab on such premises including when on the vehicle decks of the ferry. Protective headgear (hard hats) must be supplied and worn in such areas where there is likely to be danger of falling objects from above (e.g. where cranes are working). Drivers must leave the vehicle cab when parked on a straddle-carrier grid or where containers are being lifted on to or off the vehicle.

6.7 Lorry Road User Charging: LRUC

It has long been held by the anti-lorry brigade, other environmentalist groups and even government itself that lorries do not pay their way in road taxes in relationship to the amount of wear they are alleged to cause to our roads. There are many contentious aspects to this issue for which there is insufficient space to debate here; neither it is appropriate to do so because intermodal lorries, with which we are solely concerned here constitute barely a miniscule proportion of the total number of all vehicles on our roads. However, it is important to recognize that the tax to be charged for heavy lorries using strategic routes is a key issue.

In the UK, the long-standing vehicle excise duty (VED) system is to be supplemented – modernized, the government calls it – by a controversial system of lorry charging that is due to be introduced in 2007–2008. The so-called LRUC, is currently being developed by Her Majesty's Revenue and Customs (HMRC) so that both British and foreign hauliers pay an amount related to the distance that they travel on UK roads, to ensure that they contribute towards the road-wear costs they impose in the UK. Overall, the LRUC is intended to be 'revenue neutral' for hauliers who already contribute taxes through UK fuel duty. The charge will apply to all heavy goods vehicles over 3.5 tonnes that travel in the UK but not to cars or light vehicles. This will mean that vehicles will have to be fitted with an On-Board Unit (OBU) that will identify and record the vehicle location both when the ignition is switched on and when it is switched off. Roadside equipment will monitor the vehicle's travel progress and record number plate details with Automatic Number Plate Recognition (ANPR) software and there will also be mobile equipment that detects vehicle payment status to ensure full compliance. The toll charge, payable to a new Directorate established with HMCE, may vary according to lorry size, road type and possibly time of day, but there will be offsetting tax cuts through a reduction in hauliers' fuel duty which will be repaid with their LRUC bill.

Germany introduced its own LKW-Maut heavy lorry tolling system in January 2005 after two false starts in 2003 due to technological hiccups, and this is a scheme that is reputed to be less complex than the proposed UK scheme. The Maut system applies to trucks above 12-tonne gross weight which are fitted with an OBU that automatically identifies its location via satellite-based technology when joining and leaving motorways. The toll due to be paid is based on the distance travelled and this information is transmitted to the collection agency, Toll Collect, which issues the registered operator with a monthly invoice. Manual payments may be made at the toll terminals or via the Internet on a per journey basis. Tolls are charged at a rate of between €0.09 and €0.14 per kilometre (i.e. between approximately 6 pence and 10 pence in UK currency GBP at present rates of exchange; namely, £1.00 = €1.42641 as at 14 January 2005) depending on the number of axles they have and the emissions category they fall into (i.e. Euro I, II, III, or IV). Despite the best will of the UK Government, and other national governments, these tolling schemes are certain to add to road hauliers' operating costs, as well as adding to their burdens of paperwork and technological headaches to deal with. They are also bound to prove to be yet another motivation for transferring freight from road vehicles onto rail or waterway systems for the long haul, not surprisingly, just what the UK Government and the EC wants.

6.8 Road traffic accidents

A reduction in road accidents is cited as one of the likely beneficial, and very welcome, consequences of shifting more freight off the road and on to alternative transport modes. It is an unfortunate fact of life that, despite the fact that modern heavy vehicles are designed and equipped with safety matters and road accident risks very much in mind, and the best possible training of drivers, awareness campaigns and safety propaganda, road accidents do happen and heavy lorries do become involved in, or cause, a proportion of them; environmentalist campaign groups would have us believe that lorries are the evil motivator of many more accidents than statistical data shows to be the case. In fact, in the UK, heavy lorry accidents accounts for only 8 per cent of collisions on trunk roads although, unfortunately, they do, account for 26 per cent of all trunk road casualties. This is principally because they are large and take a greater distance to stop in an emergency and not necessarily because they are ill maintained or carelessly driven. But nowadays, heavy lorries are much more efficiently maintained than hitherto (under a strict legislative and law enforcement regime) and most are fitted with anti-lock braking systems (ABS), which contribute significantly to a reduction in accident rates. However, in spite of progress towards road accident reductions the EC still sees the overall situation as a disaster from the human, social, and economic points of view. The pre-May 2004 15 Member States of the EU report more than 40 000 road deaths every year and the fact that 1.7 million people are injured annually in road accidents, at a total cost estimated at €160 billion per year. And further, according to the EC, the situation is even worse in the 10 new Member States that acceded to the EU in May 2004 where about 12 000 more fatalities occur yearly. But, if positive measures such as the greater application of intermodalism can help to reduce this dreadful toll of human life, pain and suffering then the switch of traffic from road to alternative modes is very well justified.

6.9 Road haulage operations

It is a fact that most general goods begin the journey to their destination on the back of a lorry either loaded directly into the vehicle itself, or into an ISO container or swap body which is then loaded on to a road vehicle, depending on the forwarding arrangements made by the shipper. The initial choice of transport mode may be that of the shipper or they may have contracted the load movement to a road haulier who will plan the onward movement to the destination. For this the haulier may use his own vehicle for the full journey or tranship the unit load on to rail or waterway for the trunk haul and then arrange for a local road haulier at the destination to deliver the consignment to the customer. Conversely, the shipper may deal directly with a rail freight operator who will arrange the necessary road haulage links at either end. There are other intermediaries in the transport chain, such as freight forwarders who operate no vehicles of their own but who contract as necessary with the relevant intermodal operators on behalf of the shipper. The point is that the procedures for arranging intermodal shipments are simple whether via direct or indirect contact with the transport providers. In this respect there should be no inhibitions or obstacles to making the crucial decision as to how freight is shipped; namely, by the road haulage route or by more environmental friendly intermodal modes.

It may to be interesting to the reader to know that many UK road hauliers have established highly successful reputations in various aspects of intermodalism, but for our purposes here just two names in particular are cited by way of example. These firms are widely known in the intermodal context and both have been mentioned previously; they are the Potter Group and the Malcolm Group. The Potter Group claims to be the leading independent UK road and rail-freight logistics service provider, offering, among many related activities, a series of intermodal services through its connected network of distribution centres. The company's main office and base is at Ely in Cambridgeshire and it has locations at Selby in the North and Knowsley in the NorthWest which provide road and rail transfer points for conventional rail wagons, swap bodies, containers, and ISO tanks. The distribution centres are connected to the international rail network via the Channel Tunnel and are capable of receiving both full trainloads and single wagons. The centres are equipped with shunting locomotives, intermodal handling equipment, bulk discharge facilities and covered areas for unloading high-capacity freight wagons. A fleet of modern 44 tonnes, Euro II road vehicles is maintained at each distribution centre to provide customers with flexible and cost-effective transportation services.

The Scottish-based Malcolm Group, a long-established road haulage company, operates fast daily and overnight road–rail logistics services in conjunction with rail freight operator Direct Rail Services (DRS) from the Daventry International Rail Freight Terminal (DIRFT) at Crick, Northamptonshire to Mossend, near Glasgow, and from Grangemouth (Scotland) to Linwood, Aberdeen and Crick. The company operates around 450 heavy trucks, 1000 trailers and has extensive warehousing facilities with which it serves many 'blue chip' companies, including supermarket chain Asda, via its rail freight links with DRS.

Freightliner, too, although principally a major rail freight operator, has a fleet of over 180 heavy vehicles and 500 trailers, which it uses for collection and delivery of containers carried on its rail services. Well-known logistics group, Transport Development Group (TDG) also operates intermodal services, principally for BP's Grangemeouth plant. Some of the major supermarkets also have their own delivery vehicle fleets as well as using intermodal road–rail services where it best suits the economics of their operations. Well-known names, such as Asda, Argos, Marks and Spencer, Safeway, and Superdrug are among those that have found great benefit in the efficiency, reliability and speed of rail services, achieved mainly as a result of being able to avoid the constrictions of reduced truck driver working hours and the notorious bog downs of road traffic congestion which destroy the most carefully calculated delivery schedules.

7
Rail-Freight Operations

Mm Gudrun Winner-Athens, Managing Director of German forwarding company Winner Spedition, a Board Member of intermodal operator Kombiverkehr, and a well-known and respected figure in European intermodal circles, talking in 2003 about an intermodal vision for the future which was reported in the Swiss published *International Transport Journal*, expressed very clearly the advantages and disadvantage of trains in a manner that the writer cannot do better than record here. She said:

> The railways have three intrinsic advantages. Firstly, they are very competitive with large uniform loads over long distances, secondly they can reach a high speed and thirdly the railway system is planable and reliable. However, this threefold advantage is seldom realised in practice.

She pointed out that even in its core market, that is long haul, the railways only have a market share of about 5 per cent. The average length of international routes served by private European goods train operators is 800–850 kilometres. Two-thirds of this total is generated on transalpine routes, and not because of the efficiency of the rail-based services, but because of extraneous factors (i.e. topography, and fiscal and regulatory measures in Switzerland and Austria, etc.). Furthermore, she went on to explain that intermodal trains are not particularly fast. Even in the case of premium quality block trains or shuttles, mean speeds, including halts en route, are only 45–55 kilometres per hour (kph) and that is without including the time between the handover of the load units at the point of departure to their handover to the consignee at the other end. 'In comparison', she said:

> trucks manage average speeds of between 60 and 70 kilometres/hour, depending on the route and time of day. It is this difference that counts against the railways. As far as reliability is concerned, Europe's railways have really left the intermodal operators out in the cold in recent years with miserable on-time arrival rates of 40–50 per cent. Today one is relieved when punctuality rates of 90 per cent are achieved by Kombi-Netz 2000+ services and with rates of 80–90 per cent on the Brenner Pass route – thanks to intermodal operators themselves seizing the initiative with a few well-managed services. In contrast, road hauliers are achieving punctuality rates of 99 per cent and more, day in, day out.

7.1 Britain's privatized railway

A new era for Britain's railways was heralded in 1996 when privatization brought considerable change and particularly so in the sphere of freight operations, which have undergone a tremendous transformation. In place of the old nationalized British Rail (BR) network, freight operations are currently in the hands of five independent operating companies (i.e. 'Freight Operating Companies', FOCs), namely English, Welsh, and Scottish Railway (EWS), Freightliner, Great Britain Railfreight (GBRf), Direct Rail Services (DRS), and Advenza Freight each of which contracts individually with Network Rail for use of the national rail infrastructure, for which they pay agreed access charges. Such has been the success of private freight operations that, according to Network Rail, freight carryings since 1996 have increased by 41 per cent from 31 billion to 43.7 billion gross tonne kilometres.

This virtual metamorphosis was brought about by the implementation of European Union (EU) legislation in the form of Directive 91/440/EEC of 1991, the so-called 'Railway Directive', and in Great Britain (GB) the *Railways Act 1993*, which in broad terms effectively opened up the whole of the European rail network to private enterprise operations. In Great Britain, the pre-existing rail-freight operations of BR, namely its old Loadhaul, Main line and Transrail businesses, plus Rail-freight Distribution, were all sold off to an American private enterprise operator and in their new single-entity guise as EWS they became commercially independent, as did the Freightliner rail container business which has retained its original name, competing with each other, with any newcomers that may come along, and with other transport modes, particularly road and inland waterways.

The main objective of the legislation mentioned above was, and still remains, to instigate a status of independent operator on all national rail systems throughout the EU, 'so they behave in a more commercial manner adapting to market needs, and all in the interests of improved efficiency'. Its particular aim was to 'open up the railways for use by independent operators and facilitate access to rail networks throughout the EU for organizations engaged in the international carriage of goods by combined transport'. In simple terms, it means, for example, allowing a freight shipper or transport operator in the UK to hire a freight train, load it with combined transport units, then run it through the Channel Tunnel on the through-rail service and into the European rail network, direct to its destination – and all under his own control. Formerly, prior to rail privatization in 1996, this was not possible in international rail freighting, with responsibility for shipments largely, and control totally, being passed into the hands of the nationalized BR operation as soon as the loading units were on board the train – a scenario frightening enough to dissuade many consignors from shipping their goods by this means.

Under the terms of the European Commission (EC) Directive, the management of rail transport services and of the railway infrastructure had to be separated, with separate accounting systems for each; we saw this in the UK with the establishment of Railtrack (the track operator), now Network Rail, and a whole clutch of independent regional passenger train operating companies and the FOCs identified above. With this 'open-access' structure there is greater facility and incentive for private operators such as road hauliers to incorporate rail trunking within their operating plans. This in turn encourages the use of intermodal transport loading units, whether as swap body, container traffic, or as loaded piggyback semi-trailers, and consequently contributes further to the overall ideology of switching long-haul freight traffic from road to rail.

7.2 Rail operations in Europe

In Europe too, significant changes have taken place as a result of the first railway directive and the subsequent railway legislative packages described in Chapter 4. The most notable change took effect from 15 March 2003 when all railway systems within the EUs were liberalized; this date signifying the official opening of the European rail-freight market to competition from private operators and to free access across inter-state borders on the Trans-European Rail Freight Network (TERFN) for established national rail companies, and new entrants into the rail-freight marketplace. Unfortunately, not all EU Member States were geared up for such a big and bold leap forward at that time; some countries had not adopted all the necessary rules or set up bodies qualified to issue the requisite safety certificates and allocate the necessary train paths to allow open access to their rail-freight networks. But the EC was on the ball, taking the necessary measures to ensure that it did happen as quickly as possible.

Part of the reason for the delay was a number of technical hurdles that had to be overcome; not least such shortcomings as:

- differing voltage levels between rail systems;
- incompatible locomotives;
- different safety systems;
- non-standard rail gauges of some countries (e.g. Spain, Portugal, and Russia);
- current need to change train crews at borders;
- lack of compatible track-and-trace systems;
- priority being given to passenger services.

7.2.1 The legislative packages

The so-called 'revitalization' package of March 2003 comprised a series of measures enabling rail companies to provide competitive European services to meet the demands of industry and forwarders, and to secure one of the key objectives of the EU's Transport White Paper, which is to transfer a large chunk of freight from road to rail. The relevant Commission Directives are as follows:

- Directive 2001/12/EC, amending Council Directive 91/440/EEC, on the development of the Community's railways.
- Directive 2001/13/EC, amending Council Directive 95/18/EC, on the licensing of railway undertakings.
- Directive 2001/14/EC on the allocation of railway infrastructure capacity and the levying of charges for the use of railway infrastructure and safety certification.

The main points of the package concern:

- Separation of essential functions on the basis of a specific, exhaustive list of tasks that have to be assigned to an authority other than the railway undertaking to ensure the principle of non-discrimination between competing railway undertakings.
- A regulatory body, independent from any infrastructure manager, charging body, allocation body, or applicant for capacity, is to be established in each country to ensure fair and non-discriminatory access conditions for ail railway undertakings.
- Guaranteed access rights to the TERFN: guaranteeing access rights to all licensed rail operators providing international rail-freight services on the network and meeting national safety requirements. Member States wishing to do so would have the possibility of granting more extensive rights; at the latest from March 2008 onward the access rights must be extended to the whole European rail network.
- Setting of charges for the use of infrastructure: the principle of charging on the basis of marginal cost was accepted; but with the possibility of mark ups, if the market would bear them, on condition that the charges were set in a transparent and non-discriminatory manner and that the competitiveness of international freight transport was guaranteed.
- Definition of transparent and fair rules, and procedures for the allocation of train paths: this is an essential prerequisite for high-quality services of railway undertakings.

7.2.1.1 Safety issues

Under an agreement in the Directive amending EC Directive 91/440/EC, safety rules have to be made by entities other than the railway undertakings themselves and existing rail safety rules applied by all railway undertakings, thereby ensuring that the pre-existing high level of safety is maintained after the opening of international rail-freight markets. The second railway package established what was called an 'Integrated

European Railway Area' and a *European Railway Agency* 'to steer and boost the implementation of a truly European rail safety approach and the interoperability of the rail systems'. Additionally, to further back up the intended high level of rail safety, the Commission proposed a Directive – still to be implemented – on train-driver licensing, specifying training requirements and ability standards for train drivers.

7.2.1.2 Competition

With these new rules of the game, traditional railway companies have had to deal with new competitors from among themselves and from private companies such as, for example, German operator 'Rail4Chem', which specializes in the transport of chemical products: in 2003 it had 14 locomotives and had already chalked up a million tonne/miles during the lead up to open access. There were other operators on the scene too at that time, such as HGK, NetLog, Shortlines, RTC, and DLC. Swedish furniture manufacturer Ikea also set up its own rail operation and this was a particular early success story for the 'open-access' regime, operating twice-daily rail services between Sweden and Germany, crossing Denmark en route, which meant, among all the other early problems, working in three languages. Van Dieren Maritime is now operating this service for Ikea in collaboration with Intercontainer-Interfrigo (ICF). However, all this 'open competition' has not happened as smoothly as might have been expected. A report in International Freighting Weekly in March 2004 suggested that Europe's rail giants were still discriminating against newcomers a year after the international freight market was opened to competition – the newcomers citing a catalogue of unfair practices and higher access charges. For their own part, the existing national rail companies – the so-called 'giants' – have been endeavouring to better satisfy their customers' demands by improving services and especially by improving the organization of frontier crossings, the main bottlenecks in the system. The examples of Mannheim–Woippy on the French–German border, and Brenner Rail Cargo on the Munich–Verona route were frequently quoted as being forerunners of the new European approach.

7.2.1.3 EC objectives

The EC's objectives with this legislation were very clear: international freight trains should be able to cross borders without administrative and legal barriers, and benefit from the technical improvements already made by the industry. Diesel locomotives certified in most Member States or electric multi-current locomotives should be able to cross borders without worrying about voltage differences. Similarly, signalling systems should be compatible to allow High-Speed Rail Interoperability between Member States thereby allowing freight trains to cross borders without stopping, achieved by installation of the Euro-wide control system known as the European Rail Traffic Management System (ERTMS) in all Member States. Additionally, the EC requires all rail companies operating on the networks of Member States to comply with the national health and safety rules of those States and with their national rules with regard to employment law.

7.3 European Railway Agency

The EC's ambition to establish an integrated Euro-wide railway network, mentioned briefly above, took a major step forward in July 2004 when the new European Railway Agency, which is responsible for improving the safety and interoperability of Europe's railway networks held its inaugural meeting in Valenciennes (France). This meeting brought together for the first time representatives of all the 25 EU Member States, of the EC and of the railway companies (i.e. infrastructure managers, railway industry, staff representatives, passenger organizations, and railway freight customers involved) all of whom are party to the setting up of the new body. According to the EU's Europa Press Release, published on its website in July 2004, the Agency is intended to be a driving force in Europe's policy on modernizing the railway sector. It mentions the fact that the existence in the 25 Member States of technical rules and national safety rules that are mutually incompatible is a major handicap to the sector's development and that it will be the Agency's task to gradually approximate these technical rules and set common safety objectives to be achieved on all European railways. Through its work, the Agency intends to directly contribute to creating an integrated railway area that is competitive and guarantees a high level of safety. It is believed that once the new Agency is phased in, between 2004 and 2006, it will help to breath new life into the European railway sector, which is a key element of the EU's common transport policy, and should, in particular, 'improve the balance between the various modes of transport to the benefit of those that are the most environmentally friendly and guarantee a high level of safety'.

7.4 The Euro-wide railway: Railion

The changing face of European rail freighting is amply demonstrated by the EU's ambitious plans for Euro-wide intermodal operations, which can be seen coming to fruition in many quarters. A fine example of this is the German Railion operation, which calls itself 'the first European freight railway'. Formerly this was the freight division of German national rail operator Stinnes AG, the transport and logistics division of Deutsche Bahn (DB), and named at that time DB Cargo. From 1 September 2003 this operation was re-named Railion Deutschland and it has linked up with the rail-freight transportation activities of the Danish and the Netherlands' railway systems. At the time of its launch, Railion was transporting some 280 million tonnes of freight and serving 4500 customer sidings annually and crucially, for intermodal aspirations, saving an estimated 100 000 heavy truck journeys every day. According to the company, half of all its 'transportations' were already cross-border at the time of the launch and the company's intention is to extend the network right across Europe to make the rail system more competitive with road transport, especially on international routes.

As the company said in a press release in March 2003, when rail liberalization was actually beginning to happen:

> Transporting Italian tiles from Ancona via Vienna and Frankfurt right up to Copenhagen, non-stop with one freight train, without having to change locomotive or driver at the borders – soon that will no longer be just a pipedream. What has long been possible for trucks will soon be reality for the railways, too: Through co-operative agreements with railways abroad and stake holdings, Deutsche Bahn wants to break down national barriers to freight transport through Europe. Time-consuming stops for locomotive and driver changes at the borders between Germany and countries like the Czech Republic, France, Denmark, Switzerland, Belgium, and the Netherlands will soon be consigned to history. However, it is not enough to overcome existing obstacles at borders. Rather, the standard of quality that our customers expect must be guaranteed on all routes: transportations from a single source and over and beyond national boundaries.

It is pleasing to be able to comment that these ideals have become a reality. Press reports indicate that new trans-Europe intermodal services are opening up and, importantly, both reliability and punctuality targets are being met. A good example is the recently announced (IFW 24 January 2005) transcontinental rail-freight alliance operating under the name, European Bulls. This new organization promises a new era of quality rail-freight services under an integrated management structure with performance guarantees. The grouping comprises Rail4Chem (Germany), Ferrovie Nord Cargo (Italy), LTE (Austria), Comsa (Spain), and Viamont (Czech Republic). The five railway companies said:

> they wanted to create a pan-European network of 'one face to the customer' cross-border services with routes operated by one or more members but with one railway responsible

to customers for the entire route. The biggest breakthrough would be the end of time-consuming border stops to change locomotives.

7.5 UK rail-freight strategy

While the UK's rail-freight operators (RFOs) have been quietly getting on with their businesses, racking up success stories with increasing tonnages being carried and new routes being developed, fundamental changes have been taking place in the structure of the rail industry itself under provisions set out in the Government's White Paper, *The Future of Rail*, published in July 2004 and incorporated in the Railways Bill of 2004. The new structure is based on the principle of public and private partnership, recognizing rail's status as a public service specified by Government and delivered by the private sector. First, and most significantly, the Bill (later to become an Act of Parliament) allows the Government to wind up the Strategic Rail Authority (SRA) and take charge itself of setting the national strategy for the railways with the objective of streamlining the structure of the rail industry. In future, it will be for Ministers, accountable to Parliament and the electorate, to set the national strategy – which, currently, is not the case – and to determine the level of public expenditure and take the strategic decisions on what this should buy.

Among the other key measures in the Bill, Network Rail is to be given a strengthened role as operator of the network with overall responsibility for its performance – there will be no more passing of the buck for poor performance. The Office of Rail Regulation (ORR) will carry responsibility for safety, performance and cost to reduce bureaucracy, and to ensure that these issues are looked at as a whole. Responsibilities for railway safety will also transfer to ORR from the Health and Safety Commission/Executive and will remain independent of government and the industry. Other matters in the Bill concern rail passenger issues and certain powers for decision-making are to be devolved to the Passenger Transport Executives and Transport for London (TfL) on matters relating to light rail and bus integration with rail services.

7.6 Rail freight in decline

It was widely recognized, particularly prior to the 1996 UK rail privatization and the 2003 EC 'open-access' legislation, that rail freight had been in decline for many years and that even now, in 2004, it represents only about 8 per cent of the total amount of goods transported in the EU, compared with, for example, over 40 per cent in the USA. One of the reasons for this is that faced with the ever increasing requirements of customers in terms of punctuality and costs, national rail companies, in the past, have been unable to offer the quality of service required in their international services, for which the average speed was a mere 18 kph. However, Lord (Tony) Berkeley, head of the UK's Rail Freight Group, speaking at a Brussels conference in March 2003, said that contrary to declining trends in European rail freighting, the scenario in the UK was different with over 50 per cent growth in this sector being achieved over recent years, albeit from a very low base. Overall, he predicted that the open-access package would lead to a resurgence of rail freight across Europe, but it would need pressure from rail users to ensure that services and networks are developed.

7.7 The loading-gauge issue

One of the constraints on rail-freight operations in the UK, as mentioned above – unlike Europe where there is no such problem – has been the limited loading gauge on most of the domestic rail network. The loading gauge represents the maximum dimensions for rail vehicles to pass safely through tunnels, bridges and railway station platforms on the network. Track gauge (i.e. the width between the lines), on the other hand, is compatible on the Railtrack network with that of most European countries at 1435 millimetres (4 feet 8½ inches), the exceptions being Spain and Portugal (where the gauge is 1668 millimetres), Finland

(1524 millimetres), Russia (1520 millimetres), and Ireland which also has a different track gauge. The standard *loading* gauge on most of the UK network is not compatible with Europe. So while continental rail rolling stock could run on BR track, many continental railway wagons and those used in piggyback operations are too high or too wide to allow their use in GB. Hence the reason why there has been no widespread integration of European intermodal operations into Britain's rail-freight network. Traditionally, on our relatively ancient network, the clearance has not been sufficient to accommodate the passage of piggyback freight, which is described more fully below, and newer generation shipping containers which are now up to 9-foot 6-inch tall, hence the common term 'tall containers'. However, now in the early 2000s this situation is being addressed with the necessary re-construction work being undertaken and the opening up of numerous sections of the rail network to tall container traffic. In particular, these route enhancements applied early in 2004 to the Port of Tilbury, which was the first UK port to begin operational handling of 9-foot 6-inch containers on standard wagons by rail. This was, in fact, the first completed element of the £40 million Felixstowe to Nuneaton project. This was followed in October 2004 by the opening of the 9-foot 6-inch rail-freight corridor connecting Felixstowe with the West Coast Main Line (WCML); Felixstowe being the UK's largest gateway for containers – some 1.7 million annually of which 25 per cent are now 'tall' containers and by 2010 these are anticipated to comprise 50 per cent or more of all containers landing at the port.

7.7.1 Loading gauges

Code letters and numbers signify the different loading gauges, of which there are a number in use throughout Europe. Network Rail's standard gauge for the UK rail network is called W6/W6A and is the smallest gauge generally in use, but nevertheless, adequate to accommodate the passage of all rail passenger traffic and most normal freight wagons. This gauge code was adopted from the Railtrack codes and before that, those of BR, but for the rest of the freight gauge codes Railtrack changed the nomenclature in 1999. The other relevant UK gauges are the W7 gauge (previously BR W6A: for Freightliner 8-foot-high containers); W8 (also previously BR W6A: but in this case to accommodate for Freightliner's 8-foot 6-inch-high containers), and W9 (formerly BR's SB1c gauge). The remaining intermediate gauges are known as 'Berne' gauge as determined by the international rail organization, Union International de Chemins de Fer (UIC), and are used across the continent. These are: W10 (which equates to UIC's standard 'A' gauge); W10w (which also equates to UIC's 'A' gauge but with 100 millimetres extra in the width); W11 (which equates to UIC's standard 'B' gauge) and W11w (which equates to UIC's 'B' gauge but with 100 millimetres extra in the width). The largest gauges are UIC 'C' giving 4650 millimetres height clearance from track level and Eurotunnel's which at 5600 millimetres high permits its own special freight shuttle wagons and all other forms of intermodal freight to pass through the Channel Tunnel with safety. For piggyback operation, where 4-metre high road vehicle semi-trailers are carried unaccompanied on low rail wagons, 4.13 metres (4130 millimetres) headroom clearance above the track level is needed, exceeding the UK SB1 gauge, which accommodates swap body traffic within a headroom clearance of 3965 millimetres (i.e. UIC 'B' + gauge). For complete road vehicles to be carried on special rail wagons rolling motorway-style gauge UIC 'C' is necessary. A diagram showing the profile of static loading gauges for GB W6 up to the Eurotunnel gauge is shown in Figure 7.1.

7.8 Piggyback operation

Piggyback operation is a widely adopted practice in Europe, but has had limited application here in the UK, due to reasons of cost and the limitations on loading gauge described above. A great deal of enthusiasm was generated in the UK for this form of intermodalism in the early 1990s when an organization called 'The Piggyback Consortium' produced a report showing that some 400 000 heavy lorry journeys

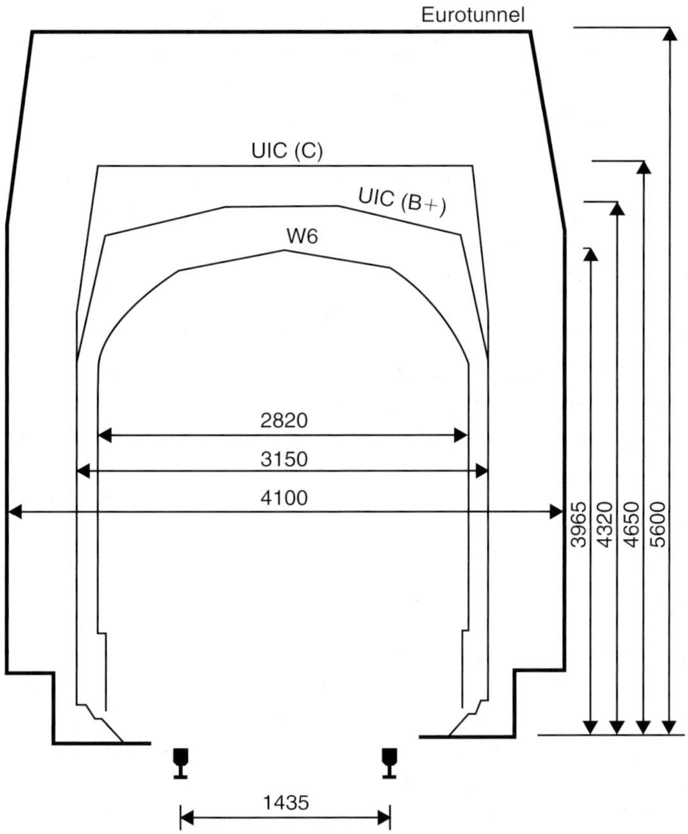

All measurements are in millimetres

Fig. 7.1 Static UK and European rail loading gauges shown in profile.

annually could be removed from the road network if the WCML and a number of other major routes were upgraded to accommodate such traffic. However, since the upgrade was painfully slow in materializing due to ever-increasing projected construction costs the enthusiasm died away. Today, apart from a number of highly specialized operations such as that of the Lafarge Cement company, piggyback in the UK is seen as a very expensive and relatively inefficient operation – after all, it is not only the load that is carried, but the complete semi-trailer on wheels which represents a lot of dead weight to be hauled and paid for.

7.9 Rolling highways

Somewhat similar in concept to piggyback transport is the so-called rolling highway intermodal system in which entire lorries and their trailers, accompanied by their drivers, are carried on drive-on low-height rail wagons: piggyback is invariably an unaccompanied service. Eurotunnel's freight shuttle service through the Channel Tunnel employs this type of drive-on/drive-off system. Another specialist in this type of operation is Swiss company Hupac renowned for its transalpine services across Switzerland for heavy trucks that are prohibited from transiting the country by road (see Figure 7.2). The particular advantages of rolling highway systems are that, generally, transit can be guaranteed in all meteorological conditions; there are no customs procedures when crossing through Switzerland; the system is suitable for ordinary trucks with no

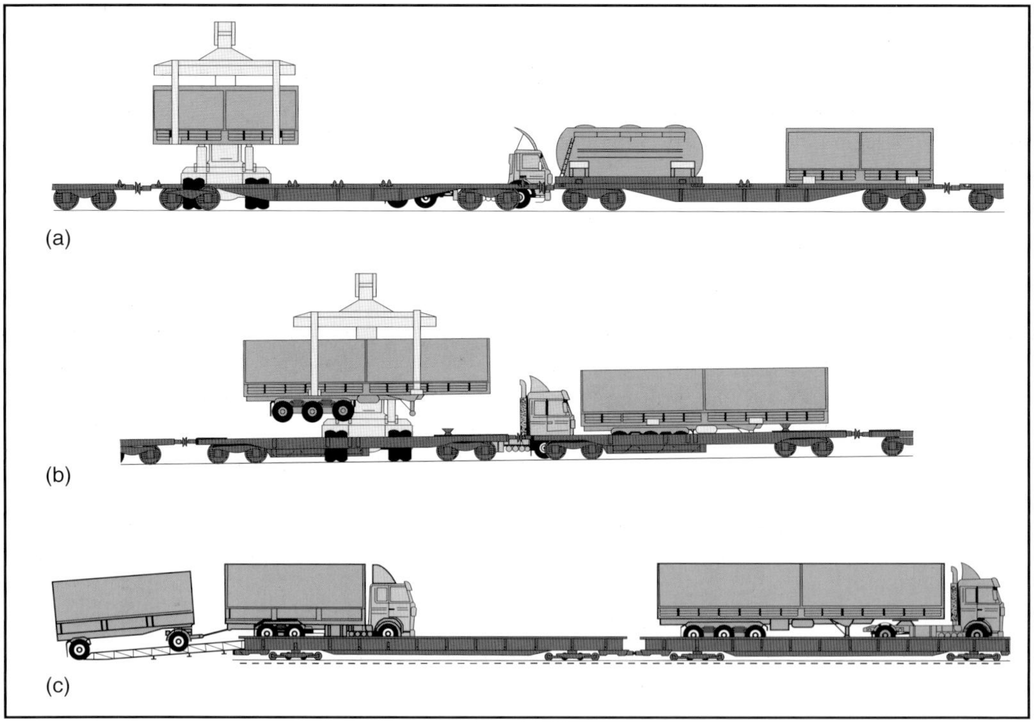

Fig. 7.2 Schematic illustration showing various intermodal systems (combined transport techniques): (a) unaccompanied swap bodies and containers (b) unaccompanied semi-trailers (piggyback/huckepack/Kangaroo) (c) accompanied vehicles on the rolling motorway. (*Source*: Hupac.)

special equipment; and the hours spent by drivers in restaurant or sleeping cars are considered as compulsory resting time under the EU drivers' hours law provisions.

7.10 The Central Railway project

The British Government finally turned its back on this project in March 2004 by refusing to support the proposed £5 billion, 400-mile Central Railway from Liverpool to France via Manchester, Sheffield, Leicester, and London, onto the Channel Tunnel and then onward to France. The project entailed modernizing and operating approximately 650 kilometres (400 miles) of railway between Liverpool and a terminal in Northern France, via the Channel Tunnel. All but 40 miles of the route could have followed existing underused or disused railways, but even where this was not possible, such as around the west of London, it would still have primarily followed existing transport corridors. The railway's core business was intended to be a high-volume, low-cost shuttle operation carrying unaccompanied lorry trailers between the UK and mainland Europe. In effect, it would have converted millions of lorry trunk hauls into block train traffic and its plans were for up to eight 1500-metre-long trains to run per hour each way, thereby, it claimed, removing more than 5-million heavy lorry journeys a year from key UK motorways, including the M25, M20, M1, M40, and M6, by the year 2015.

The project promoter, Central Railway, was a Franco-British project, supported by the French government and working closely with SYSTRA, the French railway company, Societe Nationale des Chemins de Fer's (SNCF's), engineering subsidiary, and RFF, the French rail infrastructure company, to develop the project in France and a senior SNCF executive had joined the Central Railway board of directors. Central Railway's application to the British Government was for support in the promotion of a Parliamentary Bill to authorize powers to build the £10 billion line. The Government's refusal of support was based on its view that Central Railway had not demonstrated that it would be able to secure the necessary financial backing to build and run the line and that if the project ran into difficulties the Government would be left to intervene with taxpayer's money.

7.11 Channel Tunnel Rail Link

The Channel Tunnel Rail Link (CTRL) is a 69-mile (108-kilometre) long high-speed rail line between Central London (St Pancras) and the Channel Tunnel – the first major new railway to be constructed in the UK for over a century and built as a Public Private Partnership project. While principally intended for speeding up passenger traffic between London (Waterloo) and Lille, Paris, and Brussels, knocking some 20 minutes off the previous journey times, with travel speeds up to 186 miles per hour (mph) (300 kph), the line also has a role to play in speeding freight traffic to the Channel Tunnel. According to TfL, which engaged specialist consultants to study prospects for the route, the CTRL will provide a unique opportunity to operate high-gauge Continental freight trains between mainland Europe and London when the line is fully completed in 2007, but there are currently major issues that need to be resolved in regard to the provision of suitable terminal facilities. TfL says that only two sites in the Barking (Essex) area offer significant opportunities within the M25 (i.e. the motorway ring around London), but major planning hurdles have to be overcome before a terminal can be built. It is also looking at the possibilities of providing terminal facilities in the Thames gateway area.

7.12 Eurotunnel

Eurotunnel's early success carrying significant volumes of intermodal freight through the Channel Tunnel (recorded elsewhere in this book) came almost to grief when in the latter months of 2001 and early 2002 the influx of illegal immigrants stowing away on rail wagons brought the through-Tunnel freight operations to a virtual stop (Figure 7.3). Vast numbers of immigrants were assembling at the Sangatte Red Cross centre near the Tunnel entrance at Calais ready to stowaway on freight trains at every opportunity. They broke through the security fencing then risked their lives hanging precariously on or even underneath the freight wagons – at the same time many were stowing away on lorries travelling to the UK via the Channel Tunnel freight shuttle and the cross-Channel ferry services, slitting lorry sheeting to get on board vehicles and posing threats of violence to drivers who tried to throw them off. This episode marked a significant downturn for Eurotunnel's performance. However, with steps taken at an official level to stem the flow of immigrants, the company has put that turbulent period behind it and has been boosting its freight performances, although in 2004 it launched a new commercial strategy 'Project Dare', designed to rejuvenate the financial returns from its freight businesses (i.e. the lorry shuttle and the rail-freight services). Under this new scheme, existing regular shuttle users are being asked to project their future business with Eurotunnel – on which the rates charged to them will be based and casual users of the service are to be priced out of contention. The company is hopeful that this project will be successful and will enable it reduce its mounting financial problems. As may be imagined, this new philosophy has not gone down well with the company's regular shuttle clients who see it as nothing more than a hike in rates and some of whom have already made plans to switch to exclusive use of the cross-Channel roll-on/roll-off ferry services.

Another Eurotunnel strategy to boost freight traffic through the Tunnel is for the company to start running its own international freight services, so that, in effect, it becomes another RFO along with EWS and others. Certainly there is sufficient capacity for creating more train paths on the existing infrastructure – some 50 000 paths are available. In order to achieve this objective Eurotunnel needs to apply for a Railway Undertaking Licence in France.

Fig. 7.3 Intermodal freight train emerging from the Channel Tunnel. (*Source*: Eurotunnel.)

7.13 Freight aggregators and integrators

Aggregator or integrator; what is the difference? Who are they? What do they do? These two terms have come very much into common usage with the increasing interest in and development of intermodal transport. According to the Oxford English Dictionary, the term aggregate means to 'form a whole by a combination of several disparate elements'; integrate means 'to combine or be combined to form a whole'. A study of these definitions cannot but lead to the conclusion that they virtually mean the same. Thus, by definition, anybody (any company) who fulfils the role of 'aggregator' or 'integrator' must be doing the same thing – combining, in this case, freight loads. Taking this definition business a stage further, we can refer to a report commissioned by the EC from a consortium of consultants based on a study of this function – see bibliography. Here 'Freight Integrators' are described as 'service providers who arrange full load, door-to-door transportation by selecting and combining without prejudice the most sustainable and efficient mode(s) of transportation'.

Intermodal aggregators market timetabled intermodal shuttle trains between terminals located in principal UK and continental centres (Figure 7.4). They 'buy' train services from the FOCs (such as EWS, DRS, GBRf, Advenza Freight, and Freightliner) and then sell the space on the trains to end users. Thus, in effect, they 'aggregate' demand from individual customers into whole trainloads. The principal UK intermodal aggregator is Allied Continental Intermodal (ACI, part of the Swiss group ICF).

The concept of 'freight integrator' as a developing profession for individuals was mooted in the *European Transport Policy for 2010: Time to Decide* White Paper of 2001 which emphasized the role of a transport flows organizer who was 'able to combine the specific strength of each mode at European and world wide level to offer their clients the best services'. The EC issued a consultation on the subject in October 2003, seeking public views on how such freight integrators could improve their knowledge, awareness, and understanding of intermodal transport; how intermodalism could be simplified through further standardization; how greater co-operation could be achieved between transport users; and how

Fig. 7.4 Typical view of a busy intermodal freight terminal. (*Source*: Bob Sweet/Freightliner.)

responsibility and accountability in intermodal transport could be clarified. Although nothing further has been heard of this scheme at the time of writing this text, at least the European Intermodal Association (EIA) has taken up the cudgels with its own scheme called 'Intermodal Master Class'. The course, which is being developed by European universities that have logistics specialities in their programmes, will comprise lectures and working visits.

7.14 Locomotive power

A highlight of the current rail-freighting scenario has been the revolution in heavy freight traction since the launch in 1998 of the Class 66 locomotive. This Canadian-built General Motors diesel locomotive of 3200-horsepower/92 000-pound-feet tractive effort and capable of 75 mph operation has proved to be highly successful in the UK. In fact, so much so that there are now almost 400 of these machines currently operating, predominantly with EWS, which has more than 250 of these locomotives in service. Freightliner followed suit putting a significant number into operation including some lower-geared versions with reduced speed (65 mph), but increased tractive effort (105 080 pounds/feet). Other operators include GBRf – also with some lower-geared versions for heavy operations – and DRS, which took 10 Class 66s in 2003. Following on from the Class 66 model, a high-speed Class 67 version has been developed which is capable of operation in GB at up to 125 mph.

Interestingly, the success of the Class 66 in the UK led to a number of trials in mainland Europe with German, Belgium, Swedish, and Norwegian RFOs. Here the key to success lay in the fact that as a diesel locomotive the Class 66 – there will be different European national model designations – is a quicker and more cost-effective solution for cross-border international rail haulage than the use of expensive multi-voltage electric locomotives. Thus, it has become attractive to the new band of open-access operators now running, or on the verge of running, new inter-European rail-freight services.

8

Inland Waterway, Short-Sea, and Coastal Shipping

Transport by inland waterway, short-sea, and coastal shipping is taking on an increasingly important role in the development of Euro-wide intermodal freighting operations. Many individual European Union (EU) Member State governments and the European Commission (EC) itself see these modes as offering great potential for transferring freight traffic away from Europe's grossly overcrowded road networks, especially as the waterways are currently so significantly under utilized and because they offer a truly environmentally beneficial alternative to the bane of congested roads and polluting heavy lorries. According to information about inland waterways on the EU's Europa web site (www.europa.eu.int):

> Compared to other modes which are often confronted with congestion and capacity problems, inland waterway transport is characterized by its reliability and has a major unexploited capacity. Inland waterway transport has major assets. It is particularly effective and energy-efficient; its energy consumption per tonne-kilometre of transported goods corresponds to one-sixth of the consumption [of road transport] and to half of that of rail transport. Its noise and gaseous emissions are modest. According to recent studies, the total external costs of inland navigation (in terms of accidents, congestion, noise emissions, air pollution and other environmental impacts) are seven times lower in than those of road transport. Inland waterway transport ensures a high degree of safety, in particular when it comes to the transport of dangerous goods. Finally it contributes to the decongestion of the overloaded road network in densely populated regions.

The important role of waterways is clearly demonstrated on a world scale, by such famous man-made constructions as the 80 kilometres Panama Canal linking the Atlantic and Pacific oceans, the 173 kilometres Suez Canal joining the Mediterranean Sea and the Red Sea, and the St Lawrence Seaway in Canada which combines both river and man-made canal sections through the Great Lakes linking the Canadian City of Quebec, at the mouth of the St Lawrence estuary, with Duluth, Minnesota in the USA, some 3760-kilometre inland from the Atlantic Ocean. Others, are grand on a European scale, such as the 250-kilometre-long Canal du Midi through France which links the Atlantic Ocean and the Mediterranean; and the 100-kilometre-long Kiel Canal, opened in 1895 and said to be the world's busiest artificial waterway, which effectively divides Denmark from Germany, linking the North Sea with the Baltic, and saves shipping a detour of some 250 nautical miles around the northern tip of Denmark. And in the context of inland waterway freighting, good examples are the Rhine–Rhone canal connecting with the Mediterranean Sea and the Rhine–Main–Danube link between the North Sea and the Black Sea providing, by a combination of river and man-made canal, a 3500-kilometre-long trans-Europe freight link (Figure 8.1) shows the routes of Europe's principal freight waterways. There are of course many other examples, but these few illustrate how significant waterways have become in the freighting scenario.

Waterway shipping can be divided into four main categories as follows:

1. Deep-Sea Shipping which relates mainly to international inter-continental shipping.
2. Short-Sea Shipping which relates to transport between the United Kingdom (UK) and continental Europe.
3. Coastal Shipping which is coast-to-coast national shipping.
4. Inland Shipping which uses the inland waterways for transport.

8.1 Waterway statistics

It is unfortunate that inland shipping on the British Waterways (BW) network was in steady decline between the years of 1987 and 1997. In fact, according to the *Transport Statistics Bulletin – Waterborne Freight in the UK 2003*, published by the National Statistics office of the Department for Transport in November 2004, domestic freight traffic on the inland waterways declined by 20 per cent over the decade in terms of both goods lifted and goods moved (the latter taking account of the distance travelled in kilometres). Freight volumes on coastwise traffic have also declined. The statistics show that of all modes of transport in the UK, waterborne transport carried 6 per cent of all goods lifted and 24 per cent of goods moved in 2003. The inland waterways moved a mere 1 per cent of all freight carried in the UK, but this was 3.5 million tonnes and sufficient to remove some 200 000 lorry journeys from the roads. Conversely, from an EU perspective, freight carrying by short-sea shipping during the decade of the 1990s grew by 29.6 per cent and, according to the EC, 40 per cent of all trade within the EU is now carried by sea, a total surpassed only by road transport. In 2003, 125 billion tonne-kilometres of freight were transported by inland waterways in Europe. In the European Northwest region, in the hinterland of the EU's largest seaports (e.g. such as Rotterdam and Hamburg), it is estimated that the modal share of inland waterway transport (IWT) can reach up to 43 per cent.

8.2 Inland waterways

The EC, in a leaflet published by its Energy and Transport Directorate General (DG) in 2003, – *Inland Waterway Transport: a transport solution that works* (available on the Europa web site), says:

> the inland waterways carry some 12 per cent of the EU Member States' freight, and this is an industry that is constantly growing – in fact by some 17 per cent in the 1990s – and expected to grow even more since the enlargement of the EU by 10 additional Member States in 2004. There are more than 35 000 kilometres of waterways connecting hundreds of cities and industrial regions in the 18 out of the EU's 25 Member States that have inland waterways, 10 of which have an interconnected waterway network. The modal share of river transport accounts for 7 per cent of the total inland transport in the European Union.

Inland waterway freighting is claimed to be a reliable and congestion-free mode that is cheap to use, economical to operate, environmental friendly, socially acceptable, and modern; over the past 10 years, the fleet has been continuously modernized and these days offers high safety standards and up-to-date navigational equipment, and, importantly in the intermodal context, the waterways are significantly under utilized so there is capacity to spare for expansion from a switch of freight from road haulage. With many European firms now looking to boost their 'green' credentials, switching their freight to inland waterways offers them positive results in terms of improved public and customer perception of how they conduct their business activities. In fact, an EC study into the cash spent on the socio-economic costs of various

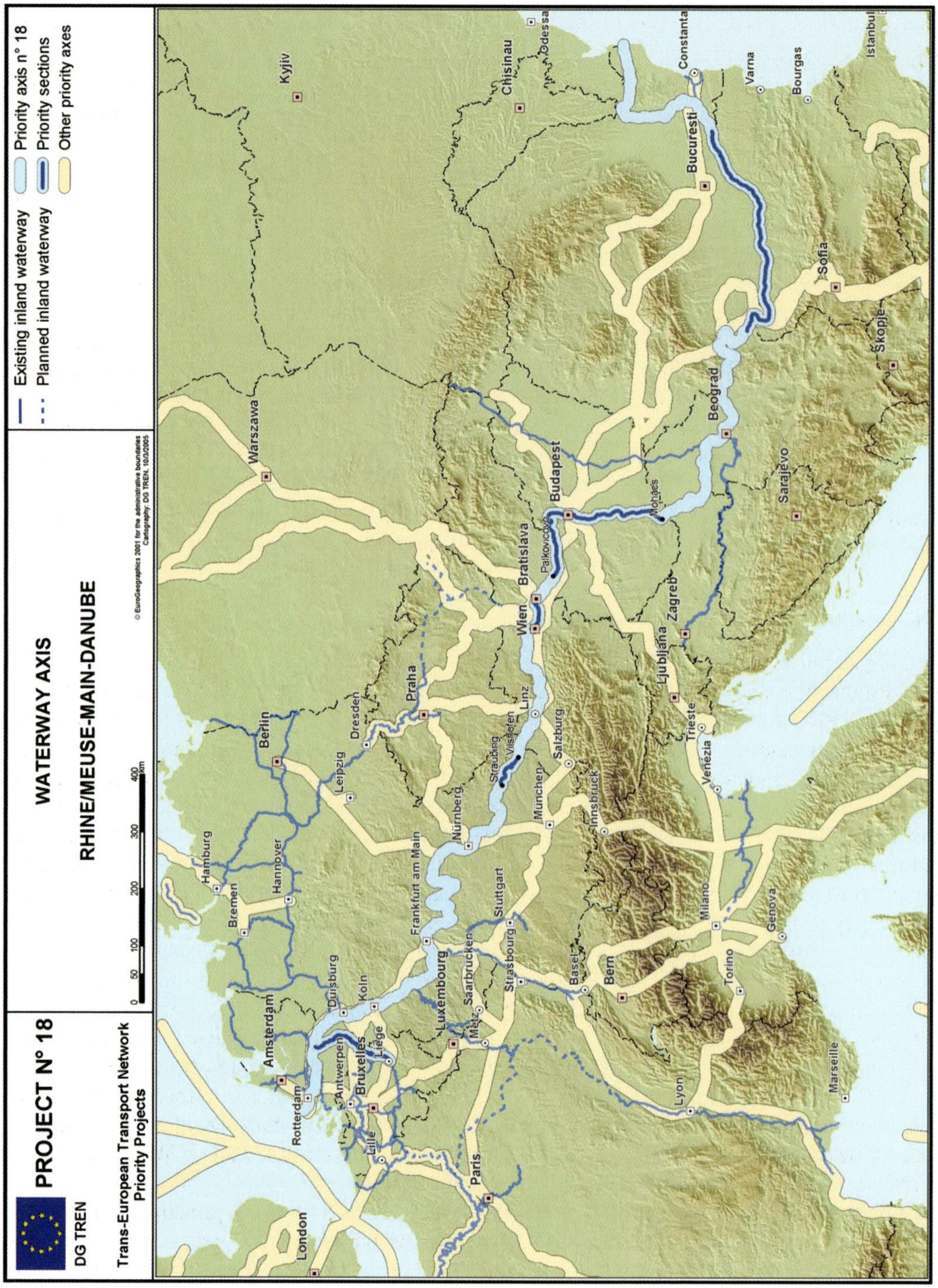

Fig. 8.1 Europe map showing the Rhine/Meuse-Main-Danube waterway axis (*Source*: EC Trans-European Transport Network from Europa website).

types of transport, (such as accidents, air pollution, climatic change, noise pollution, congestion, effects on the countryside and the urban environment) revealed that while road transport accounted for 91.5 per cent of the costs, air transport accounted for only 6 per cent and rail transport only 2 per cent, the result for inland navigation showed the tiniest of figures, namely, a mere half of 1 per cent. This together with its low fuel consumption clearly makes IWT one of the most sustainable forms of transport. Emissions in particular, in fact, are falling even further as newer vessels with more efficient engines are introduced into inland waterway fleets, resulting, again according to the EC, in emission levels on the inland waterways falling by as much as 25 per cent during the 1980s and 1990s.

Inland waterways comprise both navigable rivers and man-made canals. In both Britain and Europe many of the largest rivers have become traffic routes for goods-carrying vessels. In Great Britain, for example, the rivers Thames, Humber, Clyde, Avon, Trent, Mersey, and Ouse carry varying amounts of freight traffic on canal barges, while in Europe the inland waterways have long been major freight arteries right across the Continent, especially when one considers such rivers as the Rhine, Main, Danube, Elbe, Meuse, Seine, and Rhone and their connecting canal systems. For example, by using the Rhine-Main-Danube waterway link freight barges can travel from Rotterdam on the river Maas, down the Rhine, into the river Main and via the Main–Danube canal into the river Danube and thence to the Romanian Black Sea port of Sulina, a distance of some 3500 kilometres through 15 countries. This is a freight route of major importance and one that significantly encourages road–waterway and rail–waterway intermodalism on a very large scale and which clearly demonstrates the effectiveness of using such a system of combined transport modes. It is an unfortunate fact that, to date, freight volumes on the Danube waterway have reached only about 10 per cent of its potential capacity.

8.3 UK inland waterways

Canal building started in Britain with the 1¾-mile-long Exeter Canal built between 1564 and 1566. But it was not until the so-called 'canal age' between 1760 and 1850 that canal building really forged ahead on a large scale with industry and agriculture depending on them as the main source of transporting their raw materials and products. In fact, as Charles Hadfield points out in his well-known book, *The Canal Age*, they were more than just a means of transport; they developed towns and altered trade patterns and they were responsible for a new breed of labourer, the 'navvy', and for revolutionizing the engineering profession. By 1850, Britain's canal network, in its heyday, extended to some 4023 miles. Today, the network comprises 3200 kilometres (about 2000 miles) of river-based commercial navigations and canals providing an arterial system that reaches deep into the heart of many towns and cities, most of the rest now being largely disused. The majority of these waterways are today owned and operated by a public corporation, BW, formerly the British Waterways Board (BWB), and are mostly given over to recreational activities such as leisure boating, fishing, towpath walking, jogging, and cycling, nature conservation and water supply; just 620 kilometres of the network are maintained as commercial waterways on which BW currently carries some 3.5 million tonnes of freight annually, about one per cent of all freight carried in the UK. Other waterways are mainly privately owned and operated by members of the Association of Inland Navigation Authorities (AINA), including, for example, the Broads Authority and the Environment Agency, the Manchester Ship Canal Company as well as many small canal companies. No single organization has responsibility for the UK waterways network as a whole, BW manages about half of it and the Environment Agency a further quarter of the total.

The Manchester Ship Canal (opened by Queen Victoria in 1894) is the UK's principal inland waterway which brings deep-sea shipping 56 kilometres inland to Manchester at a current rate of 3000 ships annually carrying some 18 million tonnes of cargo. Other major canals catering for sea-going craft are the Crinan and Caledonian canals in Scotland; the former is just 14.4-kilometres-long and provides a short-cut for shipping from the West Coast of Scotland and the Western Isles (Crinan on the Sound of

Jura) to the Clyde estuary (Ardrishaig on Loch Fyne) saving the long and difficult sea route around the Mull of Kintyre. The Caledonian Canal which is partially a natural waterway (61.4 kilometres) and partially man-made construction (35.4 kilometres), stretches across the Scottish Highlands from Inverness on the Moray Firth to Fort William. It is said to be one of the world's most scenic waterways and it provides a short-cut link for both commercial and private vessels sailing from Northern European and Scandinavian ports to the West coast ports of England and to Ireland.

Besides these particular waterways, Britain also has a wealth of narrow inland canals used mostly used for leisure boating and other recreational activities. However, it is the network of broad canals that carries the majority freight traffic, most of it being bulk commodities not directly linked to road or rail carriage at one or other end of the transit, but some of it is of an intermodal nature which does involve transfer to and/or from either road vehicle or rail wagon at one end or other of the waterborne transit. The narrow canals (up to about 7-feet wide) have limitations so far as freight carrying is concerned, with a maximum gross weight capacity of only some 20/25 tonnes which, these days, is less than a single large lorry load, and a maximum speed of only 4–6 miles/hour so they are clearly not competitive with road haulage in anything except niche operations and for carrying, for example, aggregates and waste into and out of city and urban areas. But they are highly popular with leisure boaters, fishermen, and towpath walkers and joggers. The broad canals, on the other hand, are ideally suited for freight traffic being some 14 feet wide and capable of allowing passage for barges of gross weights of 40–80 tonnes.

Britain's principal commercial waterways include the River Thames ('Old Father Thames' in the famous song), the main waterway in Southern England albeit not under BW control. It rises in the Cotswold Hills and passing such historic towns as Henley on Thames and Windsor (i.e. the Royal Borough of Windsor) flows down through the City of London to the London Tideway carrying both inland waterway and an amount of short-sea shipping. On Britain's West coast, England's longest river, the Severn, provides shipping access to the industrial midlands via the 16-mile-long Sharpness Canal to Gloucester Docks and then up the river to Worcester, although the volume of freight traffic shipped by this route is limited in terms of ship capacity and volume; much of it is timber. The River Nene, which runs from the Northampton area, where it is linked to the Grand Union Canal, to the Wash on England's East Coast, carries a small volume of short-sea shipping as far upstream as the port at Sutton Bridge which can accommodate vessels of up to 5000 tonnes at its 750-metre-long quay. Much of this traffic is agricultural produce into and out of the adjacent Fenland area, but by way of example of short-sea shipping's usefulness and flexibility in a more general sphere, recently the port and river proved to be the most accessible point for receiving a number of very heavy gas turbine units for the nearby newly-built power station at Sutton Bridge. These units were transported the short distance from the port to their final destination by road.

Northwest England is where the canal revolution was born more than two and a half centuries ago and where today canals are still the focus of a great deal of waterway activity, albeit primarily by leisure boaters and fishermen among others. However, the Manchester Ship Canal (mentioned above) is a major freight waterway carrying vessels of up to 40 000 tonnes deadweight. The canal is 56-kilometre-long and has docks and berths along most of its length. It handles some 3000 ships annually and in 2002 almost 7 million tonnes of freight passed through the port. On the other side of the Pennines, the Yorkshire and North East region is home to some of Britain's most important freight waterways where the navigable rivers Humber, Trent, and Ouse link with the Aire and Calder Canal, and the Sheffield and South Yorkshire Navigation. The Humber, which provides access to the North Sea for the rivers Trent and Ouse, has deep-water docks at Immingham, Grimsby, and Hull where freight traffic is transferred to barge for onward shipment via the inland waterways. Of particular interest here is the use of the port of Immingham for berthing and discharging lighters from a Lighter Aboard SHip (LASH) mother ship on arrival from the Continent (as shown at various stages of the loading/unloading operation in Figures 8.2 and 8.3). The River Trent links with the Humber above Scunthorpe and is a busy river for freight traffic from such centres as Gainsborough, Newark, and Nottingham. The river carries coastal vessels and Rhine barges

Fig. 8.2 The LASH system in operation showing the barges inside the sunken hull of the mother ship prior to unloading (*Source*: Herfurth Shipping (UK) Ltd).

Fig. 8.3 LASH barges being towed from the mother ship (*Source*: Herfurth Shipping (UK) Ltd).

mainly from European Ports. Much of the barge traffic comprises aggregates from Trent Valley quarries being shipped to Hull and Leeds, for example. Maximum cargo weights vary between 300 tonnes to or from Nottingham and up to 1000 tonnes in the case of Gainsborough.

The Aire and Calder Navigation starts at Goole on the Humber and extends through major industrial conurbations to Leeds and Wakefield via Knottingley, Ferrybridge, and Castleford. Much of this traffic again comprises bulk cargoes such as oil and sand (and formerly coal) in barge loads of up to 700 tonnes weight as far as Leeds. The Sheffield and South Yorkshire Navigation joins the Aire and Calder Navigation west of Goole, which it links to Rotherham via Doncaster, Mexborough, and Swinton again through heavily industrialized conurbations. The River Ouse links the inland port of Goole to the Humber and is a busy waterway with barge traffic en route from the Humber up river to such destinations as Selby and York where, respectively, barges of up to 1000 tonnes and 400 tonnes capacity can be accommodated.

We have, in the UK, a valuable transport asset in our waterways system and one that could be used to relieve the mind-numbing, frustrating, costly, and time-wasting congestion that we the public, as road users, and more importantly, the lorries delivering our goods, suffer on a daily basis. It is gratifying to note that after years of decline, new life is being breathed into the system by the efforts of commercial organizations plying their trade and by Government with, at long last, its forward-looking policies for infrastructure regeneration and for encouraging modal switch from road to rail. Two of these policy initiatives are described briefly below.

8.3.1 'Waterways for Tomorrow'

In June 2000 the UK Department for Environment, Food, and Rural Affairs published a document, *Waterways for Tomorrow*, which follows on from the Government's 1998 'New Deal' White Paper in setting out proposals for the future of the inland waterways. Much of the document is concerned with the waterways' role in providing leisure, recreation, tourism, and sport facilities as well as stressing its heritage, environmental and educational benefits. Little is proposed in regard to freighting apart from describing how the system has been in decline for many years – accounting, in tonne-kilometre terms, for less than 1 per cent of domestic freight moved, other than on the larger river navigations and canals; and how the planning system (described within the document) can help the waterways fulfil their potential by facilitating the transfer of freight to water. The Government reiterates its commitment to this objective, where, it says, 'this is practical, economic, and environmentally desirable.' This document was followed in 2002 with a report, *Freight on Water: A New Perspective*, from the Freight Study Group set up by the Government to examine the scope for increasing freight traffic on the inland waterways of England and Wales (see below).

8.3.2 Short Sea and Waterways Forum: 'Sea and Water'

The Short Sea and Waterways Forum (SSWF), originally launched in July 2003 and funded by the Department for Transport to act as a central co-ordinating point for water freight interests in the UK, was re-named 'Sea and Water' in 2004. Its aims are to increase the number of freight loads carried by water in and around the UK by raising awareness of the advantages of water freight as a transport mode and by providing a direct line of communication between industry and Government. It also runs the Short-Sea Shipping Promotion Centre, which operates along similar lines to its European counterparts in promoting the benefits of water freight transport across the UK and within the EU. The Forum is chaired by Professor James McConville of London Metropolitan University and has both Inland Waterway and Short-Sea Shipping Committees.

Freight on Water, the 2002 Government report (mentioned above) which examined the scope for increasing waterborne freight in the UK recommended that the development of the water freight industry was hampered by a lack of cohesive representation and that a strong representative body for water be created. 'Sea and Water' (S&W) was established, in part, to address that deficiency by bringing the industry together and by providing a common voice for all interested parties including carriers, customers and infrastructure providers. It aims to promote the movement of freight by water as part of an integrated transport policy, which includes inland, coastal and short-sea shipping interests. S&W's central objective is to facilitate the further integration of water freight into the wider logistics chain through a combination of co-ordination and promotion. In summary, it:

- acts as the focus for all short sea, coastal and waterways interests in the UK;
- provides support to the industry in their efforts to develop the movement of freight by water;
- provides a direct link between the water freight industry and government;
- works to eliminate the obstacles to using water to move freight;
- work to integrate water freight with other modes of transport;
- raises awareness of the potential of short sea, coastal, and inland waterways freight transport with prospective customers, politicians, planners, and the public;
- provides support for prospective customers and the industry generally;
- serves as a think tank for the development of a national short-sea, coastal, and inland water transport strategy and the preparation of a rolling programme for infrastructure improvement.

8.3.3 Traffic and Terminals

While it is acknowledged that the loading and unloading of waterway vessels for some cargoes is time-consuming and consequently expensive, the application of modern handling systems and the significant moves towards the containerization of suitable traffics are changing this. Similarly, it is also acknowledged that waterways do not have a widespread role in local distribution: How could they have? But, conversely, this transport mode has proven its credentials as being ideal for the carriage of bulk materials, non-urgent and non-perishable cargoes, unitised loads and, as we have seen in a relatively new departure elsewhere in this Chapter, for the carriage of abnormal indivisible loads (AILs). Actually, the list of suitable cargoes for inland waterway shipping is quite extensive if one identifies the individual bulk cargoes that constitute much of the waterway-carrying scenario; for example, grain and other agricultural produce and products, a variety of aggregates, bagged cement, and of course coal which has always been traditional waterway traffic, but obviously much less so with the demise of the UK coal mining industry. Petroleum and chemical products, hazardous goods cargoes, feature largely on the inland waterway menu as does, increasingly, the carriage of domestic and industrial waste to disposal depots. Steel and other metals are carried, as is timber, frequently referred to these days as forest products. Plus, there is the growing container traffic carried on the waterways, much more so in Europe than in the UK, but an increasing and welcome trend here.

8.3.4 Abnormal loads by waterway

An interesting development in inland waterway shipping is utilizing its potential for carrying very large (i.e. 'abnormal') loads (AILs) such as generators and transformers, thus removing them from the road network where they cause hold-ups to other traffic, create accident risks and the risk of damage to road surfaces, under-road services and to street furniture and roadside buildings. Specialist AIL carrier, Robert Wynn & Sons, has developed an innovative transportation system involving a Multi-Purpose Pontoon (MPP) vessel, the Inland Navigator, designed to carry loads up to 1200 tonnes and a converted inland waterway barge which can carry single piece loads of up to 300 tonnes in weight. The Inland Navigator was purchased and converted at a cost of £500 000, 99 per cent of which was met by a Department for Transport 'Freight Facilities' grant that was awarded to the company in 2002. The grant for the vessel underscores the Government's policy of maximizing the use of water for the movement of these disruptive 'abnormal loads'. In order to reach the National Grid Transco substation at Staythorpe, Nottinghamshire substation from the Humber Estuary, where the first two loads of 280 tonnes each were destined in January 2004, such loads would normally have been transported approximately 70 miles along the region's road network, travelling at between 5 and 12 miles/hour and causing significant disruption to other road users over a three day period. In this instance, the use of IWT via the river Trent meant that all road movement was avoided and consequently there was no congestion or inconvenience to vehicle drivers.

In yet another example of the potential for carrying abnormal loads by inland waterway rather than by road, Wynns used another of their special vessels, the *Terra Marique*, in April 2004 to move a Concorde aircraft fuselage from Heathrow airport, near London, to a new museum home near Edinburgh. After a short road haul from Heathrow to the largely unused Thames river port of Isleworth, the aircraft was loaded on the 80-metre-long barge for its journey down the Thames and around the English East coast to its destination on the Firth of Forth.

8.4 Inland waterways in Europe

Europe's inland waterways are extensive as we have already seen and carry significant volumes of traffic over long distances and across, these days, 'invisible' national borders within the enlarged EU. The general trend is towards increased tonnages, but there are shortcomings. For instance, in 2003, which was a very hot summer in Europe, the water levels in the Rhine basin were exceptionally low resulting in a situation where many of the heaviest (i.e. deepest draft) vessels were inhibited in their operations or largely prevented from plying their trade at certain times; there were reports that barges carrying oil products, for example, were able to load to only 40–50 per cent of their maximum capacity. Germany's Federal Statistics Office, for example, reported that carrying on its waterways in the first 9 months of 2004 was up quite significantly by 5.5 per cent on the same period in 2003 which itself was down on the 2002 figure on account of the 'low water' problem. The trend towards increasing container, 'box', traffic on Europe's inland waterways was amply demonstrated by results for 2004 published in a communiqué from French authority Voies Navigables de France (VNF) which showed that box traffic on inland waterways in France increased by 37 per cent to 74 000 teu (20-foot equivalent units; that is, 20 foot ISO containers) during the first 9 months of 2004. It reported (in January 2005) particularly strong growth in the River Rhone basin and that traffic in the Rhone basin had increased by 53 per cent and that in the Rhine and Seine basins and the Nord-Pas de Calais region had exceeded 30 per cent. Overall, the communiqué stated that traffic on French waterways increased in 2004 by 5.9 per cent to 57.9 million tonnes and by 5.5 per cent when measured in tonne-kilometres. Figure 8.4 shows the class categories for European inland waterway vessels and their respective tonnage capacities as solo vessels and when operating in 'push' convoys. Figure 8.5 shows a freight barge loaded with containers on the River Elbe approaching the Port of Hamburg. Such is the growth of this type of traffic that it is expected that this river alone will carry some 200 000 teu containers in the year 2010.

8.4.1 The Prospects for Inland Navigation within the enlarged Europe (PINE) project

In 2003 the EC commissioned a study on the current and future situation in the inland waterway sector and on its future prospects in the enlarged EU (i.e. from May 2004). A consortium consisting of four experienced organizations; namely, Buck Consultants International (The Netherlands), ProgTrans (Switzerland), VBD European Development Centre for Inland and Coastal Navigation (Germany) and via Donau (Austria) carried out the project; *Prospects for Inland Navigation within the enlarged Europe* (PINE). The study dealt primarily with freight transport and particularly analysed and compared the situation in the four main IWT corridors comprising:

- the Rhine and its tributaries (The Netherlands, mid-western Germany, north of Belgium, Luxembourg, France, and Switzerland);

Class	Type motor vessel	Tonnage	Comp. push convoy	Tonnage
0	Leisuret	<250		
I	Spits	250 – 400		
II	Kempenaar	400 – 650		
III	Dortm-Eemsk Canal Ship	650 – 1,000		
IV	Rijn-Heme Canal Ship	1,000 – 1,500		1,250 – 1,450
Va	Big Rhine Ship	1,500 – 3,000		1,600 – 3,000
Vb	Push Convoy (2)			3,200 – 6,000
VIa	Push Convoy (2)			3,200 – 6,000
VIb	Push Convoy (4)			6,400 – 12,000
VIc	Push Convoy (6)			9,600 – 18,000
	Push Convoy (6)			9,600 – 18,000

Fig. 8.4 Class categories for European waterway vessels (*Source*: ECMT).

- the East–West corridor (northern and eastern Germany, Poland, and Czech Republic);
- the Danube corridor (southeastern Germany, Austria, Slovakia, Hungary, Romania, Bulgaria);
- the North–South corridor (parts of the Netherlands and Belgium, France).

Isolated waterway systems in the UK, Finland, Sweden, Lithuania, Italy, Spain, and Portugal were also analysed.

The study identified the main characteristics and performance of inland waterways, concluding that:

> In the European Union of 15 Member States (EU-15 – i.e. prior to enlargement to 25 States in May 2004), IWT ranks third in inland freight transport after road and rail with 440 million tonnes per year; representing a 3.5 per cent market share in volume and 125 billion tonne-kilometres (t/km) or a 6.5 per cent market share in transport performance. IWT (inland waterway transport) has been unable to keep pace with the rapidly expanding road sector and has therefore gradually lost its modal share since 1970. However, its transport performance in t/km has grown by 23 billion t/km or over 20 per cent during the 30-year period [i.e. 1970–2000]. The enlargement in 2004 has added about 3.5 per cent in t/km (of IWT); the [future] accession of Romania and Bulgaria would raise that figure to five per cent.
>
> The importance of IWT in the various countries and regions shows a great diversity. Its centre of gravity undoubtedly lies in the Rhine corridor and this is not expected to change. The Netherlands, Germany and Belgium on their own provide some 113 billion t/km or 90 per cent of the IWT performance in the EU-15; IWT's modal share amounts to 40 per cent, 14 per cent and 12 per cent respectively in these countries and is thus higher than the rail share in the Netherlands and some regions in Germany. A precondition for this high modal share of IWT is adequate demand and above all the availability of an efficient infrastructure in terms of waterways and ports.'

The PINE study report defines what its compilers see as the outlook for IWT and offers a series of recommendations for consideration and action by the relevant participants. Since these points are considered to be crucial to the future success and prosperity of the IWT sector the author has taken the liberty of

Fig. 8.5 A freight barge loaded with containers on the River Elbe approaching the Port of Hamburg (*Source*: Port of Hamburg Marketing).

quoting here (with due acknowledgement to the EC and the team of consultants who carried out the study) the 10 main measures recommended and the three key precondition themes, which it suggests are essential if the opportunities and potentials of IWT are to be fulfilled; namely, equal and fair competition between transport modes (i.e. intermodal competition) as well as within the sector throughout different countries (i.e. intramodal competition); harmonized framework conditions (especially fiscal and legal enforcement of the rules); and adequate infrastructure of Europe's waterways and ports to provide seamless transport with vessels of sufficient scale across the entire European transport route.

A summary of the consultant's recommendations is as follows:

1. *Legislation* improve and harmonize the legislative framework. This is particularly important in regard to technical regulations for vessels, manning requirements, social standards, boat master's licences, and fiscal aspects; in this respect, the report suggests that harmonization, updating, monitoring, and control of existing regulations and imposing penalties for violating them is more important than inventing new rules.

2. *Waterway infrastructure* improve and extend waterway infrastructure. The main objective is to improve the waterway infrastructure by proper maintenance, by removing bottlenecks and, where necessary and justifiable, by extending it.
3. *Ports* enhance performance of transhipment interfaces. The report suggests that industrial areas along waterways could be made more attractive, by adopting a co-ordinated spatial planning policy in favour of IWT. Hinterland connections of ports by rail and road should also be improved to offer new or improved logistic services.
4. *Information systems* increase safety and efficiency of IWT, by means of River Information Services (RIS), the report suggests, can significantly contribute to increasing the safety and efficiency of IWT by giving additional attention to technical framework conditions, interfaces with external systems, applications, and acceptance of user/system operators as well as implementation support.
5. *Human resources* provide sufficient supply of work force and improve skills and social standards.
6. *To overcome the shortage of skilled nautical personnel* in some European countries and the shortcomings regarding knowledge, investing in education and training programmes would be of crucial importance, it suggests.
7. *Fleet* modernize the fleets. The long lifetime of vessels hinders scale enlargement as well as the adoption of innovative technologies. To create the right preconditions, additional systematic R&D programmes are needed to develop innovative modernization measures. Vessel technologies also need to be boosted to reduce negative environment emissions by new propulsion technologies and more environmental friendly engines as well as further increasing transport safety through the implementation of double-hull technology.
8. *Market* integrate IWT into logistics chains. The consultants recognize that the success of the inland navigation system in new and more demanding markets depends on its integration into and adaptation to the requirements of the entire door-to-door logistics chain.
9. *Sector* improve co-operation between and innovation of enterprises. The major tasks to be taken up by the sector itself (i.e. both professional organizations and IWT enterprises) are concentrated around co-operation and innovation.
10. *Image* enhance awareness and acceptance of IWT. According to the report, 'strong and in some cases unfortunately rather negative perceptions exist in the minds of policy and logistics decision-makers about the performance and competitive strength of the inland waterway system.' It suggests that improving this picture through systematic and professionally processed information would contribute to a better competitive position of inland navigation.
11. *Facts and figures* create a knowledge base on IWT. Facts and figures provide vitally important information for all decision-makers the report concludes, be they (barge/ship) skippers, (freight) shippers or policy makers. 'However, in various areas of this project, a clear lack of up-to-date, compatible, and reliable data has been identified. This is not only valid at regional, country, and EU level but particularly for the four corridors and the new Member States. Clearly, better knowledge than presently available is necessary.'

8.5 Short-sea and coastal shipping

Along with the increasing interest in freighting by inland waterway, there is renewed interest too in the possibilities offered by short-sea and coastal shipping as yet another environmentally viable alternative to road freighting. There is nothing revolutionary or very new about this mode; after all, cargo shipping has a very long history. But the new wave of thinking, driven by the obvious conclusion that we must find a viable and sustainable alternative to road freighting, sees the possibilities that exist with this mode, given adequate encouragement by way of government grants to develop adequate terminal facilities, for example. Also, increasingly, a change of mind-set is surfacing among consignors that speed of transit is not the be all and end all of freight movement, along with an awakening to the fact that a 'green' agenda has its own particular merits.

In the words of the EC's 2001 White Paper; *European Transport Policy for 2010: Time to Decide*:

> For centuries, sea and river dominated goods transport in Europe. Major towns were built on rivers or on estuaries and the large trade fairs in the Middle Ages were always held at river or seaports. Nowadays, despite a slight revival, water transport is the poor relation even though it is a mode which is not expensive and does less damage to the environment than road transport.

Short-sea and coastal shipping is far from being a new phenomenon. But what, perhaps, is relatively new is the concept that this form of shipping, as opposed to the deep-sea ocean trades, can be a vital congestion-buster for our overcrowded land-based transport systems, especially road freight transport. This mode of transport has, in fact, become so important for this reason that the EC made special mention of it in its White Paper. In a section of the paper headed 'Linking up the modes of transport'; it identified the so-called 'missing link' and at the same time made some important comments about shipping in general. It says:

> Shifting the balance between modes involves, looking beyond the rightful place of each particular mode and securing intermodality. The biggest missing link is the lack of a close connection between sea, inland waterways and rail.
>
> The European fleet has shrunk to the benefit of flags of convenience and fewer and fewer people want to become seafarers. There is a growing shortage of sailors in the European Union. Since the beginning of the 1980s, the European Union has lost 40 per cent of its seamen. There is a desperate need for merchant shipping officers. Between now and 2006 the Union will be some 36 000 sailors short, If properly trained and competent, sailors ensure the safety of shipping, efficient operation of vessels, proper maintenance, and reductions in the number of accidents and victims, and in sea pollution.
>
> Intra-Community maritime transport and inland waterway transport are two key components of intermodality which must provide a means of coping with the growing congestion of road and rail infrastructure and of tackling air pollution. Up until now these two modes have been underused, even though the Community has huge potential (35 000 km of coastline and hundreds of sea and river ports) and virtually unlimited transport capacity. The way to revive them is to build motorways of the sea …' (see below).

According to the EC:

> Short-sea transport grew by 29.6 per cent between 1990 and 1999. A total of 40 per cent of all trade within the European Union is now carried by sea, surpassed only by road transport. But its potential is much greater. Still needed are improved links to inland networks and the establishment of 'sea motorways' between key ports. Promoted in this way, short-sea shipping could take substantial volumes of goods traffic off Europe's congested roads and ease major road and rail bottlenecks.

In a briefing note on short-sea shipping in 2002, the Commission, as follows, identified its role, the challenges, the goals and the Community action needed to stimulate growth, which it defines as a priority.

- Short-sea shipping
  - an efficient and environmental friendly transport mode,
  - often, the most cost-effective means to shift long-distance traffic off Europe's roads,
  - an essential link to islands and outlying regions.

- The challenges
  - to promote short-sea shipping's reliability and frequency,
  - to ensure that its reliability, quality, and safety are recognized,
  - to remove unnecessary costs and delays at ports.
- The goals
  - to integrate shipping more fully in door-to-door freight transport services,
  - in conjunction with rail and inland waterways, to absorb predicted increases in EU goods transport.
- Community action
  - active promotion of short-sea shipping;
  - administrative simplification of shipping logistics;
  - targeted investment in infrastructure and support for R&D (i.e. research and development) where possible.

A strong proponent of the short-sea shipping concept is Geest North Sea Line BV, a mainstream intermodal operator that provides coastal shipping as well as road-borne services. Figure 8.6 shows one of the company's vessels 'Geest Trader' setting sail with a load of its own brand containers.

8.5.1 EU 'motorways of the sea' project

The EC's White Paper of 2001, *European Policy for 2010: Time to Decide*, proposed the development of 'motorways of the sea' under the general heading 'Linking up modes of transport' (i.e. in Part One of the White Paper). As the document states:

> Short-distance shipping has been around for a very long time: there are thousands of wrecked vessels around the Mediterranean dating back to Roman times. Short-sea shipping

Fig. 8.6 Geest North Sea Line vessel 'Geest Trader' setting sail with a load of the company's own brand containers.

> carries 41 per cent of goods traffic within the Community. It is the only mode of goods transport with a growth rate between 1990 and 1999 (plus 27 per cent) approaching that of road transport (plus 35 per cent). In millions of tonne-kilometres [t/km], the volume of trade carried between 1970 and 1998 increased by [a factor of] 2.5 representing 44 per cent of the total volume and 23 per cent of the total value of the goods transported within Europe. There are examples of efficient services between southern Sweden and Hamburg, between the ports of Antwerp and Rotterdam, and between south-east England and the inland port of Duisburg. But the current volume of traffic in Europe is well below potential capacity. Sea transport is not just a means of carrying goods from one continent to another; it is a real competitive alternative to land transport. For this reason, certain shipping links, particularly those providing a way around the bottlenecks in the Alps and Pyrenees, should be made part of the trans-European network, just like the motorways or railways. At national revel, shipping routes between European ports will have to be chosen to create networks, for example between France and Spain or between France and the United Kingdom. Similar routes will also have to be encouraged between Poland and Germany. However, these lines will not develop spontaneously. Based on proposals from the Member States, they will have to be 'sign-posted', notably by granting European funds (from the Marco Polo Programme and the Structural Funds) to encourage start-ups and give them an attractive commercial dimension.

Following the White Paper, the EC issued, in July 2004, a consultation document on the 'Motorways of the Sea' concept seeking views from interested parties. At the time of closure of the consultation period no less than 65 submissions had been received from national, regional, and local authorities, trade associations and single entities such as maritime and port operators and research institutes. Based on the comments received, and further consultation, the EC plans to prepare the actual implementation of the project and issue further documentation in 2005, the point reached as this book closed for press.

However, in the meantime it is useful to consider here the basics of the project to understand how it will be moved forward in the future. This is adequately explained in a document published by EUROPA on its web site in November 2004 as follows:

> In its Transport White Paper of September 2001, the Commission proposed the development of 'Motorways of the Sea' as a 'real competitive alternative to land transport.' To help these lines develop, the White Paper states that European funds should be made available. These 'motorways of the sea' should be part of the Trans-European network (TEN-T) – see Chapter 11 for a more detailed description of the various TENs projects. The 'motorways of the sea' concept aims at introducing new intermodal maritime-based logistics chains in Europe, which should bring about a structural change in our transport organisation within the next years to come. These chains will be more sustainable, and should be commercially more efficient, than road-only transport. Motorways of the sea will thus improve access to markets throughout Europe, and bring relief to our over-stretched European road system. For this purpose, fuller use will have to be made not only of our maritime transport resources, but also of our potential in rail and inland waterway, as part of an integrated transport chain. This is the Community added value of motorways of the sea. The adoption of Article 12a of the TEN-T Guidelines of 29 April 2004 gives a legal framework for funding the 'motorways of the sea' … and three main objectives for the projects (as follows):
>
> 1. freight flow concentration on sea-based logistical routes;
> 2. increasing cohesion;
> 3. reducing road congestion through modal shift.

Four corridors have been designated for the setting up of projects of European interest:

- Motorway of the Baltic Sea (linking the Baltic Sea Member States with Member States in Central and Western Europe, including the route through the North Sea/Baltic Sea canal) (by 2010).
- Motorway of the Sea of Western Europe (leading from Portugal and Spain via the Atlantic Arc to the North Sea and the Irish Sea) (by 2010).
- Motorway of the Sea of southeast Europe (connecting the Adriatic Sea to the Ionian Sea and the Eastern Mediterranean, including Cyprus) (by 2010).
- Motorway of the Sea of southwest Europe (western Mediterranean, connecting Spain, France, Italy, and including Malta and linking with the Motorway of the Sea of southeast Europe and including links to the Black Sea) (by 2010).

These corridors provide one essential part of the projects: the 'floating infrastructures' of our European seas. However, it is up to industry, Member States, and the Community to implement financially and operationally sound projects to use these maritime resources better for new intermodal maritime-based transport systems.

To make motorways of the sea a success, three conditions must be present for each project.

- First, in order to obtain the necessary concentration of freight flows, choices have to be made concerning ports and intermodal corridors and services.
- Second, all actors in the supply chain have to be committed to these projects.
- Third, motorways of the sea need to feature the best available quality throughout the chain in order to be attractive for users.

By 2010, a fully fledged network of motorways of the sea should be established throughout Europe on the corridors mentioned above.

8.5.2 Coastal highways

An imaginative concept designed to encourage further development of coastal shipping as a means of relieving the UK's heavily congested road network was launched in 2004 under the Coastlink brand name: Coastlink is an association of shipping companies, intermodal operators, ports, stevedores, logistics specialists, and shippers under the chairmanship of noted shipping expert David Cheslin. The idea (outlined in an article published in *International Freighting Weekly* (26 April 2004) under the heading; *UK Coastal Highways – An Impossible Dream*) was that a high-frequency integrated coastal shipping service linking key ports should be established. Particularly, UK east coast ports such as those on the river Forth, those on England's northeast coast, Humberside, Felixstowe, and those in the Thames/Medway area could ideally all be linked with high-frequency services, as could the west-coast ports on the rivers Clyde, Mersey, and the Bristol Channel, with cross-Irish Sea links to Belfast and Dublin. Large volumes of regular traffic would clearly be needed to sustain such services and there would be a need for much greater co-operation between port operators and between shipping lines and for a standard loading unit, essentially the shipping container.

But is this all a pipedream? It would seem to be so because by the end of 2004 there was a realization that the scheme was unlikely to work for two basic reasons; first that at present, both UK and European ports are stifled with container traffic causing severe bottlenecks and consequent delays to both shipping and to road and rail traffic delivering into and trying to clear containers out of the ports; and secondly that the costs and delays associated with all the handling, especially on short-distance traffic, kill off the economics of such a scheme. Basically, the conclusion must be that rather than keep lifting containers on and off road vehicles and enduring the delays associated with this, better to leave the container on the lorry and let the driver press on to its final destination. Scotland to London, northeast England to southern England and vice versa can be achieved in relatively few hours and almost certainly within the same day; a performance not likely to be achieved if it is loaded on to a ship. In this connection, there is a view that the introduction of the German LKW Maut lorry toll scheme in Germany from 2005, and similar schemes due to be implemented in the UK by 2008, will drive up road haulage costs and remove some of the speed and flexibility associated with this mode of transport in favour of both inland waterways and short-sea shipping. But without improvements to port infrastructures and particularly turnaround performance, this is unlikely to occur.

Having largely deflated the idea that short-sea shipping is about to boom, it is important to note that all is not lost on this front. At the end of 2004, a classic example arose to illustrate how short-sea shipping across the North Sea can be successful. Shipping line Sea wheel, which specializes in providing full load, door-to-door and quay-to-quay freight services between the UK, Ireland, and Continental Europe, has launched a daily lift-on/lift-off (LO-LO) container service between Rotterdam and Goole (northeast England) using 150–200 teu ships. It has also launched a Duisburg (Germany) to Goole service using its 'unique' river-hulled sea-going vessels, which, with their relatively shallow draft (3.6 metres) and the capability to lower their bridge housings, can sail 250 kilometres up the River Rhine to Duisburg and 40 kilometres down the Humber to Goole. This shipping line's philosophy illustrates how clever thinking can open niche markets; its strategies include using smaller ports with good transport links which are close to important markets as is the case with Goole. Similarly, in Europe, Duisburg is well away from the congestion at Rotterdam and is an important rail centre with connections offering overnight services to all main German railheads as well as daily connections to Italy, Austria, Switzerland, Poland, Croatia, and Hungary.

8.5.3 Roll-on/roll-off ferry services

Roll-on/roll-off (RO-RO) ferry services have been with us for many years; in fact, the development of this system for transporting road vehicles on short-sea crossings, such as those across the English Channel, has been one of the most significant advances in modern transportation. It was following the Second World War that the concept adopted for military tank landing craft of the type used for the D-Day landings was applied in the construction of merchant ships in the late 1940s and early 1950s. Thus commercial application of the RO-RO principle was borne. It proved to be extremely popular on cross-Channel routes, encouraged by the increasing demand for international road transport journeys between the UK and the Continent. For both the shipper and the ship operator, RO-RO shipping offers a number of advantages over traditional LO-LO cargo shipping; most notably the speed at which the crossing can be accomplished with modern ships and the rapid turnaround in port.

This mode of road–cross–Channel shipping is a classic example of intermodalism at its most efficient. Freight is loaded on a lorry at an inland location, is then driven to the port where the vehicle rolls straight on board the ferry ship via a loading ramp/bridge and is secured in the hold for the crossing. The driver meantime, takes his break in the comfortable lounge or restaurant facilities provided. On arrival at the destination port, the vehicle is driven off the ship invariably within a few minutes of docking, thereby dramatically cutting loading and unloading times, and the driver continues his journey to the final destination. The significant advantage of this system is that the load travels undisturbed from the time of original loading and securing right through to the consignees' premises. Also, if the vehicle travels accompanied by its driver through to the final destination, as most do, the consignor can take confidence from the fact that the driver will take proper care to ensure that it arrives at the right place, hopefully at the right time, and with no, or only the minimum, of interference apart from any inspection by officialdom or in the event of unfortunate accident.

Modern RO-RO ferry ships are fast, travelling at up to 25 knots, and the largest can carry up 2000 passengers and literally hundreds of cars. Specialist freight-only services are operated by such companies as

Sea France, Stena Line, Norfolk Line, and P&O Ferries, the largest operator with some 30 per cent of the cross-Channel market (of which about 50 per cent of its net revenue comes from freight traffic, its only competitor from a UK perspective is Eurotunnel with 40 per cent of the cross-Channel market share). Many of the ferry ships on the cross-Channel routes can accommodate 80–120 heavy trucks, turning round in port in no more than 1 hour in many instances, and providing such amenities as; electrical plug-in for temperature controlled transport; hazardous cargo arrangements, including on-deck cargo space; special arrangements for wide loads, demountable, and unaccompanied traffic; as well as a range of comfortable driver facilities.

8.6 Container shipping

Deep-sea container shipping is largely outside the remit of this book, but nevertheless the author feels that it is important to recognize that ISO containers, shipped worldwide in their thousands, almost invariably start and end their global journeys on the back of a lorry maybe in China, Japan or Australia at one end of the route and to an inland destination in the UK at the other end; this is intermodalism. It is important too to recognize the enormity of this business. According to the Institute of Shipping Economics and Logistics (ISL) *Shipping Statistics and Market Review* (SSMR), container traffic amounting to some 219.5 million teu (13 per cent up on 2002) was recorded as being shipped in 2003 through the world's top 62 ports, of which 28 were Asian, 17 European and 15 were American, each having a throughput of more than one million teu. Remarkably, in 2003, approximately 63 per cent of world container traffic was attributable to Asian ports alone, Europe accounted for 20.2 per cent and America 15.4 per cent. Also in 2003, 172 new fully cellular container ships with a combined capacity of 570 000 teu were added to the world container-ship fleet. Most of these ships followed the trend towards ever-greater capacities, on average their capacity was 3320 teu, but 26 of them were of 6000 teu capacity or above – around 75 per cent of the world's merchant fleets comprise container vessels. The very largest of container ships in service at the beginning of 2004, according to SSMR, were four vessels operated by (Orient Overseas Container Line OOCL, registered in Hong Kong) with capacities in excess of 8000 teu.

On a smaller scale, both in the UK and in Europe, smaller ships plying short-sea and coastal trades and, increasingly, many river/canal barges are constructed or adapted for carrying ISO containers, many of them carrying no more than 100 or so units. In his excellent little book, *Short Sea Shipping: 2003/2004*, published by Ships in Focus, Preston, England, author Gilbert Mayes notes an increasing number of short-sea vessels engaged in the container liner trades with a corresponding increase in slot capacity. While their capacity is recorded in 20-foot equivalent units (i.e. teu) in the conventional shipping manner, many of these smaller ships, he notes, in fact now carry the bulk of their loading in 40-foot containers or even 45-foot units. He also comments that:

> Feeder container vessels have also shown a small increase though the main change has been in the additional ports that are now visited, often to pick up and set down a small number of containers, again reducing road traffic.

9

Environmental and Economic Issues

Environmental pressures are driving the Euro-wide freight industry to take a close look at how it conducts its business. Both in the UK and in Europe there is increasing concern about the detrimental (some would say catastrophic) effects on the environment of a growing road-freight industry; traffic congestion, air pollution, lorry noise and vibration, high levels of energy consumption, and road accidents being among the major worries. But against a scenario of ever-increasing volumes of long-distance heavy-lorry traffic moving across Europe as a result of expanding Single Market trade there are viable alternatives.

9.1 An environmental solution

Switching long-distance freight from direct door-to-door delivery by road haulage to a system which combines the respective merits of road, rail and/or water transport is a sensible and operationally acceptable alternative. Rail freighting, which links many individual consignments into single 1000/1200-tonne train loads, and causes much less of an adverse effect on the environment while doing so, is ideally suited for undertaking the long-haul trunk leg of freight journeys between terminals, while lorry transport has the flexibility to service these terminals with efficient local collection and delivery of individual freight consignments. A relatively similar scenario applies with waterways intermodalism where many individual heavy-lorry loads may be transferred to river or canal barges capable of carrying anything from as few as 30 ISO-type containers to in excess of 400, 20-foot equivalent units (i.e. teu), thereby reducing the adverse effects on the environment of road freighting and the consequent traffic congestion. Together, these combined road–rail or road-waterway modes provide the most environmentally sound solution to the problem of moving large volumes of long-haul freight. In the circumstances, bearing in mind that for short-distance operations neither rail nor waterway is viable anyway, intermodal transport in this form is the most efficient mode, and it is the solution that the public at large would invariably see as the most desirable solution to many road traffic congestion problems.

9.2 Freight by road, rail, or waterway?

It is widely recognized and accepted that lorry traffic is a heavy pollutant of the environment; far more so than rail freighting. But, as we have seen over so many years, commercial demand favours road haulage to the point where more than 80 per cent of all freight moved in the UK goes by road, and a broadly similar situation applies in Europe. There are many reasons for this, not least the past inabilities and failures of the railways to provide the flexibility and levels of service, to say nothing of acceptable pricing structures, required to meet the demands of a sophisticated consumer-driven, Just-in-Time (JiT) society.

However, a straight switch to rail of all freight currently carried by road, as the anti-juggernaut brigade advocates, is just not practicable. Supermarkets, High Street stores, DIY warehouses, and the many other retail outlets where people shop for basic needs and expect to find their favourite product brands on the shelves are not and are never likely to be rail connected. Similarly, most other commercial and industrial premises that receive and despatch goods are not rail connected either and never could be for economic reasons and because they have no space to accommodate rail sidings and the loading and unloading paraphernalia that goes with them.

For freight movements in and out of such premises, whether the goods are for delivery locally or to distant domestic or even international destinations, only a road vehicle is suitable, provided it is of a size and has sufficient manoeuvrability to enter the invariably cramped loading bays. Road transport has proved the efficiency and flexibility with which it can provide a freight service to the smallest of commercial enterprise trading under a railway arch or to the largest of manufacturing premises. Rail is both physically incapable of handling this business and does not want it; it is not its operating forte.

Collecting and delivering individual consignments, responding to urgent, short-notice demands, adhering to the strict timetables so often vital, for example, in supermarket delivery operations, providing guaranteed (next-day or timed) deliveries of important consignments, and meeting crucial JiT delivery schedules to keep continuous manufacturing processes supplied, are all part of the everyday business of road transport which rail cannot accommodate. The case of bulk traffics, of course, is a different matter, rail being more suited to handling large-volume flows of coal, although much less these days than hitherto, chemicals, petroleum, aggregates, quarried materials and such like, but then much of this traffic, where it is feasible and economic to do so, is already moved by rail.

So, no matter how powerful the arguments and the environmental pressures to get rid of the heavy lorry, it is destined to remain a major feature of the freight scenario. But this is not to say that we should have to tolerate the worst excesses of its polluting potential. Considerable progress has been made over recent years by both vehicle and component manufacturers, under the pressure of an increasingly stringent legislative regime, a more environmentally aware market place and their own social consciences, to reduce exhaust emissions and the noise emanating from heavy vehicle power units, drive trains and tyres. Furthermore, it is a specific statutory offence for drivers to cause or allow their vehicles to make excessive noise which could be prevented, or to belch out smoke and fumes as a result of mechanical defects or poor maintenance. Most lorries do not do this these days, although it is recognized that there are still rogue operators who persist in cutting maintenance corners and drivers who could not care less.

9.2.1 The alternative solution

Between the two extremes of everything by direct door-to-door road delivery or all freight by rail or waterway, a middle road is possible, the alternative solution referred to above. This is, as we have seen above, the use of the heavy lorry for local collections and deliveries and rail or waterway for the long haul where the economics and practicalities begin to make sense; in other words a viable combination of modes, namely, intermodal transport. This is not only a practical solution but an environmentally sound solution too, exploiting what the individual modes are respectively good at and at the same time eliminating a significant proportion of the worst environmental effects of lorries: the pollution, the noise, the traffic congestion, and the use of much less scarce energy resources into the bargain. There is a widening recognition of this option as a modern solution to the 'all-by-road' or 'all-by-rail' prejudices one so often hears.

As this book has already shown, intermodal transport is fast developing and expanding and one of the major reasons for its doing so is the environmental pressure emanating from the European Commission (EC), from national governments and from the public at large. This pressure is forcing freight shippers to include an environmental factor in the formula they use for devising by what means and by whom their freight is to be consigned. Increasingly, they are considering the possibility that road haulage combined with rail or waterway transport can provide an environmentally acceptable solution while at the same time meeting economic criteria and demanding service schedules.

It has to be said that intermodal transport is not a panacea for all transport ills and woes; for example, it will not eliminate all environmental blights, nor indeed can it cope with all traffic. For some traffic intermodal transport is just not suitable, for others it does not offer the level of service required; although, it is worth pointing out that despite the fixed mind-set with which many shippers conceptualize their freight movements as always being of the utmost urgency, in many instances this is just not the case. Furthermore, for some freight movements intermodal services do not operate on the routes required and for others it will just not be economically viable. And there is still the case that staunchly anti-rail or pro-road consignors will *never* be convinced of the suitability or reliability of the road–rail/water option, and nowhere on the horizon is there any suggestion of forthcoming legislation compelling them to make a mandatory switch of freight from road to either of these alternative modes. A keystone of the ECs campaign to promote the development of intermodal transport is that there should be a choice for the user – and this does not just mean a choice of road–rail services, but also a choice as to whether or not to use this combination of modes at all – there are other options, such as the use of inland waterway transport, or short-sea shipping.

9.2.2 Environmental benefits

There is absolutely no doubt whatsoever that intermodal transport promises significant benefits in regard to environmental improvements. In fact, as a general comment, it could be said that greater use of this form of transport holds out the prospect of reducing, albeit only relatively marginally, the worst elements of environmental despoliation caused by heavy road-freight traffic. To understand the benefits that a progressive switch from door-to-door road delivery to intermodal transport would bring, one has to consider the specific effects which road traffic has on the environment. For every one of the following environment-blighting penalties described below, a transfer of more freight traffic to other modes would produce a marginal benefit. Given a sufficient conglomeration of such marginal improvements we can but hope that, in due course, really tangible benefits will accrue.

9.3 The EC's view

According to the EC in its document, *The Future Development of the Common Transport Policy*, published in 1992, the growth in transport activity throughout Europe in the 1970s and 1980s not only led to pressure on capacity (and consequent problems for the infrastructure), but also raised broader environmental issues; for example, the effects on transport of economic and technological developments have exacerbated the areas of conflict between transport and the environment. These areas of conflict take the form of energy consumption, operational pollution, land-intrusion, traffic congestion, and the risks inherent in the carriage of dangerous goods. The Commission said at that time that since the beginning of the 1970s, transport as a whole (road, rail, waterway, sea, and air) had become a major consumer of non-renewable energy. In particular, road transport's energy consumption had risen by 103 per cent, representing an annual growth rate of 3.8 per cent per annum, greater even for that of air transport at 3.6 per cent per annum. The available data for atmospheric pollution caused by road transport showed a substantial increase over the previous two decades for a number of pollutants. The most important of these is the emission, via the vehicle exhaust, of carbon dioxide (CO_2), the main man-made 'greenhouse gas' that contributes to the global warming effect, which increased by 76 per cent between 1971 and 1989; an average annual increase of 3.2 per cent.

Other pollutants include:

- nitrogen oxide (NO_x), which contributes indirectly to the 'greenhouse effect' and directly to acid rain and the build-up of tropospheric ozone, which increased by 68 per cent;
- particulates that are detrimental to human health which increased by 106 per cent;

- hydrocarbon (HC) emissions which contribute to the 'greenhouse effect' and the build-up of tropospheric ozone and may have potential carcinogenic (i.e. cancer inducing) effects, which increased by 41 per cent.

The Commission also pointed out further detrimental effects of road transport. It causes water and soil pollution and is a source of noise and vibration while transport infrastructure (i.e. roads and road-related infrastructures) causes land-intrusion with a permanent and often irreversible impact on the landscape and the urban environment. Increasingly, too, road transport is the cause of traffic congestion. For example, the Commission stated that the volume of road traffic (in vehicle/kilometre) had doubled between 1970 and 1989 for both car and freight vehicles with an average annual growth 3.7 per cent. One of the main contributory factors for this growth, it cites, was the development of car ownership in the European Union (EU) (and on the basis of trends since 1975, the number of cars in the Union is expected to increase between 25 and 30 per cent during the period 1990–2010). Finally, road transport is also, unfortunately, the cause of physical damage to persons and property, and of fatal accidents. As the Commission said:

> The resulting economic costs are impossible to quantify but despite progress made in the prevention of accidents in recent years, they remain a substantial burden, financial and otherwise, to be borne by society as a whole.

9.4 The environmental impact of transport

As we have seen above, road freighting impacts on the environment in various ways, but it is useful to consider these differing impacts in more detail, as the Commission's report also confirms.

9.4.1 Energy consumption and pollution

All power-driven transport consumes energy and causes operational pollution. The energy consumption of the transport sector (of the economy) represents 30 per cent of the total final energy consumption in the EU virtually equivalent to that of industry. In the case of road transport, operational pollution mainly takes the form of gaseous emissions such as CO_2, HC, NO_x, carbon monoxide (CO), volatile organic compounds (VOCs) and sulphur dioxide (SO_2) plus noise, vibration and visual intrusion. Road transport is the worst offender in terms of energy consumption and operational pollution; it consumes over 80 per cent of the total energy used in the transport sector and contributes over 75 per cent of its total CO_2 output.

9.4.2 Congestion

Traffic congestion is a major problem causing delays to essential journeys, driver frustration and excessive costs by way of increased fuel usage and additional driver costs (e.g. for overtime payments), and it adds significantly to the amount of pollutants being discharged into the atmosphere thus exacerbating environmental inefficiency. Largely, traffic congestion occurs where infrastructure capacity, due to old infrastructures or poor highway planning, is insufficient to cope with demand, particularly at peak periods. It is most prevalent in the core regions and around large conurbations. Apart from improving road use planning, solutions to the problem lie mainly in devising schemes that either discourage traffic in particular areas or on specific routes or schemes that encourage a diversion of traffic to alternative modes. The former can be achieved by the application of tolling systems whereby users of key routes must pay more for the privilege of doing so; the latter, as we are seeing, by governmental persuasion and financial incentive. Not all road tolling schemes add much to the trend towards intermodalism, especially those devoted to driving traffic away from city centres (these are largely aimed at car and light van users) but the motorway toll initiatives that affect heavy lorries may certainly make a contribution in this direction.

9.4.2.1 Congestion charging schemes

One of the most significant measures being adopted to reduce road congestion is to impose charges on vehicles for the use of the road infrastructure; in central London, for example, or on the motorways in Germany; or, in due course, on parts of the UK road system.

London congestion charging Daily congestion charges are payable by vehicle drivers entering a central area of London that covers routes crossing the River Thames by Lambeth, Westminster, Waterloo, Blackfriars, Southwark, or London Bridges; but not included in the charging area are Euston Road, Tower Bridge, Elephant and Castle, Vauxhall Bridge, Victoria, and Park Lane. The charge applies only on weekdays only between the hours of 7.00 a.m. and 6.30 p.m. Cameras, located at the boundary intersections and at random points within the zone, record vehicle registration numbers which are cross-checked by computer with records of charges paid. Vehicles running on gas or electricity are exempt after paying an initial registration fee. All other commercial vehicles are included in the scheme, unless they are emergency service vehicles or are vehicles on municipal duty, not competing with private firms. Drivers who regularly enter the charging zone may pre-register and pay a fee otherwise payment of the daily fee may be made by credit card (a call centre processes credit-card numbers) or by buying carnets from newsagents and convenience stores and from ticket terminals in car parks, hospitals, and other local amenities. Any driver who has not paid prior to entry into the charging zone may pay later that day. Failure to pay the daily charge results in evaders incurring a penalty, issued to the vehicle owner (not necessarily the driver), but this is halved if paid within 2 weeks. There is an appeals procedure. The scheme is predicted to reduce traffic by a modest amount of only about 10–15 per cent, but it is anticipated that at least 34 other towns and cities across the UK may follow London's lead; interestingly, the citizens of Edinburgh rejected such a scheme early in 2005. The way for making such congestion charges was paved by the UK Government's 1998 New Deal for Transport White Paper and its Transport Plan 2010 published in July 2000, but will it help to divert more freight traffic from the roads on to rail or waterway? In the author's view this is highly unlikely.

Proposed Lorry Road User Charge While the London congestion charge may not divert freight traffic to alternative modes, the Lorry Road User Charge (LRUC) being proposed for the UK to take effect from 2008 very likely will make a contribution towards this aim. The toll charges will undoubtedly add to road haulage costs, irrespective of the Government's best intentions that it should be a 'fiscally neutral' tax scheme, and this will be the incentive that encourages lorry owners to seek alternative means of moving freight over longer distances. Although at the time of writing this text (early in 2005) the scheme is a long way short of finalization. The outline is that it should apply to all lorries over 3.5-tonne gross weight, regardless of their nationality, using all UK roads (with potentially higher tolls payable for motorway use) and that they should contribute on a fair and equal basis towards the costs they impose. The charge will relate to the distance travelled as monitored by vehicle-mounted sensors via satellite links and for UK-based road hauliers the fuel duty they pay will be reduced by an amount proportionate to the road user charges they have to pay, hence the concept of fiscal neutrality … hopefully.

LKW Maut (Germany) Germany has a scheme for charging heavy lorries over 12-tonne gross weight for use of the country's 12 000 kilometres of highways. After a couple of false starts the scheme finally took-off from January 2005. It applies to relevant trucks, equipped with 'on-board' units (OBUs), which are tracked by global positioning satellite (GPS) from outer space. Road users must register with the toll operator, Toll Collect GmbH, in advance of road usage and fit the vehicle OBU, which calculates the toll due based on the road being used and the distance travelled. A manual back-up system is available which ensures that non-registered users and non-OBU-equipped trucks still pay the toll.

9.4.3 Land use

Transport affects the environment not only in terms of energy consumption and operational pollution, its infrastructure, through land-take and -intrusion, has a permanent and often irreversible impact on the environment. Road transport is again the worst offender with a land-take that considerably outstrips that of the railways and other modes. Within the Community the total land area used by road transport is calculated to amount to 1.3 per cent against 0.03 per cent for the railways. Measured by unitary means on the basis of the infrastructure needed to move one transport unit (in persons or tonnes of freight) over a given distance, the EC says that road transport is the least efficient. There is continuing concern that growing road traffic volumes will result in ever-increasing demands for new road building, but this is unsustainable. We cannot keep concreting over the countryside simply to provide infrastructure for more lorries to carry more goods and forever greater numbers of private cars to go wherever they will. We have railways, we have waterways, and we have a very long coastline, we must use these to much better effect by transferring more freight onto rail or waterway transport modes.

9.4.4 Dangerous goods

The carriage of dangerous or, as they are sometimes called, hazardous, goods carries a potential risk for the environment as well as for human life. Road transport accounts for a large proportion of such carriage, so the risks exist for accidental pollution of the environment and, since this form of transport invariably operates within relatively close proximity to the public, harm to people as a result of spillage incidents and more serious lorry accidents. While the road carriage of dangerous goods is very strictly controlled by a large body of complex Euro-wide legislation that is stringently enforced on the ground by the police and other agencies, this in itself is not sufficient to eliminate the risks mentioned above. The preferable solution must be that as much of this traffic as possible should be conveyed by rail and waterway; both modes being in much less close proximity to the general public.

9.5 'Something must be done'

As the Commission makes clear in its 'Future Development of the Common Transport Policy' document, mentioned above, not all of the environmental problems described above can be laid at the door of road-freight transport alone, indeed far from it. But it does highlight the fact that transport is generally detrimental to the environment, and that leaves us with the thought that anything that can be done in one sector alone (namely the one we are principally concerned with here, road-freight transport) will help to ameliorate the problems outlined. Hopefully, of course, those more concerned with other transport sectors, and the car-using public at large, will all be playing their respective parts in reducing harm to the environment. But waiting and hoping is no good. Some positive action needs to be taken because forecasts of growth in transport demand show that in what the Commission calls a 'business as usual' scenario, with a reasonably favourable economic climate, the expansion of the road sector is likely to be buoyant; – in fact, it predicts a near doubling of demand for passenger and freight transport in the foreseeable future; although technological progress (on vehicle exhaust emissions for one thing) and measures already taken will, it says, mitigate the environmental impact. However, in the absence of additional policy actions, significant worsening of the situation is still likely as regards pollution, notably CO_2 emissions, traffic congestion and road accidents. And as a final warning, the Commission says that even if a lower economic growth slows the rate of deterioration for a time, the risk of the development of the transport sector being unsustainable in the medium to long term due to its broad environmental impact remains real. In other words, the writing is on the wall: 'something must be done'. But, improvements in the environmental performance of transport, in terms of reducing pollution, will only be achieved by the effective participation of all those involved. This means the necessity for rationalizing and managing transport demand by means of a shift towards more environmentally friendly modes and collective transport, as well as better utilization of existing capacity. The development and promotion of intermodal transport systems are seen as the best solution for inducing such a modal shift for freight transport.

9.6 Vehicle exhaust emissions

It is an inescapable fact that all vehicles emit pollution into the atmosphere. Most of this is attributable to carbon emissions that result from the vehicle engine combustion process. In the complete combustion of hydrocarbons (HCs) the emission products are CO_2, water, and unaffected nitrogen, but where the combustion of HCs is incomplete, as is often the case with motor vehicles, the emissions comprise unburned HCs, nitrogen oxides, CO, CO_2, and water. It is these emissions that are responsible for carbon monoxide poisoning and for damage to the global environment. In fact, among these emissions it is CO that presents the greatest threat to human health (and life) and it is CO_2 (mainly resulting from chemical processes as the CO emissions pass through the vehicle's catalytic converter) which blights the atmosphere and causes the 'greenhouse' effect.

As mentioned earlier, vehicle and component manufacturers are striving to achieve solutions to the environmental harm done by heavy-goods vehicles, partly as a result of government regulation. In fact, since 1 April 1991, all newly registered diesel-engined goods vehicles exceeding 3.5-tonne gross weight have been required to comply with EU Directives which set stringent gaseous emission (i.e. exhaust emission) limits. The EU has a three-part legislative programme (under Directive 91/542/EC) to reduce the amount of NO_x, CO and unburned HC discharged into the atmosphere from vehicle exhausts. The first stage of this programme (the so-called Euro-I standard) applied to all new goods vehicles from 1 July 1992; tougher Euro-II standards applied to new diesel-engined goods vehicles over 3.5-tonne gross weight from the end of 1996 and even more stringent heavy-lorry emission controls came into effect from 2001, with the Euro-III standard. The Euro-IV standard is to be brought into force from 1 October 2006 and Euro-V from 1 October 2008 (see table at Figure 9.1).

9.6.1 Comparative analysis: energy consumption and CO_2 emissions

A scientific study; *Comparative Analysis of Energy Consumption and CO_2 Emissions of Road Transport and Combined Transport Road/Rail*, published by the Institut für Energie und Umweltforschung, Heidelberg

	Date and category	Test cycle	CO	HC	NO_x	PM	Smoke
Euro-I	1992, <85 kilowatts	ECE R-49	4.5	1.1	8.0	0.612	
	1992, >85 kilowatts		4.5	1.1	8.0	0.36	
Euro-II	1996.10		4.0	1.1	7.0	0.25	
	1998.10		4.0	1.1	7.0	0.15	
Euro-III	1999.10, EEVs only	ESC & ELR	1.5	0.25	2.0	0.02	0.15
	2000.10	ESC & ELR	2.1	0.66	5.0	0.10 0.13*	0.8
Euro-IV	2005.10		1.5	0.46	3.5	0.02	0.5
	2008.10		1.5	0.46	2.0	0.02	0.5

* for engines of less than 0.75 dM^3 swept volume per cylinder and a rated power speed of more than 3000/min.

Fig. 9.1 Summary of diesel engine emission standards and their implementation dates (*Source*: EC).

GmbH (IFEU), and Studiengesellschaft für den kombinierten Verkehr e.V.(SGKV), in 2002 and jointly commissioned by the International Road Transport Union (IRU), and Bundesverband Güterkraftverkehr Logistik und Entsorgung e.V. (BGL), has determined that combined transport, using both rail and road, is *not* inherently superior to pure road transport in energy consumption and CO_2 emissions. The study was carried out by the two German research institutes: IFEU, which looks at energy and environmental issues, and SGKV, which specializes in combined transport. Dr Ulrich Höpfner of IFEU said: 'our research shows that shifting freight from trucks to trains can reduce primary energy consumption or CO_2 emissions but not every time and not automatically'. Hermann Grewer, Vice President of the IRU and President of the BGL, commented:

> The ECs Transport White Paper aims to shift freight from road to rail and other modes which it claims are more 'environmentally friendly'. But the Commission did not provide evidence to support this claim. Our study, the first ever in-depth analysis in this field, shows that shifting freight from road to rail will not really reduce the overall environmental impact of transport.

The study shows, in fact, that combined transport only offers a significant environmental advantage over all-road transport for two reasons.

> One is, paradoxically, where the electricity used to power the trains comes from nuclear power stations and other non-fossil-fuel sources. The other is where feeding and distribution by road are over short distances or in the same direction as the overall journey.

This study is particularly interesting because, among many diagrams and sets of tables, it illustrates, taking an example of 19 key routes, comparisons of primary energy consumption and CO_2 emissions between combined road–rail transport and pure road transport. In the case of energy consumption, a variation of between three instances where consumption was up to 15 per cent higher for combined transport than for pure road transport and two instances where it was more than 40 per cent lower. In eight instances combined transport consumption was up to 20 per cent lower than road transport and in a further six cases the consumption for combined transport was between 20 and 40 per cent lower than for road transport. When it comes to CO_2 emissions, the study showed that in only two instances was combined transport inferior to road transport – by producing 3 per cent higher levels of CO_2, whereas in the other 17 instances combined transport out-performed pure road transport by between 20 per cent and more than 50 per cent. It is important to point out, as stated in a footnote to the study, that an underlying factor of these results lies in the fact that the higher the nuclear component in the electricity generation mix used to power trains, the lower the overall CO_2 emissions generated by combined transport; it suggests that a unit of electricity from a fossil-fuel power station results in CO_2 emissions more than 12 times greater than the same unit produced in a nuclear power plant.

9.6.2 CO_2 reduction

Yet another report on the same subject; *CO_2 Reduction Through Combined Transport* was published in July 2003 by the International Union of Combined Road–Rail Transport Companies (UIRR). In its introductory letter to the survey the UIRR stated:

> In the light of the commitments on reducing the emission of greenhouse gases by 2008–2012 incumbent upon the EU within the context of the Kyoto protocol, the UIRR had undertaken a detailed survey within the framework of the Pilot Actions for Combined Transport (PACT) programme (see Chapter 10). This survey consisted of measuring the savings made in terms of energy consumption and the reduction of CO_2 emissions when using combined rail–road transport instead of transport wholly by road. Reductions in CO_2 emissions, to quote just these, going from an average of 18 per cent when using the rolling road technique to 55 per cent and over when unaccompanied combined transport

> is used, must encourage all decision-makers to fully support the widest possible deployment of combined transport services.

This survey report, like the one described above, uses a variety of illustrations and graphs to make clear the points it is making. Of particular interest are some of its conclusions. Set against the background of the 1997 Kyoto Protocol commitment by the EU to achieve emissions (i.e. greenhouse gases) reductions, including CO_2, by 8 per cent by 2008–2012 compared with the 1990 level, it noted that by 2002 the CO_2 level had been maintained at the 1990 level (but had not reduced) and in fact that the global warming rate over the last 25 years had accelerated. It also noted that transport is becoming less and not more environmentally sustainable; that transport is the fastest growing energy consumer in the EU; and that, as we already know, CO_2 emissions from transport are a major contributor to the greenhouse effect. The report concludes with a number of key points and recommendations, namely:

> Basically, all savings in CO_2 emissions realized by intermodal transport can be traced back to three sources:
>
> - Intermodal transport consolidates smaller loads into large volumes which can be moved with less energy consumption per unit.
> - Combined transport shifts traffic from road to rail. The rolling friction of steel wheels on rails is less than that of rubber lorry tyres on road.
> - Rail transport, at least in long-distance traffic on the main axes, uses mainly electric traction in Europe. The relative savings in CO_2 are higher than those of energy, where electric energy is generated by waterpower or nuclear power.
>
> There are a number of further parameters to reduce the CO_2 emissions:
>
> - Encourage and promote all technical, operational and commercial solutions that increase the average capacity use per intermodal transport train.
> - Encourage and promote all infrastructure upgrades, which will allow the operation of longer and heavier trains.
> - As the saving of energy is achieved mainly during the rail carriage: construction of terminals and promotion of rail infrastructure (alpine base tunnels) in order to reduce rail distance and grade.
> - Set fair and equal framework conditions for all modes including external costs, then competition will drive operators to minimize distance, energy input and to maximize the capacity use.'

The table at Figure 9.2 shows the levels of CO_2 emissions by each transport sector.

Means of transport		*Share in total CO_2 emissions*	
1.	Road: total	79.7%	
	– private car		55.4%
	– goods vehicles		22.7%
	– buses + coaches		1.6%
2.	Aviation	10.9%	
3.	Rail: total	3.9%	
	– passengers		2.8%
	– goods		1.1%
4.	Inland waterways	0.7%	
5.	Other transport	4.3%	

Source: Commission of the European Union/TNO Policy Research 1991.

Fig. 9.2 Levels of CO_2 emissions by transport sector (*Source*: EC).

9.7 Environment reports

It is useful to mention here two further reports that have considerable bearing on the subject of heavy lorries and the environment. The first was an account of an international seminar; *Freight Transport and the Environment*, ECMT, Paris (1991), which examined in detail, against a background of growing freight transport volumes and worsening environmental problems, how it would be possible to obtain the best balance between the wealth-creating characteristics of freight transport and its welfare-damaging effects on the environment. Broadly, the seminar ended with three main conclusions:

- That action must be taken against excessive noise and gaseous emissions, in particular that a convention should be established for certifying international haulage vehicles as meeting state-of-the-art technology performance (see below, Section 9.7.1).
- That action must be taken against excessive speed. (This, of course, has already happened with the introduction of legal requirements for goods vehicles to be fitted with speed limiters (in the UK over 7.5-tonne vehicles, in Europe over 12-tonne vehicles).)
- That action should be taken against excessive distance travelled by road freight (by introducing a regime of higher prices – road pricing). (The German LKW Maut and the UK's proposed LRUC schemes are evidence that this is happening.)

The second report was based on a study by NEA Transport Research and Training in the Netherlands; *The Transport of Goods by Road and its Environment in the Europe of Tomorrow* (summarized original version), for the IRU, Geneva (1992). Its purpose was to 'throw new light on the means to be employed to achieve the objectives of transport policy', which are based on two ideals:

- that the road transport of goods is pervading our infrastructures and harming our environment;
- that rail transport, and more particularly, combined transport, offers a clean alternative to the transport of freight by road; by virtue of this fact appropriate measures must be taken to shift long-distance road transport to the railways.

The reports conclusions are that:

- Technological developments (by regulatory measures) are one of the principal means of reducing emissions; it projects overall reductions of around 34 per cent by the year 2010.
- Large capacity goods vehicles (over 16-tonne gross weight) have a lower level of pollutant emissions, therefore the use of these should be encouraged in preference to smaller vehicles.
- The transfer of goods on journeys over 1000 kilometre from road to rail would reduce NO_x emissions.
- Empty journeys, despite regulatory measures (i.e. freeing cabotage, etc.) will remain a major factor (in creating NO_x emissions) comprising some 37.5 per cent of vehicle/kilometre travelled in Europe on inter-regional transport.
- The adaptation of road infrastructures to the level of demand for road transport (thus avoiding the problems of congestion and stop-go traffic) will permit a reduction of two-thirds of the pollution emitted by goods vehicles.

It adds two riders:

- that while the problem of congestion is considerable in terms of energy costs, it can triple the levels of emissions on sections of problem roads;
- harmonization of vehicle weights will have a positive effect (on environmental pollution) if it results in the authorization of larger capacity vehicles.

9.7.1 Environmental plaques on heavy trucks

In response to required action against lorry noise and gaseous emissions, the ECMT has devised a scheme for marking environmentally acceptable heavy lorries (the so-called 'green lorry' scheme). Under this scheme, vehicles are marked on the front with a circular plaque, and carry an environmental certificate of conformity, denoting that they meet specified minimum standards for noise emissions and chemical exhaust emissions for diesel engines as well as certain other safety requirements. Compliance with these standards allows for the issue of an additional quota of multilateral quota road haulage permits as well as the plaques indicating whether they are a 'green' or 'greener and safe' lorry, the former with Euro-I specification engines and the latter with Euro-II specification engines. Vehicles with Euro-III engines are required to display a plaque showing the number '3'. These requirements are set out in the ECMT's document *Guide for Government Officials and Carriers on the Use of the ECMT Multilateral Quota*. This document can be viewed and/or downloaded from the ECMT's website at: http://www1.oecd.org/cem/pub/pubpdf/MQguide.pdf. Also accompanying the *Guide* is a document called, *Documents to be on board a Tractor Vehicle.*

The plaques (referred to in the *Guide* as stickers) fall into the following categories:

For 'green' lorries
A green circular plaque of 200 millimetre-diameter with a 20 millimetre-wide white border and displaying:

- a white letter 'E' (Environment) or white letter 'U' (Umwelt/environment) denoting vehicle with a Euro-I engine.

For 'greener and safe' lorries
A green circular plaque of 200 millimetre-diameter with a 20 mm-wide white border and displaying either;

- a white letter 'S' (Sicher/safe) denoting a vehicle with a Euro-II engine and safety features; or,
- a white figure '3' denoting a vehicle with a Euro-III engine and safety features.

The safety features referred to above include; rear under-run protective devices, lateral protection, rear view mirrors, lighting and light signalling devices, tachograph, speed limiters, rear marking plates, anti-lock braking systems and steering, all of which must comply with relevant EC/EEC/ECE/UN Directives and Regulations.

Austria and Germany have their own domestic plaques to indicate vehicles that meet low-noise standards as follows:

'L': denoting Lärmarm Fahrzeug (i.e. low-noise vehicle) Austrian version of the symbol for 'quiet' diesel vehicles, which are;

- small and medium-sized lorries of no more than 150 kilowatts producing noise levels no greater than 78 decibels; and
- heavy lorries exceeding 150 kilowatts with noise levels no greater than 80 decibels.

'G': denoting Geräuscharm Fahrzeug (i.e. noiseless/quiet vehicle) German version of the Austrian 'L' plate for use only on diesel vehicles meeting German low-noise requirements – similar to those for Austria as above.

Emissions standards
ECMT specified noise emission standards for so-called 'green' vehicles are as follows:
Noise emissions standards for 'green', 'greener and safe' and 'Euro-III safe' lorries

- For vehicles with maximum power output up to 150 kilowatts = 78 decibels
- For vehicles with maximum power output over 150 kilowatts = 80 decibels

	'green lorries' (gram/kilowatts)	'greener and safe lorries' (gram/kilowatts)	'Euro-III safe lorries' (gram/kilowatts)
CO	4.9	4.0	2.1
HC	1.23	1.1	0.66
No_x	9.0	7.0	5.0
Particulate*	0.4	0.15	0.10

* gram/square kilowatts.

Fig. 9.3 Chemical exhaust emission standards for diesel engines (*Source*: ECMT).

ECMT chemical exhaust emissions standards for diesel engines are shown in Figure 9.3.

9.8 The economic issue

By comparison with the environmental augments in favour of intermodalism, which are very powerful, the economic case is weak. The drive towards intermodalism, in fact, is seen by most participants to be much more of an environmental issue than it is an economic issue but they do recognize that the economics of any switch of modes for reasons other than of pure cost must, nevertheless, be financially sustainable. In many instances, it is inevitable that a switch to intermodal freighting invariably adds cost as we have seen elsewhere in this book. For example, if piggyback road trailers are loaded on to a rail wagon, the costs stack up significantly. First, there is the additional cost of the trailer itself, which has to be specially built or strengthened to withstand being lifted under load from the road on to the rail wagon. The built-in extra strength may well add to its unladen weight that will reduce its payload potential within maximum legal vehicle weight limits. Then there are the actual lifting costs, accomplished by heavy-duty cranes or lift trucks, which have high levels of fixed and operating costs to be recovered. And the rail operator has to bear the costs of freighting loads, which comprise not only the goods in transit, but also the semi-trailer with its heavy frame and running gear (i.e. wheel-axle sets).

Compare these costs with those likely to be incurred by the original road vehicle taking the load directly through to its final destination and the benefits of intermodalism begin to fade away. So it needs another incentive to boost justification for a switch to rail – for the marginal extra costs, the fact that the train will carry many trailers in one shipment and remove that number of heavy lorries on long-haul journeys from our roads is a clear and positive environmental benefit. If the reader thinks that the piggyback example is extreme, then even with standard ISO containers carried on normal road vehicles (i.e. with no additional strengthening), there are still the lifting and loading/unloading costs and time delays which dissuade many road hauliers from even contemplating switching traffic from lorry to rail or canal barge. Basically, they take the view that once the container is on the lorry, it may as well stay there until the journey is complete whether this is merely from Birmingham straight up the M6 motorway to Glasgow or from Manchester to somewhere distant in Spain or Italy, for example. It will take a lot of persuasion and a significant change of mind-set to really overcome the perceived obstacles to switching from road to intermodal freighting.

10
Grant Aid and Government Support

The subject of grant aid for intermodal transport operations was previously outlined in Chapters 3 and 4, but in view of its significance this chapter is devoted to a rather more detailed discussion of the important grant schemes that have already aided intermodal developments and those currently in operation, and the projects they are designed to facilitate.

Grant aid (known by various names and designations) is a principal aspect of European Union (EU) and UK Governmental policy designed to generate environmental benefits by encouraging a switch of freight to more sustainable modes of transport, such as from road to rail or to inland waterway and short-sea modes. An important point to note in regard to such aid is that the European Commission (EC) is, and will continue to be, concerned to ensure that any funding granted is necessary, that it must not lead to unacceptable distortions of competition and that it must comply with Community rules, which, basically, forbid the propping up by governments of so-called 'lame duck' enterprises.

Since 1996 the EU has developed a number of grant aid policies. Firstly, with the Pilot Actions for Combined Transport (PACT) programme, which ran from 1996 to 2001, and more recently with the Marco Polo programme, with a budget of €100 million (i.e. euros) for the 25-Member State EU, plus three non-EU members; Iceland, Lichtenstein, and Norway – running from 2003 to 2006. A follow-up programme, Marco Polo II, is scheduled to run from 2007 to 2013 with a budget of €740 million. In this case, the programme is extended also to cover countries bordering the EU. The general purpose of these aid programmes is described as being to help achieve a modal shift of the expected progressive increases in international road freighting to rail, inland waterway, and short-sea shipping. The UK Government, for its part, has also developed various grant aid schemes for essentially the same purposes.

10.1 UK Government grants

UK operators wishing to invest in intermodal transport systems that effectively divert freight from road-borne transport to rail or waterway modes, may be eligible for a grant from the UK Government which is anxious to provide positive encouragement to switch freight from road to rail. Four different types of grant are available at the time of writing this text. But it is essential to note that the award of any grant is subject, of course, to satisfying relatively onerous criteria. These public funds are not doled out willy-nilly, they have to be justified and 'transparent', in other words, they must be capable of being seen, under detailed scrutiny if necessary, as having been applied to good and purposeful effect in the public interest.

The four types of grant available are as follows:

1. *Rail Freight Facilities Grant (FFG)* provided to part-finance the capital costs of providing rail freight facilities.

2. *Waterborne Freight Facilities Grant (WFFG)* made available to part-finance the capital cost of providing waterborne freight facilities at ports and wharves.
3. *Track Access Grant (TAG)* which helps offset the track access charges levied on rail freight operators for use of the rail network in return for generating environmental benefits by switching freight traffic from road to rail.
4. *Company Neutral Revenue Support (CNRS) scheme* which is relatively new, being launched only in February 2004 with a budget for 2004/2005 of £22 million and which is designed to further growth in the movement of deep-sea, short-sea, and domestic rail freighting of intermodal containers. It will replace around 80–85 per cent of the TAG.

These schemes have resulted in varying amounts of aid being awarded to qualifying projects; for example, at their peak in 2001, £32.7 million was awarded for rail projects and £15 million for inland waterway and coastal and short-sea shipping projects. There was something of a hiatus in the grants scheme in 2003 when the awarding authority, the Strategic Rail Authority (SRA), suspended the FFG, and the TAG because budget constraints meant that it no longer had sufficient discretionary funding to support new schemes. However, in February 2005 the UK Government changed tack with its grants scheme by announcing a new approach to reducing the impact of freight transport on the environment and congestion and to improving safety. This follows on from its July 2004 White Paper, *The Future of Transport* wherein it stated an intention to start administering grants to focus more on the overall objectives of reducing congestion, pollution and accidents, rather than promoting individual modes. Under the new arrangements it plans for the following:

- Merge the water freight grants and sustainable road haulage budget into a single pot from April 2005.
- From April 2007 extend the single pot to include rail freight grants including FFG. Subject to the passage of the Railways Bill, the administration of rail and water freight grants will be brought together in to a single unit.
- Prioritize schemes that offer the best value for money in terms of their impact on the environment, safety, and congestion.
- Set a minimum value for money threshold whereby every pound spent on grants will 'buy' benefits valued at £1.50 or more.

To give the reader some idea of the value of these grants, it is projected that taking the years 2005–2006 and 2006–2007 together, a total of £50.4 million is to be made available for rail freight grants and £19.2 million for water and road freight schemes. The rail figures are to include an additional £2 million available to support new applications. In 2007–2008, when rail freight grants are incorporated into the new arrangements, the total budget for all modes was provisionally set at £22.6 million.

10.1.1 FFG scheme

The FFG scheme, originally established in 1975/1976, is designed to help fund the capital costs of providing new freight handling facilities or re-investing in existing facilities, its aim being to encourage a switch of freight traffic from road to rail or waterways where it is not currently using these modes, or to retain existing rail and waterways freight traffic that may be lost to road. In recognition of the environmental and other benefits which accrue from such a switch, the grants are intended to help offset the high infrastructure investment costs involved in developing suitable rail and inland waterway transfer facilities. The grants are available to private freight service operators (e.g. road hauliers), as well as to freight customers and consignors. They cover all relevant forms of capital expenditure and equipment, including for the purchase of locomotives and freight-specific rail track and infrastructure. The basis of grant assessments is widened to include an allowance for lorry traffic avoided on inter-urban dual carriageways and motorways as well as on rural and urban roads. Grants are also available to assist in situations where goods are currently travelling by road, for the development of new rail traffic flows which would otherwise go by road, and for re-investment where the traffic would revert to road if there was no further capital investment in the rail facility.

The amount of grant available in any particular case is determined by three criteria:

1. Normally up to 50 per cent of the capital cost of rail freight equipment is available, although in cases of exceptional environmental benefit a higher percentage may be payable.
2. The amount paid may be constrained by how much environmental benefit arises from a particular scheme. The Government has stated its willingness to pay up to £1.50 for each lorry-sensitive mile saved.
3. Sufficient grant is paid to tip the balance in favour of rail (or inland waterway) when compared with alternative road transport. No grant is payable where projects can be commercially justified in their own right.

Most facilities needed to either handle freight or carry freight are eligible for grant aid including, for example:

- design and consultancy costs;
- rail siding infrastructures (e.g. track, ballasting, signaling, etc.);
- loading and unloading equipment such as cranes and conveyors;
- associated land, buildings, and weighbridges;
- access roads, screening, landscaping, etc. for environmental purposes;
- tractors and trailers, containers, and pallets;
- locomotives, rail wagons;
- refurbishment costs for existing or second-hand equipment and facilities.

The authority to pay FFGs is contained in the Transport Act 2000. Administration of the grant for waterways is the responsibility of the Department for Transport (DfT) in England and by the Scottish Executive and Welsh Assembly in Wales. Administration of rail grants in England currently rests with in the SRA, which also advises the Scottish Executive and the Welsh Assembly on this matter. (The SRA is scheduled to be disbanded by the Government in 2005 and its functions transferred back to the DfT.)

The key requirement for the payment of any grant aid is for the Secretary of State to be satisfied that if the proposed facility were not provided, the freight which would have used it would in fact be carried by road. The amount of FFG that can be paid depends on the value of environmental benefits and the need for grant, as determined by a financial appraisal comparing rail or inland waterways with the road alternative. In preparing the financial case for FFG, estimates of tonnage flows for each year of appraisal together with annual costs and revenues associated with the project are considered. Cash flows, expressed in present-day prices, are analysed for both road and rail/waterways options. The net cash flow of the rail/waterways option minus the road option is discounted at a real discount rate of 10 per cent, and is calculated to give the Net Present Value (NPV) of the scheme. A negative NPV indicates that the rail/waterways option will not be financially viable without the grant and the amount of financial support is calculated on the basis of the shortfall. In calculating the environmental benefits of rail, the assessment of externalities is based on the criteria of 'sensitive lorry miles' (SLMs). This is a method by which the benefits of using rail for the transport of freight is based on the monetary value for each mile that would have been travelled by lorry according to the type of road used (i.e. SLMs). The value of environmental benefits is calculated by applying specific rates for each road lorry mile over the qualifying road routes (see below). Coverage of the FFG scheme has now been extended to coastal and short-sea shipping. However, because of the EU regulations regarding state aid, the grant will not be available for the purchase of vessels and any grant awarded is limited to 50 per cent of the capital cost of the project.

10.1.2 TAG scheme

TAGs, introduced under provisions contained in the Railways Act 1993, are intended to assist with the charges raised by the rail track operator Network Rail for access to the rail network in the interests of switching freight flows from road to rail for environmental reasons. Under the railway 'open access' scheme, Network Rail (the track operator) can accept freight traffic from private companies (e.g. road hauliers operating intermodal services) provided that each traffic flow at least covers the costs it directly imposes on the rail network, such as track wear. In cases where a traffic flow genuinely cannot pay even these marginal track charges, the Government is prepared to pay the TAG to encourage the use of rail instead of freight going by road. Such grants are available to cover up to 100 per cent of the track charges where necessary in order to attract traffic to rail and where it is justified by wider environmental and other benefits arising from the avoidance of heavy lorry movements on the roads. The TAG scheme operates in a similar way to the FFG scheme described above in that a fixed budget is set and applicants have to be able to demonstrate:

- that in the absence of a grant, the traffic in question would travel by road; and
- that there are sufficient wider benefits to justify the required level of grant.

Grant is available in amounts up to:

- a level necessary to tip the balance in favour of rail,
- the value of benefits which the grant will secure,
- the total track access charge payable to Network Rail, whichever is the lower.

It is a requirement that grant applicants demonstrate that they have obtained competitive quotes for road haulage and, wherever practicable, rail freighting. Benefits are assessed on the basis of lorry miles avoided and are based on the road values for SLMs given below. Although the grant is payable only to rail freight train operators, applicants may be operators, customers, or consignors who contract with a train operator to provide train services, or to a consignor who intends to operate train services on his own account.

10.1.3 Waterborne Freight Grant scheme

Taking freight off congested roads and moving it by water can have environmental and wider social benefits, but it can be more expensive. Thc FFG described above is also, therefore, available to assist with the extra costs generally associated with moving freight by water by offsetting the capital costs of providing water freight handling facilities (Figure 10.1 shows the amount of waterways grant awarded from the year 1994 to 2002). It is also available to help companies re-invest in existing water freight facilities. Any company wishing to move freight within Great Britain (GB) by water may apply for the Waterborne Freight Grant (WFG) in much the same way that those conveying freight by rail may apply for such grants, as already described above. Most proposals for capital expenditure on the facilities needed to handle or carry freight by water is likely to be eligible for a WFG, but in recognition of EU State Aid rules, the grant will not be payable for the acquisition or modification of ships (i.e. defined as self propelled vessels which require certification to operate outside domestic smooth water limits). Nevertheless, grants may be considered for the acquisition or modification of vessels where the greater part of the freight movement, which is the subject of the grant application, is on an inland waterway.

The amount of grant offered will be the *lower* of the value of the environmental benefits or the amount needed as demonstrated by the financial case. In maritime cases there is a further cap of 50 per cent of costs eligible for grant. However, in a case where there are exceptional environmental benefits and financial analysis indicates that a figure in excess of 50 per cent grant is required, it may be possible to secure EU agreement to a higher percentage award. As we have seen, the grant scheme is all about producing benefits

	1994	1995	1996	1997	1998	1999	2000	2001	2002
Inland waterway projects									
Number of grants	–	–	1	1	3	8	7	7	6
Value of grants (£ million)	–	–	0.1	0.1	1.4	7.5	4.5	3.6	1.7
Lorry miles saved (million)	–	–	0.02	0.02	0.11	0.49	0.28	12.74	2.70
Coastal and short-sea shipping projects									
Number of grants	–	–	–	–	–	–	–	3	4
Value of grants (£ million)	–	–	–	–	–	–	–	11.4	3.9
Lorry miles saved (million)	–	–	–	–	–	–	–	1309.14	7.99
All projects									
Number of grants	–	–	1	1	3	8	7	10	10
Value of grants (£ million)	–	–	0.1	0.1	1.4	7.5	4.5	15.0	5.6
Lorry miles saved (million)	–	–	0.02	0.02	0.11	0.49	0.28	1321.88	10.70

Fig. 10.1 Freight facilities grants for inland waterway and coastal and short-sea shipping projects 1994–2002 (*Source*: DfT).

(public, environmental, and social) from freight being moved by water rather than by road. Broadly, the benefits are calculated by taking the tonnage that is committed to water over an agreed number of years and by working out how many lorry journeys this will save over that period of time.

The main financial criterion for a WFG is that when compared with the road alternative, the proposed facilities would not be financially justified without grant. The DfT calculates how much grant is required using a discounted cash flow appraisal. The financial appraisal compares the costs of moving the same tonnage of freight between the same points using water and road. There are seven main sections to the appraisal:

A Economic assumptions
B Tonnage transported per annum
C Analysis of the road option
D Analysis of the water option without WFG
E Calculation of financial deficit
F Analysis of the water option with WFG
G WFG calculation

10.1.3.1 Financial appraisal of water freight grants

The financial appraisal of a WFG is similar to the financial appraisal for a FFG. It takes the costs incurred and revenues received for the water-based scheme being proposed, and compares these with the costs and revenues of a road-based alternative. A DfT Guidance Note describes how the amount of grant is calculated that would be needed to make the water-based scheme as financially attractive as road. It says that the financial appraisal should be prepared by the applicant based on information on tonnages, costs, and revenues associated with the project provided in the financial summary table of the application form. The appraisal is presented as a year-by-year breakdown of the revenues, capital and operating costs under the water and road alternatives with all cash flows expressed in real (present day) prices to exclude the effects of general inflation in future years.

The Guidance explains that the capital costs of assets owned by the organization promoting the project, and specifically purchased for either the water or road-based alternative scheme should be recorded as and when incurred and not shown as depreciation or interest. Capital allowances that can be set against corporation tax should be recorded as appropriate, but because they are calculated on a current price basis, the effect of allowances has to be converted to constant prices to be consistent with the rest of the appraisal. The details of this calculation are the same as with FFG, but it should be noted that for WFG, the grant is not netted off the capital expenditure. If assets are purchased through a Hire Purchase (HP) agreement or other loan-based agreement the principle element of each payment should be recorded as capital expenditure, but the interest element should not be included. Operating costs for the road alternative should include road haulage costs, handling costs, and other recurring costs whereas operating costs for the water alternative should include the different elements of the shipping costs such as crew, fuel, and handling costs as well as other costs. Operating costs may also include annual costs that are related to capital items that are hired, leased, or chartered from other organizations. Corporation Tax should be calculated as a proportion of the cash flows, after including any allowances, at the corporation tax rate appropriate for the company. In accordance with current Corporation Tax guidelines, 50 per cent of the Corporation Tax liability should be calculated as being paid in the year in which the cash flow appears, and the remaining half in the following year.

The final step of the financial appraisal is the real after tax cash flow of the water alternative minus the cash flow of the road alternative. This net cash flow is then discounted at a real discount rate of 8 per cent and summed to give the NPV. A negative NPV indicates that the water alternative is not financially viable without grant. WFG will be paid as a rate per tonne actually moved, and paid in arrears. Thus the appropriate calculation of the actual grant payable is the undiscounted net cash flow of the water alternative compared with the road alternative over the 3-year period. However, grant payments are also subject to two other constraints. They cannot exceed:

- two-thirds of the value of environmental benefits,
- 30 per cent of the operating costs over 3 years of the water alternative.

10.1.4 CNRS scheme

The CNRS scheme is a new way (i.e. since April 2004) of supporting the movement of intermodal containers in GB. It provides revenue support and aims to secure growth in the deep-sea, short-sea, and domestic intermodal rail freight sectors. According to the SRA:

> this grant is an appropriate way of supporting the movement of containers by rail because it ties the grant support to traffic moved, rather than to any individual operator, on a non-discriminatory basis. This impartiality should enhance competition amongst freight operators and enable services to be developed in response to market needs. Clear and transparent rates enable all parties to recognise the grant that is available, which will be paid to whomever the contracting parties propose as taking the financial risk of running the service. In almost all cases this would be the freight operating company (FOC), though it could be a terminal operator, customer or shipping line.

The CNRS scheme is intended to play an important role in supporting the SRA's strategy for growing rail freight within the context of the Government's 10-year Plan both by reducing the number of lorry miles in GB and by shifting the traffic to rail:

> it will also deliver an economic and environmental dividend from which we all benefit, and against which revenue support can be justified. It also satisfies the SRA' s requirements for a transparent and more equitable allocation of resources and greater value for public money.

The EC has approved the scheme for a period of 3 years to 31 March 2007 with a budget of approximately £22 million in 2004/2005, as stated above. Funding for the years 2005/2006 and 2006/2007 is expected to be broadly similar at around £20/£25 million per annum. Around 80–85 per cent of the scheduled TAG (see above) will be replaced by the CNRS scheme. Prior to the end of the scheme, its success (or value) will be evaluated to ascertain whether an extension will be pursued and whether there is likelihood of approval by the EC. Should this be the case, it is envisaged that the CNRS rates will reduce in value as the cost of rail is expected to become more competitive when compared to the road alternative.

CNRS is available to all traffics carried in standard intermodal units (containers, swap bodies, or piggyback trailers) on any Network Rail owned route, with the exception of Channel Tunnel traffic and those commodities highlighted below. The CNRS rates are the same for all intermodal units that are 20 feet or more in size. Examples of services eligible for CNRS grant are:

- a service conveying deep-sea containers/tank-containers from a port to customers in an inland city;
- a service conveying supermarket products in containers between two distribution depots within GB;
- an intermodal units carried on a 'wagonload' service (a service of mixed traffic for a range of customers).

CNRS is not available for the movement of non-intermodal wagons. CNRS is also not available to companies moving bulk commodities in full or part trainload services where the party has chosen to use an intermodal unit. This is because the cost of conveying such traffic is usually different to the generalized costs used in the formation of the CNRS rates. These services are also normally viable by rail and CNRS is not therefore required. Traffic through the Channel Tunnel is not eligible for CNRS at present, although the SRA will consider the potential for the scheme to apply to this traffic in future. The scheme divides GB into 18 regions and eligible flows attract a fixed grant rate relating to each container moved between two specific regions, whether empty or full. There are two sets of rates; domestic rate and port rate, presented in an 18 × 18 grid square format.

10.1.4.1 Domestic rate

A domestic rate applies where units are delivered by road to a rail terminal, trunk-hauled by rail, and then delivered by road to the final customer. These services share the common characteristic of two road legs.

10.1.4.2 Port rate

A port rate applies when units are loaded straight to rail at a port, trunk-hauled by rail, and then delivered by road to the final customer. These services share the common characteristic of one road leg. In circumstances where there are intermediate stages to the rail journey (e.g. the container is loaded onto rail at a port then transferred to another rail service before reaching its final rail terminal), then grant will be paid against the actual origin and destination of the rail journey.

The rates are based on the principle that a traffic flow is entitled to grant if, the following conditions occur:

- *The environmental benefits justify it*
  Environmental benefits measure the effect of removing freight from Britain's roads. The benefits are calculated using SLMs (see below), which quantify the value of taking lorries off GB roads. The distance between regions, measured from the economic centre, is used to determine the value. The individual locations of the rail freight terminals used at either end of the rail-based flow have an environmental impact and an adjustment is made to the CNRS rates based on the location of the individual sites.

- *The cost of using rail is greater than the cost of road*
  'Financial need' is defined as the difference between the door-to-door cost of using road and the door-to-door cost of a rail-based solution. These have been calculated using standardized costs derived from industry models. The grant cannot exceed the financial need of any flow of traffic.

10.1.5 Road values

For the purposes of both the FFG and the TAG schemes, the DfT will assess the environmental benefits (i.e. the value) of so-called SLMs saved by applying the following rates per lorry mile avoided over the qualifying routes:

1.	Rural single carriageway roads	£1.00
2.	Urban single carriageway roads	£1.50
3.	Urban non-grade-separated dual carriageway roads	£1.50
4.	Motorways, rural dual carriageways and urban grade-separated dual carriageway roads	£0.20

Non-public roads such as those providing access to a power station or industrial complex are excluded.

10.1.6 Benefits of grants

These grants, each in its own particular way, provide encouragement and, more importantly, direct financial assistance in the field of combined road–rail transport. There is no doubt that without the grant facility, a number of developments in this area would not have taken place. It is acknowledged, of course, that at first glance the grants appear to be more specifically aimed at switching short-distance bulk traffic from road to rail where immediate environmental benefit is there for all to see, and not least by the local community who's neighbourhood was previously pounded by heavily loaded lorries. However, major haulage groups in the UK have also seen the potential for aid in establishing rail-side distribution operations to allow for the trunking-in and -out by rail of bulk stock in swap bodies and containers.

10.2 EC grants

The Marco Polo programme is the EC's aid programme designed to succeed the former PACT programme (see below) which came to an end on 31 December 2001. Based on the positive experiences with the PACT programme, and established under provisions contained in Regulation 1382/2003/EC, the Commission proposed a further programme to supports its fight against congestion in the road freight sector, and its ambitious goals to improve the environmental performance of the transport system as a whole. The Marco Polo programme was adopted on the 22 July 2003. Its main goal is to reduce road congestion and improve the environmental performance of the whole transport system by shifting freight from road transport to short-sea, rail, and inland waterway transport; thus aiming to support the essential transport policy direction outlined in the Commission White Paper, *European Transport Policy for 2010: Time to Decide*.

The specific stated objective of the Marco Polo programme is to:

> to reduce road congestion and to improve the environmental performance of the freight transport system within the Community and to enhance intermodality, thereby contributing to an efficient and sustainable transport system. To achieve this objective, the Programme supports actions in the freight transport, logistics and other relevant markets. These actions should contribute to maintain distribution of freight between the various modes of transport at 1998 levels by helping to shift the expected aggregate increase in international

> road freight traffic of 12 billion tonne kilometres per year to short sea shipping, rail and inland waterways or to a combination of modes of transport in which road journeys are as short as possible.

The programme supports commercial actions in the market for freight transport services and is therefore different from the support given through research and development programmes and the Trans-European Network programme (see Chapter 11). It fosters modal shift projects in all segments of the freight market, not only in combined transport and can also fund actions involving third countries, but taking into account the principle of 'subsidiarity', it will focus principally on international, rather than national, projects.

Marco Polo is a broader programme both in scope and budget, and is more ambitious; also the number of participating countries has been doubled. It intends to foster modal shift projects in all segments of the freight market, not only in combined transport as was the case under the previous PACT programme.

There are three main types of Marco Polo 'Actions' as follows:

1. *Modal Shift Actions*, where start-up support is given for new non-road freight transport services, which should be viable in the mid-term.
2. *Catalyst Actions*, where support is given for launching freight services or facilities of strategic European interest.
3. *Common Learning Actions*, where co-operative behaviour in the freight logistics market is stimulated.

The programme runs from 2003 to 2007 with a budget of €115 million for the 25 Member States of the EU (i.e. the EU25) and including other countries, such as Norway, Iceland, and Lichtenstein which have joined the programme and which also contribute to the available budget. The first call for proposals for projects under this scheme was published on 11 October 2003 and closed on 10 December 2003, and the 13 (out of 92 proposals received by the EC) successful projects concluded a contract in autumn 2004. Interestingly, these 13 projects involve private investments of about €360 million (without the spend on infrastructure) and are likely to result in some 13.6 billion tonne/kilometres of freight traffic being switched from road freighting to short-sea shipping, rail, and inland waterways. In terms of environmental efficiency, these projects are estimated to save for society €15 of external costs for every 1 euro of subsidy spent. A second call for proposals was published on 15 October 2004, with the deadline for submission set for 15 December 2004. Those projects that are successful under this call were expected to have a contract by approximately mid-2005.

10.2.1 Marco Polo II (2007–2013)

Following on from the successful launch of the first Marco Polo programme, on 15 July 2004 the Commission presented a proposal to establish a second, significantly expanded, Marco Polo programme effective from 2007 onwards. Marco Polo II, as it is designated, includes new actions such as motorways of the sea and traffic avoidance measures. The programme, which has a budget of €740 million for 2007–2013, has been extended to countries bordering the EU. The Commission estimates that every one euro in grants to Marco Polo will generate at least 6 euros in social and environmental benefits. The reader should be aware that the final form of Marco Polo II depends on the outcome of the negotiations with the European Parliament and the Council during 2005–2007.

10.2.2 The PACT programme

The former EU support scheme called PACT was successful during its currency from December 1992 to December 2001 in supporting pilot schemes to promote combined transport, including initiatives by road hauliers wishing to establish road–rail services. During this time the programme achieved, for 34 out of the 63 operational actions approved, traffic shift from road to rail amounting to at least 3.5 billion tonne/kilometres, and this despite the lack of data on the other projects. According to the PACT scheme Evaluation Final Report, produced in November 2000 by UK consultants AEA Technology Environment, about *half* the projects achieved carbon dioxide (CO_2) savings for less than 20 euro per tonne of CO_2 avoided (and this accounting only for the amount paid in subsidy by the EC to the beneficiaries) and *two-thirds* of the projects achieved CO_2 savings for less than 50 euro per tonne of CO_2 avoided. This analysis suggests that projects funded under PACT generally performed well in the wider context of greenhouse gas abatement options. Taking into account other environmental effects (e.g. reduced emission of other pollutants, of noise, of accidents, and of congestion) and the possibility of project replication would further improve the cost-effectiveness of PACT projects for environmental improvement.

It is useful to record here details of the PACT programme despite its cessation in 2001 because it is regularly mentioned in intermodal circles, because its effects are still largely being felt in terms of reduced CO_2 emissions and increased intermodalism, and because, for the reader who is studying transport subjects, it does have some historic merit. According to the Commission, these pilot projects were aimed at establishing competitive combined transport services in terms of price and levels of service. The Commission said at the time that the situation in the freight market did not take account of the real costs of transport for each mode and hence it saw itself as being justified in providing aid for combined transport in proportion to its benefits to society by way of savings in external costs.

PACT projects, according to the Commission, were first and foremost intended to improve the quality of combined services and were not investment programmes for transport infrastructure, nor were they for research and development. They had to be established on the basis of the existing infrastructure and should not rely upon unproven technology. Nevertheless, they could have been linked to transport infrastructure investment measures, or to a demonstration programme for bimodal systems, or to other new systems provided that the characteristics and financing of the PACT project, the associated infrastructure measures and the demonstration projects were separated.

The specific objectives of an individual PACT project could vary on the basis of the existing level of service, the service that could technically be offered and the real quality requirements. For each project it was necessary to identify the points that were important using a complex check list of some 60-odd specific items under eight headings given in an appendix to the Commission's paper on PACT, namely; *Pilot Actions for Combined Transport (PACT). Directorate-General VII (Transport), European Commission, Brussels, 23 March 1994.* It was on the basis of the identified items in this list that the specific objectives, in quantified terms, and the cost estimate for their attainment would have been fixed.

10.2.2.1 PACT application

Projects submitted for grant consideration could cover any part of the EU in geographical terms, and even some specific routes to Central and Eastern European countries. Furthermore, grants were not restricted to enterprises solely within the EU; private and public enterprises established outside the Union could also apply for and receive financing if they were partners in a project of European interest. Mainly, the Commission gave grant priority to projects concentrating on road, rail, and inland waterway modes, but applications relating to the improvement of maritime (i.e. shipping) links could have been considered if they were concerned with connecting the peripheral areas to the centre of the EU by only the most economic or most efficient route. In making PACT submissions for grants, information covering eight specific points had to be given as follows:

1. The current and potential flows of traffic, if possible by type of goods. (An analysis of the potential traffic that could be captured by combined transport would have been of value.)
2. The quality criteria that was relevant to the project (selected from a list in the appendix to the Commission's paper, see above).

3. The likely parties involved in the project (e.g. administrations, public or private firms, and other bodies).
4. The quantified objectives, for each of the criteria a precise objective and a time scale should have been given.
5. The measures needed to complete each of the objectives.
6. The cost of each of the measures and the allocation of costs between each party involved.
7. Estimated receipts of the project (cost coverage of the project).
8. The approach to have been adopted to control and monitor the project.

The amount of Commission support for each project was established on the basis of the balance sheet of costs/receipts for each participant in the project within the following limits:

- Thirty per cent of the total cost of the project for innovative actions and direct intervention.
- Fifty per cent of the cost of the feasibility study for the project (i.e. on a specific route).
- Hundred per cent of the cost in the very particular case of a general preliminary study.

The support was calculated on the basis of the minimum that would be needed to ensure that the project was launched. The Commission said that any additional support provided directly by EU Member States had to conform to the competition rules of the Treaty of Rome.

At the end of each project, none of which was allowed to exceed 5-year duration (*Note*: bearing in mind, anyway, that PACT ended in 1997) an evaluation, made on the basis of the initial technical description of the project by the Commission and its steering committee, was prepared taking account of:

- the relevance of the initial objectives,
- the amount and the method of financing,
- the degree to which the project had been completed,
- the traffic,
- the profitability of the axis,
- the possibility of generalizing the results.

Besides the formal evaluation at the end of a project, throughout its operation the partners were required to meet frequently with the Commission to assess progress and to solve any problems; financial as well as legal, administrative, or organizational.

Such was the success of the PACT programme that by September 1995, the EU's European Transport and Trans-European Network Commissioner was describing it as an original initiative which, since its launch in 1992 with two projects, by 1995 had more than 20 such projects in operation despite a stagnating budget. With such success and continuing demand, the Commission intended to submit a proposal to the Council of the EU to 'transform this useful tool into an official programme provide with greater means'. As we have seen above, this aim has been achieved with the successive Marco Polo and Marco Polo II programmes which now extend as far into the future as 2013. A description of the key routes (i.e. 'Actions') developed under the PACT programme is to be found in Chapter 11 where the subject of the Trans-European Transport Networks (TEN-T) is discussed in greater detail (see also the Europa web site: www.europa.eu.int for further information).

11

Intermodal Networks and Freight Interchanges

The development of an extensive and efficient system of combined road–rail and road-waterway/short sea transport involves, basically, a two-pronged attack: first, identification of the transport network to provide maximum geographic coverage across the whole of the European Union (EU) and beyond; and second, massive investment in freight interchanges and other infrastructures which, particularly since the early 1990s, have begun to take shape and come on stream.

So far as network developments are concerned, the European Commission (EC) has been working for some years on its Trans-European Transport Networks (TEN-Ts) plan for road–rail networks, and more recently on waterway networks, its 'Motorways of the Sea' project, and the Galileo, satellite navigation and positioning system. Its aim is to integrate the national transport systems of each of the (now 25) EU Member States into a cohesive Euro-wide network, removing bottlenecks, creating vital missing links, joining remote and outlying regions (as well as the European Free Trade Area (EFTA) countries, i.e. Iceland, Liechtenstein, Norway, and Switzerland) to the network, and developing connections between different transport modes: the Channel Tunnel, linking the UK and the Continent, is described by the Commission as symbolic of its TEN-Ts. The development of TEN-Ts is underpinned by a series of important measures including better use of existing networks, by modernizing equipment and by an improved flow of information between systems by using electronic data interchange (EDI) and telecommunications. There is also a greater research and development effort, concentrated on interfaces between modes (so-called intermodal or freight interchanges) and on technologies to improve the quality of service.

A diversity of investment sources is involved in these Euro-wide transport projects with greater reliance on private capital (as we have seen with a number of UK road, bridge, and rail development projects). The EC sees its financial role as being that of stimulus for projects, getting things started with feasibility studies and such like, and integrating regional and national projects. On the investment front, nothing on an individual project basis compares on a financial scale with the £10 billion poured into the Channel Tunnel; a project which will not within foreseeable time-scales, if ever, achieve payback. Nevertheless, so far as UK trade with mainland Europe is concerned, the Channel Tunnel forms the backbone of international road–rail operations; without it, this particular freight revolution would not have got off the ground. But the Channel Tunnel alone is not the only major capital project destined to enhance the development of international combined road–rail transport. Investment on a grand scale is also going on elsewhere, with the development of both freight and high-speed rail routes across Europe; for example, 30 000 kilometres of new high-speed track to be constructed by European railway companies over the next 25 years; the new Gotthard and Lötschberg base tunnel rail routes through the Swiss Alps currently under construction;

and major intermodal freight hubs and intermodal 'villages' being developed in the UK and in many other European states.

This chapter is devoted to an examination of the three key elements of intermodalism in greater detail, namely routes, communications and freight interchanges (i.e. terminals). First it identifies the EU's role in establishing TEN-Ts and developing other transport and communications infrastructures; and second by examining the positive progress being made in the development of intermodal interchanges.

11.1 The TEN-Ts

Implementation of the Maastricht Treaty on EU in 1993 and adoption by the Council of the European Communities of its Decision 93/628/EEC of 29 October 1993 on 'the creation of a trans-European combined transport network' gave a new impetus to the evolution of the Common Transport Policy (CTP), particularly the provisions on TEN-Ts which provided a new basis for the EU to contribute to the establishment and development of the transport infrastructure. The EU Commission's 1992 report *The Future Development of the Common Transport Policy* identified the fact that while transport demand has grown, investment in inland infrastructure in Europe expressed as a percentage of gross domestic product (GDP) actually declined between 1975 and 1980 from 1.5 per cent to 1.2 per cent but this decline was halted in the early 1980's and the investment share remained at about one per cent throughout that decade. As the Commission said at the time, growth in transport demand and relative decline in inland transport infrastructure investment have increased pressure on the road and rail network capacity which had reached saturation at many points.

The report identified the development of TEN-Ts for different transport modes and their progressive integration as being one of the important measures needed to improve the quality of intermodal services. In particular, it pointed out the necessity for identifying those corridors where the maximum potential for transfer (between road and rail) exists and those flows of goods that particularly lend themselves to transfer. A start had been made, it said, with the Pilot Actions for Combined Transport (PACT) grants scheme described in Chapter 10. Again, in the Commission's words, terminals are important because they are the vital interfaces between modes, which in large part determine the competitiveness and utility of intermodal systems. To the extent that they become more effective, the average distance at which combined road–rail transport journeys become competitive with road transport, which at that time (i.e. 1992) was estimated to be in the region of 700 kilometres, was expected to decrease. Given the reality that most freight transport takes place over shorter distances (66 per cent within 50 kilometres and a further 20 per cent within 50–150 kilometres), the future contribution of intermodal transport depends critically on improvements in efficiency which would allow it to compete effectively in those shorter distance markets.

The EC saw the problem, insofar as TEN-T development was concerned, as being that until this time (i.e. 1992), transport networks had been designed largely from a national point of view, with emphasis frequently being placed on the development of particular modal networks rather than on the relationships between them, much less their integration as a transport system. This traditional approach had led to problems such as the absence of adequate interconnections between national networks, missing links and bottlenecks as well as obstacles to inter-operability entailing huge inefficiencies. Examples of this exist in rail operations, such as differing loading gauges and electric power supply, and in combined transport with incompatible approaches to the technical specifications for equipment. Additionally, there were differences in the geographical situation and economic histories of EU Member States that resulted in considerable divergence in the availability and quality of transport infrastructures. For example, the centre and north of the EU was much better equipped in this respect than the periphery and the south. The Commission said that general stagnation in investment in transport infrastructure over the 1980s had not provided an environment favouring the early elimination of these problems and imbalances. Overcoming them was a major aspect of the work, not just through investment but also in planning, to ensure the integration of the EU's transport system through the completion and combination of networks, taking particular account of the needs of the more geographically isolated regions. The objective was, and remains today, to link islands, landlocked and peripheral regions with the central regions of the EU.

Prior to publication of the 1992 report, the EU was already committed to giving financial support to a number of priority road-links, including two motorways and tunnels through the Pyrenees between France and Spain, a motorway connection between Lisbon and Madrid and better links between England and Ireland. Rail projects concerned, essentially, two high-speed networks, one serving the north of the Community and the other the south. The northern network is designed to connect London, Paris, Brussels, Amsterdam, and Cologne, with onward links to other destinations, while the second provides a Mediterranean backbone running from Seville via Madrid, Barcelona, Lyons, Turin, and Milan to Venice. Among the original 14 priority Trans-European Network (TEN) projects (see list below) five were directly aimed at furthering combined transport solutions, these being: the so-called *Brenner axis* which links Berlin with the Italian city of Napoli (the final sections of this key route are due to be opened in 2005); the 160-kilometre *Betuwelijn maritime* (i.e. container) freight railway across the Netherlands linking the port of Rotterdam to the German rail network at the Dutch/German border; the *combined transport line* across Ireland from Cork in the south to Belfast; upgrading and renewal of the 850-kilometre *UK West Coast rail Main Line* (WCML) from London to Glasgow due for completion in 2007; and the *Lyon to Turin link* between the high-speed rail networks of France and Italy involving 750 kilometres of new lines and a 57-kilometre long Gotthard tunnel through the Alps, one of the longest rail tunnels in the world. The remainder included such projects as high-speed passenger train links between major European cities (e.g. Paris, Brussels, Cologne (Köln), Amsterdam, London, the so-called 'PBKAL' network); various cross-border motorway links and road corridors; a new International airport for Athens at Sparta (now in service); the Øresund link between Copenhagen in Denmark with Malmö in Sweden comprising tunnel, road and rail links and including the world's longest cable-stayed bridge for road and heavy rail traffic (completed in 2002); and yet another long sea crossing by way of a 19-kilometre long-fixed bridge/tunnel link across the Baltic between Denmark and Germany across the Fehmarn Strait, brought forward in the TEN-T programme and now scheduled for completion by 2013; various inland waterway upgradings, an air traffic management system for Europe and the multimodal global navigation and positioning satellite system known as Galileo (see below).

The 14 priority TEN-T projects adopted in 1996 were as follows:

1. High-speed train/combined transport north–south
2. High-speed train PBKAL
3. High-speed train south
4. High-speed train east
5. Conventional rail/combined transport: Betuwe line
6. High-speed train/combined transport, France–Italy
7. Greek motorways, Pathe and Via Egnatia
8. Multimodal link Portugal–Spain–Central Europe
9. Conventional rail link Cork–Dublin–Belfast–Larne–Stranraer
10. Malpensa airport Milan
11. Øresund fixed rail/road link between Denmark and Sweden
12. Nordic triangle rail/road
13. Ireland/United Kingdom/Benelux road link
14. West Coast Main Line (UK rail project).

All the proposed rail, motorway or waterway links were chosen according to a range of selection criteria: economic importance for the region, employment creation, benefit for industry and viability, potential for private investment and financing, community interest like trans-frontier links, or interconnection of networks. All had to pass the environmental impact scrutiny. It was up to the public authorities of the

Member States, the regions or municipalities, and the private partners to decide on the implementation of these projects. The prime aim being to connect national networks and build the necessary motorway links, many of which were in the outlying regions of the Community (i.e. in Ireland, Greece, Portugal, and Spain). The networks in these countries then only accounted for only 20 per cent of the Community total. Another important objective was to relieve congestion on the roads with users being made to meet the real traffic costs, for example, by wider application of road tolls as we are seeing now in the early 2000s.

11.1.1 TEN-T 2002 progress report

By 2002, 6 years on from when the majority of these initial projects were proposed or commenced, the EC reported that only 20 per cent of the work had been completed. Public investment in transport infrastructure had fallen from 1.5 per cent of GDP in the 1980s to less than 1 per cent in the 1990s and the resulting delays affected cross-border projects in particular. The EU's transport commissioner Loyola de Palacio said, in launching an update report on progress with the TEN-Ts in 2002:

> At the present rate, and without additional financing, a further 20 years will be needed just to complete the work planned for 2010. At the same time, by 2010 the economic growth rate envisaged by the European Council is likely to generate increases of 38 per cent in freight traffic compared with 1998. As the Commission's September 2001 White Paper on European transport policy for 2010 demonstrates, without a major effort to rebalance traffic growth this means a rise of 50 per cent in road freight. This growth and the delays in building the TEN-T, demand a new transport policy covering improved regulation of competition, the promotion of intermodal transport and the shift of traffic from the roads, and better targeting of investment. Community action should supplement national plans, and must be guided by research to identify priorities with a real value to Europe as a whole. The 1994 European Council (meeting) in Essen initiated this concentration of effort by selecting a first series of priority projects. With the same objective, in October 2001 the Commission proposed a revision of the guidelines for the TEN. This proposal, called for by the (European) Council, strengthens the priority given to the first series of projects, takes stock of progress, and responds to new challenges with plans for six new priority projects including deployment of the Galileo satellite system and the crossing of the Pyrenees by rail.

The six new priority projects (and two extension projects) proposed by the EC in 2001 were as follows:
New projects (numbered by the EC to follow on from the first 14 projects listed above)

15. Global navigation and positioning satellite system Galileo
16. High-capacity rail link across the Pyrenees
17. Eastern European combined transport/high-speed train
18. Danube river improvement between Vilshofen and Straubing
19. High-speed rail interoperability on the Iberian Peninsula
20. Fehmarn Belt: fixed link between Germany and Denmark.

Two further extensions to existing TEN-T projects were also proposed, namely:

Project 1. High-speed train/combined transport north–south (Verona–Naples and Bologna–Milan)
Project 3. High-speed train South (Montpellier–Nîmes).

11.1.2 New 2004 TEN-T guidelines

In April 2004, yet another step forward was taken to revitalize the TEN-Ts when the European Parliament adopted new Guidelines for the 30 priority projects (which include the 20 projects listed above) declared to be of European interest, these projects are listed in detail at the end of the chapter. According to the EC's Press statement, the extension of major European axes to the future Member States should help to make enlargement a success and provide the Union with a new opportunity to reduce congestion, improve accessibility, and encourage intermodality. The new Guidelines, which follow on from a report made in 2003 by the High-Level Group on the TEN-T chaired by Mr Karel Van Miert (i.e. the so-called 'Group Van Miert Report') have a strong focus on the enlargement of the Union and the need to integrate the networks of the 10 new Member States that joined the EU in May 2004, namely Cyprus, Czech Republic, Estonia, Hungary, Latvia, Lithuania, Malta, Poland, Slovakia, and Slovenia. The Press statement announcing the Guidelines is reproduced here in full (with acknowledgement to the EC) because of its significance in identifying both the objectives and the projects contained therein:

> The list of projects also aims at ensuring modal shift and more sustainable mobility patterns by focusing investments in rail and waterborne transport. Strong focus is directed towards cross-border projects as these are said to be 'typically the most difficult ones to implement'. The estimated cost of carrying out these 30 projects will be around €225 billion by 2020. Some 20 per cent of the total amount could be raised from the private sector and the rest will have to come from the national and Community budgets, notably within the framework of the financial perspectives after 2006. The total cost of completion of the TEN-T, including the projects of common interest not identified as priority projects, will be €600 billion. Implementation of the priority projects should produce time savings for international transport, help to reduce the growing pollution due to transport and contribute to more balanced spatial development. These benefits would boost the growth potential of the Union up to 0.3% of GDP and create up to 1 million new jobs according to recent research.
>
> A new mechanism for supporting Motorways of the Sea is one of the new priority projects aiming at concentrating freight flows on a limited number of sea connections to ensure their financial viability while reducing road traffic. The importance of the development of Motorways of the Sea is to ensure that transnational maritime links between countries isolated for geographical reasons, or affected by road congestion, will be treated with the same importance as land links. The objective is to concentrate freight transport for some key links on a limited number of ports to increase the viability of these links. Member States will be encouraged to jointly establish transnational maritime links through tendering procedures to avoid distortion of competition. The new decision will also allow Community financial support through startup aid. The Motorways of the Sea provide an alternative route to existing bottlenecks (e.g. the congested Pyrenean crossing), and connect Europe with peripheral countries and island regions and states (e.g. in the Baltic Sea region).
>
> ***Declaration of European interest***
>
> The Guidelines declare the 30 priority projects to be of European interest. This declaration aims at ensuring the timely completion of the projects by focusing on them the resources available from the various community financial instruments for the networks. These declarations of European interest would open the way to co-ordinated evaluation and public consultation procedures and allow the carrying out of a single transnational enquiry in the case of certain cross-border sections. To solve the difficulties caused by Member States' separate national evaluation procedures, the Guidelines will make it possible for the Member States to co-ordinate their evaluation and public consultation procedures prior to the authorization of projects. For certain cross-border sections, such as bridges or tunnels, the Member States should aim at conducting a transnational enquiry. To enable this

process, the Commission will work on methodologies and tools for socio-economic and environmental evaluation of projects.

Concentration of aid from the Community

The fact that the priority projects will be declared to be of European interest will make it easier to focus on them the aid for the TENs, from the Cohesion Fund and the Structural Funds. The guidelines are complemented by the new rules for granting of aid in the field of TENs, which are due to enter into force in the coming months. In particular, these new rules increase the co-funding rate from 10 per cent to 20 per cent (compared with 30 per cent in the Commission proposal) for the sections of the priority projects, which crosses borders and natural barriers. This rate should provide a greater incentive for implementation of projects including the setting up of public–private partnerships. These new rules should also allow multi-annual commitments, which would provide greater flexibility in the financial commitments to promoters of TEN-T projects.

Closer international co-operation

Delays in completion of links reduce the return on the investments made by neighbouring Member States on the same axis. To encourage better synchronization of investment, the Commission will designate, case-by-case, a European co-ordinator for projects or groups of projects located on the same European axis. The co-ordinator will also give advice on the financial package for the projects.

The package will stimulate the European economy

The investments in the TEN priority projects would reduce congestion on roads by 14 per cent and the monetary value of time savings to international traffic is approximately €8 billion per year. These benefits would stimulate the economy of the enlarged Europe and increase the GDP between 0.14 per cent and 0.3 per cent according to recent research (i.e. TIPMAC and IASON projects funded under the 5th Framework Transport Research Programme). This would translate into half a million to 1 million new jobs.

11.2 Trans-European Rail Freight Freeways

In 1996, the EC's transport White paper called for higher priority to be given to the Trans-European pathways for freight with fair charges and open infrastructure access. Following on from this the Commission developed the Freeway concept as a means of promoting competitive international rail freight to counter its long-term loss of market share, mainly to road. Freeways combine unitary route planning and management with the development of faster trains paths offered by a single sales point and a range of complex commercial and legal issues are involved. The Trans-European Rail Freight Freeways (TERFFs) study, carried out by international consultants Arup, was concerned with the development of measures to assist international rail freight in general and to establish extension proposals for freight freeways across Europe. The study involved the gathering of wide-ranging market intelligence, discussions with state railways, review of rail policy, road haulage and international rail freight across seven countries, forecasts of market potential and a review of the progress with Freeways in the West, including the Glasgow–Sopron (Hungary) scheme. Proposals for demonstration projects and international co-operation on rail freight, which complement International Union of Railways (UIC) and United Nation/Economic Commission for Europe (UN/ECE) initiatives, were made. Several critical areas for actions in international rail freight were identified and studied in depth, these include: wagon and load tracking/tracing systems, border crossing delays for rail and road, wagon condition and replacement and the needs of shippers and forwarders.

The Freight Freeways idea has two main elements according to The Railway Forum in a Fact Sheet published in September 2000:

> the first is the identification of high-quality route capacity for long-distance freight haulage and the offering of a single end-to-end price for the use of the infrastructure for international services. The second is the opening up of the freight market to all licensed railway companies within the EU Member States.

Four Freight Freeways have presently been identified: Germany through Switzerland to east coast Italy; Germany through Switzerland to west coast Italy; Holland and Germany to Austria; UK to Hungary (i.e. Glasgow to Sopron). As well as these Freight Freeways, certain national railways have co-operated to introduce the Belgium/France/Italy/Spain 'Belifret' Freightway. This accords with one of the elements of the Freight Freeway concept – the identification of high-quality route capacity for freight services with improved transit times and more productive train operations. The availability of Belifret capacity is promoted through a single point of contact for the customer but the price is not made public and the new capacity will be utilized by alliances of existing railways.

Commercial development of the Freeway and Freightway concept had been slow (up to the year 2000), though some rail services now operate along parts of the Belifret corridor. Building on this experience, the idea of a Trans-European Rail Freight Network (TERFN) has emerged in debate on the infrastructure package of liberalization legislation in the European Parliament. The TERFN would define key freight routes in each EU Member State, linked up to form a network on which there would be freedom of access for international freight traffic. As we have seen above, some of the key rail freight routes conceived as part of the Freeway and Freightway network have now become part of the EC's TEN-T funded projects. A key date in the development of the TERFN occurred on 15 March 2003 when the network was thrown open for access by rail freight operators across the whole of the EU as described more fully in Chapter 7.

11.3 Motorways of the Sea

We have already considered the subject of the EC's 'Motorways of the Sea' project in Chapter 8. However, it is useful just to briefly reprise it again here in the context of the TEN-T programme; in fact, it features as TEN-T project number 21 in the list on p. 148. The concept of ' Motorways of the Sea' was first publicized in the EC's Transport White Paper of September 2001, where it proposed the development as a 'real competitive alternative to land transport.' To help these lines develop, the White Paper stated that European funds should be made available and that these 'Motorways of the Sea' should be part of the TEN-T. In an 'overview' published in November 2004, the EC said:

> The concept aims at introducing new intermodal maritime-based logistics chains in Europe, which should bring about a structural change in Europe's transport organization within the next years to come. These chains will be more sustainable, and should be commercially more efficient, than road-only transport. Motorways of the sea will thus improve access to markets throughout Europe, and bring relief to our over-stretched European road system. For this purpose, fuller use will have to be made not only of our maritime transport resources, but also of our potential in rail and inland waterway, as part of an integrated transport chain. This is the Community added value of motorways of the sea.

The TEN-T Guidelines (see p. 136) gives three main objectives for the 'Motorways of the Sea' projects:

- freight flow concentration on sea-based logistical routes;
- increasing cohesion;
- reducing road congestion through modal shift.

Four corridors, for completion by 2010, have been designated for the setting up of projects of European interest:

- Motorway of the Baltic Sea (linking the Baltic Sea Member States with Member States in Central and Western Europe, including the route through the North Sea/Baltic Sea canal);
- Motorway of the Sea of western Europe (leading from Portugal and Spain via the Atlantic Arc to the North Sea and the Irish Sea);
- Motorway of the Sea of south-east Europe (connecting the Adriatic Sea to the Ionian Sea and the Eastern Mediterranean, including Cyprus);
- Motorway of the Sea of south-west Europe (western Mediterranean, connecting Spain, France, Italy, and including Malta and linking with the Motorway of the Sea of south-east Europe and including links to the Black Sea).

11.4 Infrastructure developments in retrospect

While the foregoing text concentrates mainly on current and future projects, it is interesting and useful from a background and historical view point to look back at some of the massive and very significant infrastructure works that have been completed under the TEN-T funding umbrella, starting with the building of the Channel Tunnel. This was a massive capital investment project and the crucial link in the development of UK–Europe combined road–rail transport operations. However, this is only part of the transport infrastructure development that has taken place, and continues to take place, in both the UK and Europe. There has been nothing short of a revolution in recent years in the Euro-wide transport scene resulting from, among other catalysts, the great pressure on environmental issues and the programme to privatize and regenerate the railways).

As we have seen in the foregoing chapters, there has been a significant move to switch freight from road to rail, especially via combined road–rail operations, and this has meant considerable investment in terminals and transfer facilities. In the second case, privatization of rail has meant that a fresh look has been taken at what rail freighting has to offer and how services and facilities can be improved. And in any case, who would have thought just a few years ago (i.e. in the mid-1990s) that, to take just one example, UK state rail operator, British Rail's traditional wagon-load freight business would be divided up into three completely separate entities (Main Line, Transrail and Loadhaul), and then sold off to American rail operator Wisconsin Central, which now operates in the UK as English, Welsh, and Scottish (EWS) Rail? If that was not revolution, what was? Taking another example, as we have already discussed, but the context makes it worth repeating here, the building and opening of the Channel Tunnel revolutionized thinking, on this side of the Channel particularly, about the ways in which international freight can be consigned. Why did nobody have such thoughts before? Quite simply, the answer is that there was no facility to generate such thought; by and large international freight had to either go on a lorry and the lorry had to cross the Channel on a roll-on/roll-off (RO/RO) ferry ship, or it had to be loaded into a container and shipped on a short sea lift-on/lift-off (LO/LO) operation. Now the combined road–rail alternative for containers and swap bodies on a through rail service from freight interchanges in the UK to destinations in Europe presents a whole new concept of choice. But to achieve this option has taken significant development in infrastructure, particularly rail networks and terminal infrastructures, as the following few examples indicate.

Network Rail (and its predecessors), for example, invested £50 million in a programme of infrastructure works to boost capacity on the Channel Tunnel freight network to allow it to carry almost any container or swap body that will fit on a standard road trailer. This included cosmetic gauge work on the top corner of tunnels and bridges to accommodate ISO containers up to 9-foot 6-inch-high and 2.77-metre × 2.5-metre swap bodies on low platform (i.e. 945 millimetre) intermodal wagons. To create the SB1 (swap body 1) gauge, over 100 bridges and tunnels on the routes between London and the Channel Tunnel had to be rebuilt, with some additional work on routes between London and the major international intermodal terminals in the UK. This meant that well over 90 per cent of all registered loading equipment operating within Europe would be able to use BR's enhanced freight network, but not piggyback traffic. On the other side of the Channel too, where loading gauge dimensions have not been a problem for most container and swap body traffic, nevertheless the railways have invested heavily in infrastructure improvements, terminal developments and upgrading of freight networks, and the building of vital missing links, as well as in the 30 000 kilometres of new high-speed track across Europe mentioned previously.

In Germany, unification of the separate western and eastern rail systems (DB and DR) into a single system from 1 January 1994 involved massive reorganization and plans for a new High-Quality Rail Network (HNS) comprising direct freight trains between major economic centres. Investment, particularly on the freight side, was intended to encourage combined container, swap body and semi-trailer traffic that was seen as 'the great hope'. The plan was to increase the volume of direct train, terminal-to-terminal movements and to develop a new hub-and-spoke system, with freight terminals and container and swap body trains running return services between terminals and the hub.

Italian railways, Ferrovie dello Stato SpA (FS), also had ambitious plans to develop a new and ambitious freight project, described in the company's brochure as 'a system-changing project aimed at transformation and growth towards an ever-increasing use of intermodal processes'. However, besides the system changing, there were infrastructure developments including track improvements, such as the Alpine line through the Brenner Pass (see TEN-T project 1 on p. 147), double-track laying, loop lines and junctions (plus incentives for private sidings), and gauge improvements. Part of the scheme involves investment in rolling stock, such as intermodal and bimodal wagons, shunting locomotives, and introduction of the latest generation 800-tonne capacity, 120 kilometre/hour, freight trains operating shuttle services within Italy and internationally into Germany.

11.4.1 Swiss 'AlpTransit' tunnels

Britain is not alone in having a major transit tunnel development. In Switzerland, by 1996, work was already well advanced on the 'AlpTransit' tunnels project which will provide for a switch of heavy goods traffic from the congested Alps transit road network to rail, as well as a high-speed passenger transportation system. Although not part of the EC's TEN-T funding programme (since Switzerland is not an EU Member State), it is the country's answer to growing European and domestic demand for efficient railway links for the next century, and it should be stressed, these are key north–south transit routes for heavy lorry traffic carrying many thousands of vehicles daily. This massive 25-year construction project involves two routes with the 57 kilometres Gotthard (set to become the world's longest rail tunnel) and the 42 kilometres Lötschberg twin tunnels excavated in places to depths exceeding 2300 metres. The new Gotthard route, starting near the lake of Lucerne, will run 131 kilometres to Lugano near the Italian border, cutting the travelling time from Zurich to Milan by half. The new 60 kilometres Lötschberg route will be from south of the Swiss capital Bern to the Rhone valley, providing a direct link between two national highways. Once in operation, the new routes will provide twice the present freight traffic capacity through the Alps (including facilities for the piggyback carriage of goods vehicles over 28-tonnes gross weight which are not permitted on Swiss roads), and a high-speed line for passenger trains. It is claimed that these projects will protect the Alps from further environmental impact, saving intact the landscape and helping to maintain a clean atmosphere; and their low-altitude base and near-level lines are designed also to provide a large saving in energy.

By February 2005 41.8 per cent of the total excavation work on the Gotthard project had been completed; the Lötschberg tunnel is due to be completed and open to traffic by 2007. Cost benefit studies carried out in 1991 show that all investment should be repaid within 60 years of commencing operation early in the twenty-first century–possibly 2004/2007. The total costs amount to Swiss Francs (SF) 14 billion on the basis of the cost estimate in 1991 although by early in 2005 the St Gotthard final cost forecast had risen to some CHF 7.6 billion (i.e. SF) which equates to approximately €5 billion.

11.4.2 Europe/Scandinavia links

Other great infrastructure projects are the now completed links between mainland Europe and Scandinavia. Most important of these was the TEN-T project 11 on the list on p. 148. This is a 17-kilometre long bridge and tunnel development across the Danish Straits between Denmark, just south of Copenhagen and Malmo in Sweden. The Øresund Bridge was opened to traffic in July 2000, and comprises a four-lane motorway running above a double-track railway. The link consists of a 4-kilometre tunnel under the sea, a 4-kilometre-long artificial island and the 7.5 kilometres bridge, the world's longest cable-stayed bridge for road and heavy rail. Another bridge project, the so-called Fehmarn belt fixed link spanning the 19-kilometre wide Fehmarn Strait between Germany and Denmark, broadly follows the same route as the new Øresund crossing and will involve both road and rail links. It is listed as project 20 on the TEN-T list (see p. 148) and is due for completion in 2014–2015.

11.5 Freight interchanges (terminals)

A network of strategically located and fully equipped transfer terminals (variously called, in industry jargon, freight interchanges, Euro-terminals, international freight villages/hubs, multimodal transport/distribution parks, or some combination of these terms) served by adequate access routes for heavy lorries, and linked to main-line rail, in the UK and across Europe, is critical to the development of international combined transport operations. Additionally, with increasing transfer of freight to the inland waterways as an alternative to rail, there is need also for similarly strategic waterway terminals. These terminals are the principal interface between the modes, where road meets rail or waterway. Essential to the whole operation at this point is the availability of five basic ingredients:

- Strategic location to serve key industrial and commercial catchment areas.
- Adequate road access for heavy freight traffic.
- Direct links into the international rail freight network.
- Secure space for container/swap body storage, and vehicle manoeuvring.
- Suitable equipment for effecting quick and efficient transfer between road and rail, and road and waterway vessel.

Britain's rail network operator Network Rail has published as part of its 'Guide to Railfreight' a definition of common types of rail terminals. These include: private sidings and common-user terminals (Figure 11.1).

11.5.1 Private sidings

Rail terminals are a loading pad, warehouse, or factory connected to the national rail network. This enables rail freight services direct access to the point of despatch or receipt, removing the need for additional road movements, which may significantly increase the overall transport cost. Such savings must be balanced against the costs of developing, operating, and maintaining a private siding.

11.5.2 Common-user terminals

These may share some of the features of private sidings, but are distinguished by their accessibility to more than one customer. A number of these terminals are available within Great Britain and continental

Fig. 11.1. Typical combined transport terminal: Cologne, Germany (*Source*: German Railways DB).

Europe, and include rail-linked warehousing and intermodal transfer facilities. These are typically operated by the freight operating companies or third-party logistics providers.

In the UK a number of regional freight interchanges are the focal points of the rail freight network. These are complemented by private company sidings and automotive distribution centres. Additionally, a number of private enterprise intermodal terminal projects are at various stages of planning and development. European States such as France, Italy, and Germany, among others, are investing heavily in both new intermodal infrastructures and expanding and upgrading existing, traditional, rail freight facilities to cater for combined road–rail traffic. Germany, for example, in a long-term programme intended to drive more freight traffic from the roads and on to rail, is developing a country-wide network of 44 intermodal freight terminals, all with direct access to the national rail network and international services. While many of these involve improvement of existing rail freight facilities to intermodal standard, others are new developments including eight completely new sites in eastern Germany. These will be mainly partnership or co-operation projects between the state railways Deutsche Bundesbahn (DB), via subsidiary Deutsche Umschlaggesellschaft SchieneStrasse (DUSS), and intermodal centre management organizations.

Switzerland has a new (opened 2004) intermodal terminal at Wiler, near Utzenstorf, south-east of Solothurn mainly intended to handle unaccompanied intermodal transport units such as semi-trailers and swap bodies transported without tractive unit or driver. Here the objective is to remove some 1.2 million truck kilometres annually from Swiss roads and for road hauliers, by switching to rail, they can avoid the Swiss LSVA (mileage-based truck tax) as well as the LKW Maut motorway tax in Germany and a similar road tax in Austria.

In France, state railways, Societe Nationale des Chemins de Fer (SNCF) (Fret, the freight division), with state aid, invested heavily in equipping terminals and preparing suitable rail trunk routes for combined transport, believing that the volume of traffic consigned by this means will grow rapidly in the coming years. Together with other European railways SNCF is planning to equip 30 main routes, all with strong road–rail traffic potential, to form a true European intermodal network. Typical of these developments is a relatively new terminal 20 kilometres south of Lille in Northern France where the 'Delta 3 multimodal handling platform' went online in December 2003. This so-called trimodal logistics complex is claimed to form an efficient interface between road, rail, and water transport. Delta 3 has its own access point on the A-1 Paris–Lille motorway in the immediate vicinity, which crosses the A-21 close by. This is the main route between the Channel Tunnel and Brussels via Calais, Douai and Valenciennes. The terminal is directly connected to the main north–south line between Paris and Lille and the east–west Lens-Ostricourt line, and is located close to their intersection. The site also borders the Valenciennes–Dunkerque Canal, which allows the passage of ships of up to 3000 tonnes.

In March 2004, the UK Strategic Rail Authority published its Rail Freight Interchange Policy which, in its Foreword, it said:

> The success and growth of rail freight can only be sustained if there are enough Rail Freight Interchanges to enable modal shift. Rail freight interchanges have an important role to play whether they are terminals for aggregates or waste, sub-regional interchanges or other rail served industrial facilities and should be encouraged in the planning process, in accordance with Government policy. However, this Policy is concerned with a particular class of interchange, the Strategic Rail Freight Interchange. They are needed in relatively small numbers to serve major conurbations and are key to delivering growth of rail in the general freight market. These interchanges are long-term strategic infrastructure investments with operating lives going beyond 2020 with rail infrastructure, container handling and rail-connected warehousing, on a sufficient scale to enable critical mass for consolidation of trainload freight. By creating facilities on a scale, which effectively creates a rail connected distribution park or 'village', a wide range of businesses in the general freight market will be encouraged to locate their logistics operations, or production, where they have the option of rail or road transport. This also has the economic benefit of reducing the viable distance of the rail trunk movement, improving competitiveness with road, by locating businesses with direct access to rail, taking out the 'last mile' double handling and transport cost. In the longer term these interchanges will make a major, essential, contribution to developing the national strategic rail freight network linking, rail freight interchanges of all types, ports, and the Channel Tunnel, connected by a range of competitive rail operator services.

The policy document gives details for a number of examples of UK freight interchanges that are included in the following list:

- Glasgow (Mossend)
- Birmingham (Landor Street)
- Wakefield Euro-port
- North-East (Cleveland)
- London (Willesden)
- Port of Liverpool
- London (Stratford)
- Manchester (Trafford Park)
- Daventry International Rail Freight Terminal (DIRFT) (adjacent to junction 18 on the M1 motorway)

- The Midlands Channel Tunnel Rail Freight Terminal, within the Hams Hall National Distribution Park (near Coleshill, Birmingham)
- Direct for Europe Railport (Doncaster)
- Cardiff International Rail freight Terminal (CIRFT)
- Potter Group (Selby).

11.5.2.1 Wakefield Euro-port

International combined road–rail services from the Wakefield Euro-port via the Channel Tunnel began in January 1996. This terminal is a joint venture between the rail operator, EWS, and the local council (City of Wakefield Metropolitan District Council), which invested £18 million in the scheme (£3.5 million of which was from EU funding). Being strategically located near to both the M1 and M62 (adjacent to junction 31) motorways and the A1 trunk route, close to the Leeds/Bradford conurbation, it serves Yorkshire, Humberside and north-east England. The international intermodal rail terminal is incorporated within a 137 hectares (338 acres) freight village which will provide a range of facilities such as offices, retail premises, an hotel, truckstops, leisure, and support facilities (such as international freight forwarders). With an annual handling capacity of up to 175 000 containers and swap bodies, the terminal provides direct access to the European rail network.

11.5.2.2 Manchester: Trafford Park

A joint venture between the rail operators (EWS and Freightliner) and Trafford Park Development Corporation, this terminal opened in 1993 on a 20-acre site 6 miles to the south-west of Manchester close to the M63 motorway which is directly linked with the M62, M6, M56, and M61 motorways. It has an annual throughput capacity of 100 000 containers and swap bodies.

11.5.2.3 Birmingham: Landor Street

This Freightliner 110 000 annual capacity terminal is located just 2 miles east of Birmingham city centre, on the inner ring road with direct access to the A38 trunk route linking with the M6, M5, M42, and M40 motorways.

11.5.2.4 Glasgow: Mossend

Scotland's Euro-central rail freight terminal at Mossend in Lanakshire is a new (1994) £8.6 million development equipped to handle up to 74 000 unitized loads each year. It has direct links to the A8 and the M8 motorway and is only 10 kilometres from the M73 and M74 motorways.

11.5.2.5 London: Willesden

Developed from the old Freightliner terminal with a £10 million investment this terminal located at Willesden in north London has capacity to handle 184 000 containers/swap bodies annually. This terminal houses a purpose-built facility for HM Customs and Excise where suspect wagons, separated from trains, will be inspected for contraband.

11.5.2.6 Daventry International Rail Freight Terminal

The DIRFT operated by Tibbett & Britten is a Channel Tunnel multimodal rail terminal with 147 hectares of rail-related space to accommodate associated distribution developments. It is located adjacent to junction 18 on the M1 motorway (to which it is linked by a dual carriageway road), just south of the M1/M6 motorway intersection and the A1/M1 link road, on the electrified West Coast Main rail freight Line (WCML), which links the north-west and midlands to the Channel Tunnel, and which forms part of the EU's defined combined transport network. The terminal has a capacity to handle up to 1 million tonnes of freight annually, with five reception sidings and accommodation which allows for the loading/unloading of two full-length (i.e. up to 40 wagons) intermodal trains simultaneously.

11.5.2.7 Midlands Channel Tunnel Rail Freight Terminal: Hams Hall

Contained within the Trafalgar House/Powergen sponsored National Distribution Park at Hams Hall near Coleshill, Birmingham, the Midlands Channel Tunnel Rail Freight Terminal provides intermodal facilities for Midlands firms shipping to continental European markets. The site has good access to the Midlands motorway network, being close to the M42 motorway (junction 9) and the Birmingham Northern Relief Road (BNRR), and just 19 kilometres from Birmingham city centre. It is on the Birmingham to Nuneaton rail line, with 9-foot 6 inch-high container links to the port of Felixstowe, close to the West Coast Main Line.

11.5.2.8 Doncaster International Railport

Operated by Tibbett & Britten, Doncaster International Railport, which opened in December 1995, serves manufacturers and importers across Yorkshire, Humberside and the North Midlands, and offers direct access to the East Coast Main Line and Channel Tunnel. Trains take around 36 hours (two nights and a day) to reach the industrial centres in the south of France, east of Spain and north of Italy. The comprehensively equipped multi-user Railport boasts a total of 3 kilometres of track and has two reception sidings and four terminal sidings, allowing two trains each 600 metres long to be handled one after the other.

11.5.2.9 Cardiff International Rail Freight Terminal

Specialist distribution company Helios Properties is developing a site at Wentloog near Cardiff for its £25 million, CIRFT scheme. The proposed development of the 400 000 square feet project, situated adjacent to the existing rail freight terminal, is due to commence in the summer of 2005.

11.5.2.10 UK's European Freight Operating Centres

The UK's rail freight terminals are controlled from two European Freight Operating Centres located at Wembley in north London and at Dollands Moor, near Folkestone in Kent. Wembley fulfils a key role as a hub for Channel Tunnel freight. Here trains received from the UK regional terminals, most with loads for several countries, are resectioned into direct international block trains for direct transit to their European destinations, an operation normally taking no more about 1 hour. Where sufficient traffic volumes are generated to make it viable, single-destination block trains may be formed at regional terminals to run direct to a European location. The Wembley site can currently handle up to 70 (35 each way) international timetabled, full-length Channel Tunnel trains (up to 750 metres long), on a daily basis. It comprises three main operational areas: a group of 12 arrival and departure sidings, six secondary sorting sidings, seven train assembly sidings and four locomotive and customs sidings. There are 16 kilometres of track, 70 sets of points and 60 sets of signals, all controlled from a new signalling control centre that oversees all movements within the site. The 40-acre high security complex at Dollands Moor, located 3-kilometre west of the Channel Tunnel portal at Folkestone, controls through-Tunnel rail freight traffic. Trains pause here to await their allocated 'path' through the Tunnel and to undergo any necessary safety and security checks. This site, which has sophisticated intruder warning systems and closed-circuit TV surveillance (with more than 100 closed-circuit TV cameras), can handle up to eight international freight trains an hour (four in each direction) in sidings equipped with inspection roadways linked by an access subway.

11.6 The EC's 2004 list of 30 TEN-T projects

The indicative date for completion is in between brackets.

1. Railway axis Berlin–Verona/Milano–Bologna–Napoli–Messina–Palermo
   - Halle/Leipzig–Nürnberg (2015)
   - Nürnberg–München (2006)
   - München–Kufstein (2015)
   - Kufstein–Innsbruck (2009)
   - Brenner tunnel (2015), cross-border section
   - Verona–Napoli (2007)
   - Milano–Bologna (2006)
   - Rail/road bridge over the Strait of Messina–Palermo (2015)
2. High-speed railway axis Paris–Bruxelles/Brussel–Köln–Amsterdam–London
   - Channel tunnel–London (2007)
   - Bruxelles/Brussel–Liège–Köln (2007)
   - Bruxelles/Brussel–Rotterdam–Amsterdam (2007)
3. High-speed railway axis of south-west Europe
   - Lisboa/Porto–Madrid (2011)
   - Madrid–Barcelona (2005)
   - Barcelona–Figueras–Perpignan (2008)
   - Perpignan–Montpellier (2015)
   - Montpellier–Nîmes (2010)
   - Madrid–Vitoria–Irun/Hendaye (2010)
   - Irun/Hendaye–Dax, cross-border section (2010)
   - Dax–Bordeaux (2020)
   - Bordeaux–Tours (2015)
4. High-speed railway axis east
   - Paris–Baudrecourt (2007)
   - Metz–Luxembourg (2007)
   - Saarbrücken–Mannheim (2007)
5. Betuwe line (2007)
6. Railway axis Lyon–Trieste–Divaca/Koper–Divaca–Ljubljana–Budapest–Ukrainian border
   - Lyon–St Jean de Maurienne (2015)
   - Mont–Cenis tunnel (2015–2017), cross-border section
   - Bussoleno–Torino (2011)
   - Torino–Venezia (2010)
   - Venezia–south Ronchi–Trieste–Divaca (2015)
   - Koper–Divaca–Ljubljana (2015)
   - Ljubljana–Budapest (2015)
7. Motorway axis Igoumenitsa/Patra–Athina–Sofia–Budapest
   - Via Egnatia (2006)
   - Pathe (2008)
   - Sofia–Kulata–Greek/Bulgarian border motorway (2010), with Promahon–Kulata as cross-border section
   - Nadlac–Sibiu motorway (branch towards Bucuresti and Constanta) (2007)
8. Multimodal axis Portugal/Spain–rest of Europe
   - Railway La Coruña–Lisboa–Sines (2010)
   - Railway Lisboa–Valladolid (2010)

  - Railway Lisboa–Faro (2004)
  - Lisboa–Valladolid motorway (2010)
  - La Coruña–Lisboa motorway (2003)
  - Sevilla–Lisboa motorway (completed 2001)
  - New Lisboa airport (2015)
9. Railway axis Cork–Dublin–Belfast–Stranraer (2001)
10. Malpensa (completed 2001)
11. Öresund fixed link (completed 2000)
12. Nordic triangle railway/road axis
  - Road and railway projects in Sweden (2010)
  - Helsinki–Turku motorway (2010)
  - Railway Kerava–Lahti (2006)
  - Helsinki–Vaalimaa motorway (2015)
  - Railway Helsinki–Vainikkala (Russian border) (2014)
13. UK/Ireland/Benelux road axis (2010)
14. West Coast Main Line (2007)
15. Galileo (2008)
16. Freight railway axis Sines/Algeciras–Madrid–Paris
  - New high-capacity rail axis across the Pyrenees
  - Railway Sines–Badajoz (2010)
  - Railway Algeciras–Bobadilla (2010)
17. Railway axis Paris–Strasbourg–Stuttgart–Wien–Bratislava
  - Baudrecourt–Strasbourg–Stuttgart (2015) with the Kehl bridge as cross-border section
  - Stuttgart–Ulm (2012)
  - München–Salzburg (2015), cross-border section
  - Salzburg–Wien (2012)
  - Wien–Bratislava (2010), cross-border section
18. Rhine/Meuse–Main–Danube inland waterway axis
  - Rhine–Meuse (2019) with the lock of Lanaye as cross-border section
  - Vilshofen–Straubing (2013)
  - Wien–Bratislava (2015) cross-border section
  - Palkovicovo–Mohàcs (2014)
  - Bottlenecks in Romania and Bulgaria (2011)
19. High-speed rail interoperability on the Iberian Peninsula
  - Madrid–Andalucía (2010)
  - North–east (2010)
  - Madrid–Levante and Mediterranean (2010)
  - North/North–west corridor, including Vigo–Porto (2010)
  - Extremadura (2010)
20. Fehmarn Belt railway axis
  - Fehmarn Belt fixed rail/road link (2014)
  - Railway for access in Denmark from Öresund (2015)
  - Railway for access in Germany from Hamburg (2015)
  - Railway Hannover–Hamburg/Bremen (2015)
21. Motorways of the Sea
  - Motorway of the Baltic Sea (linking the Baltic Sea Member States with Member States in Central and Western Europe, including the route through the North Sea/Baltic Sea Canal) (2010)

  - Motorway of the sea of Western Europe (leading from Portugal and Spain via the Atlantic Arc to the North Sea and the Irish Sea) (2010)
  - Motorway of the sea of south-east Europe (connecting the Adriatic Sea to the Ionian Sea and the Eastern Mediterranean to include Cyprus) (2010)
  - Motorway of the sea of south-west Europe (western Mediterranean), connecting Spain, France, Italy and including Malta, and linking with the motorway of the sea of south-east Europe (2010)

22. Railway axis Athina–Sofia–Budapest–Wien–Praha–Nürnberg/Dresden
    - Railway Greek/Bulgarian border–Kulata–Sofia–Vidin/Calafat (2015)
    - Railway Curtici–Brasov (towards Bucuresti and Constanta) (2010)
    - Railway Budapest–Wien (2010), cross-border section
    - Railway Breclav–Praha–Nürnberg (2010), with Nürnberg–Praha as cross-border section
    - Railway axis Prague–Linz (2016)
23. Railway axis Gdansk–Warszawa–Brno/Bratislava–Wien (2013)
    - Railway Gdansk–Warszawa–Katowice (2015)
    - Railway Katowice–Breclav (2010)
    - Railway Katowice–Zilina–Nove Mesto n.V. (2010)
24. Railway axis Lyon/Genova–Basel–Duisburg–Rotterdam/Antwerpen
    - Lyon–Mulhouse–Mülheim (2014), with Mulhouse–Mülheim as cross-border section (2018)
    - Genova–Milano/Novara–Swiss border (2013)
    - Basel–Karlsruhe (2015)
    - Frankfurt–Mannheim (2012)
    - Duisburg–Emmerich (2009)
    - 'Iron Rhine' Rheidt–Antwerpen, cross-border section (2010)
25. Motorway axis Gdansk–Brno/Bratislava–Wien
    - Gdansk–Katowice motorway (2010)
    - Katowice–Brno/Zilina motorway (2010), cross-border section
    - Brno–Wien motorway (2009), cross-border section
26. Railway/road axis Ireland/United Kingdom/continental Europe
    - Road/railway axis linking Dublin with the North (Belfast–Larne) and South (Cork) (2010)
    - Road/railway axis Hull–Liverpool (2015)
    - Railway Felixstowe–Nuneaton (2011)
    - Railway Crewe–Holyhead (2008)
27. 'Rail Baltica' axis Warsaw–Kaunas–Riga–Tallinn–Helsinki
    - Warsaw–Kaunas (2010)
    - Kaunas–Riga (2014)
    - Riga–Tallinn (2016)
28. 'Euro-caprail' on the Brussels–Luxembourg–Strasbourg railway axis
    - Brussels–Luxembourg–Strasbourg (2012)
29. Railway axis of the Ionian/Adriatic intermodal corridor
    - Kozani–Kalambaka–Igoumenitsa (2012)
    - Ioannina–Antirrio–Rio–Kalamata (2014)
30. Inland waterways Seine–Scheldt
    - Navigability improvements Deulemont–Gent (2012–2014–2016)
    - Compiogne–Cambrai (2012–2014–2016)

12
Intermodal Road and Rail Vehicles and Maritime Vessels

We have seen throughout this book, how road haulage vehicles form the backbone of most intermodal freighting operations. Almost all of the initial and final legs of intermodal journeys are undertaken by standard road haulage vehicles, comprising either articulated or road–train (i.e. drawbar) combinations fitted with platform or skeletal-frame bodywork specially equipped with quick-action twist-lock devices for securing containers and swap bodies. Besides these vehicle types, there are other variations from the standard vehicle theme. First, for piggyback operations, specially strengthened and equipped semi-trailers are necessary in order to withstand the 'lifting-under-load' stresses incurred during loading and unloading; and second, the combined road–railer semi-trailer, or bimodal system, comprising a road-going semi-trailer that converts to run on rail bogies (see Figure 12.1). A number of such systems are in use or have been trialled in the UK and Europe, most notably the American RoadRailer system.

12.1 Road vehicles

The development of combined road–rail/waterway transport operations places great emphasis on the road vehicle link between locations where goods are loaded and unloaded, and the terminal where they are transferred to and from the rail or waterway networks. However, the choice of road vehicle that may be used for this purpose is restricted by legislative requirements. In particular, such vehicles must be able to carry fully laden swap bodies and ISO containers within the constraints of the 44-tonne legal limit on maximum gross weight, bearing in mind that most will be loaded up to the maximum or very near to it. For example, most Category A swap bodies weigh around 34-tonne gross while 40-foot ISO containers generally weigh around 30 tonnes when fully laden, hence the need for a 44-tonne vehicle gross weight, the minimum that would allow fully-laden swap bodies and ISO containers to be legally and safely carried on public roads.

12.1.1 Road vehicle weights and dimensions

Maximum weights and dimensions for UK registered road vehicles are specified in the *Road Vehicles (Construction and Use) Regulations 1986* (as amended) and the *Road Vehicles (Authorised Weight) Regulations 1998*. European Union (EU) limits are specified in European Commission Directive 85/3/EEC as amended by Directive 96/53/EC. Current road vehicle maximum permitted length, width, height, and weight insofar as they are relevant to intermodal transport operations, are as follows:

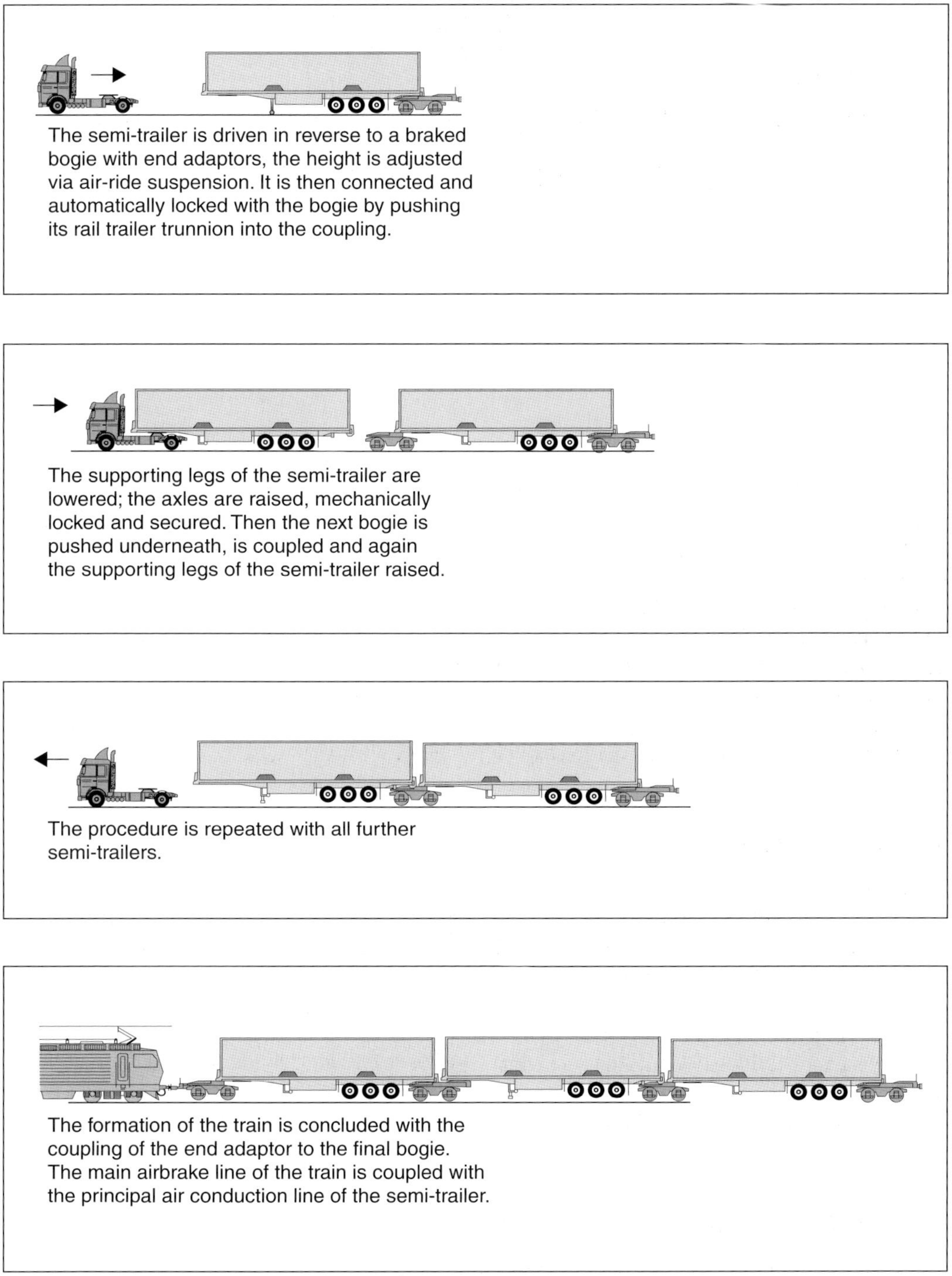

Fig. 12.1 Schematic representation of Kombirail's bimodal system showing semi-trailers being loaded to rail (*Source*: Kombirail).

12.1.1.1 Maximum overall length

- articulated vehicles 16.5 metres,
  - maximum semi-trailer load space 13.6 metres,
- drawbar combinations: 18.75 metres,
  - maximum load space, drawing vehicle, and trailer 15.65 metres (i.e. 7.825 metres each).

When determining the overall length of a vehicle or trailer, account must be taken of 'any receptacle which is of a permanent character and strong enough for repeated use'. This clearly implies that swap bodies and ISO containers are to be included in the overall length of vehicles and trailers.

12.1.1.2 Maximum overall width

- heavy vehicles 2.55 metres.

12.1.1.3 Maximum overall height

- Europe 4 metres,
- UK not restricted (but must be indicated to the driver if over 3.66 metres).

12.1.1.4 Maximum gross weights

- articulated vehicles and road–train combinations 44 tonnes.

  Note: This weight limit is subject to technical conditions regarding the minimum number of axles (i.e. six), maximum axle weights, minimum axle spacings, the fitment of 'road-friendly' suspension, and the use of a Euro-2 designated engine.

12.1.1.5 Carrying capacity

Intermodal vehicles built to the maximum permitted length and plated (i.e. approved) for 44-tonne operation can carry swap bodies and ISO containers as follows:

- Articulated vehicles
  - 1 × 7.15- or 7.45- or 7.82-metres swap body
  - 1 × 12.2- or 13.6-metres swap body
  - 2 × 20-foot ISO containers
  - 1 × 30- or 40-foot ISO container
  - 1 × 45-foot ISO container (subject to the container having tapered Euro-type corner castings)
- Road–train combinations
  - 2 × 7.15 or 7.45 or 7.82 metres swap bodies
  - 2 × 20-foot containers

12.1.2 Road vehicle specifications

For operation at 44-tonne gross weight as mentioned above, road vehicle combinations comprising articulated tractive units and semi-trailers, and drawing vehicles and drawbar trailers, must comply with statutory requirements as to their construction, the number of axles, their 'plated' weights, and their safe operation, besides the operational restrictions specified under UK law.

12.1.2.1 Construction and use of vehicles

In constructing goods vehicles and trailers, manufacturers and bodybuilders must observe requirements regarding the specification and standards of construction of components and the equipment used in the manufacture. While some of these items are covered by the European 'Type Approval' scheme, the majority of them are included in the UK's domestic Road Vehicles (Construction and Use) Regulations 1986 (Statutory Instrument 1078/1986) and subsequent amendments to these regulations which, collectively, are commonly referred to as the Construction and Use Regulations. Once a goods vehicle or trailer has been built and put into service, it is the operator as the 'user' of the vehicle who must then ensure that it complies fully with the law regarding its construction and use when on the road. It is worth pointing out that where a vehicle on the road is found to contravene the regulations, it is the operator who will be prosecuted and, if convicted, liable to meet the penalties imposed – it may even jeopardize his 'O' licence. It is no defence or excuse to say that the fault rests with the vehicle/trailer manufacturer, bodybuilder, or even the supplying dealer.

12.1.2.2 Authorized weights

In addition to the maximum vehicle weight limits set out in the Construction and Use Regulations outlined briefly above, operators may follow the statutory limits set out in the *Road Vehicles (Authorised Weight) Regulations 1998*, as amended.

Note: Both of these sets of regulations are complex and too extensive to detail here. The reader is recommended to refer directly to the legislation (available from The Stationery Office, TSO) or to the author's annual publication; *The Transport Manager's and Operator's Handbook* (available from Kogan Page Ltd, London) which explains all these in great detail.

12.1.2.3 Plated weights

The UK law demands that most goods vehicles display a manufacturer's plate showing for the vehicle and/or trailer and its individual axles the maximum weights at which it is designed to operate, and an official (Ministry) plate showing the permissible maximum gross combination/train and individual axle weights (pmw) for the vehicle which must not be exceeded on a road in Great Britain (see below for definitions).

12.1.2.4 Definitions

- Gross vehicle weight (gvw) refers to the maximum laden weight of a rigid (load-carrying) vehicle.
- Gross combination weight (gcw) refers to the maximum weight of an articulated combination comprising of a tractive unit and a semi-trailer. When the semi-trailer is uniformly loaded, at least 20 per cent of the weight of the load must be imposed on the drawing vehicle (i.e. tractive unit).
- Gross train weight (gtw) refers to the combined maximum weights of a load-carrying drawing vehicle and a drawbar trailer.

Note 1: In all cases above where the word 'weight' is used, it is common also to see the word 'mass' used instead, as in, for example, gross mass.
Note 2: In the case of articulated vehicles and road–train combinations, it is important that the respective gross weights of both drawing vehicle and trailer are compatible for the combination's maximum load-carrying potential to be legally realized.

12.1.3 Typical vehicles and specifications

A wide choice of heavy-goods vehicle and trailer makes and models, with many differing individual specifications, is available to meet the particular requirements of the intermodal/combined road–rail transport operator, even allowing for the legislative restrictions mentioned above. It is useful to consider, as a typical example, the prototype road–train combination described below. An articulated vehicle designed to operate at 44 tonnes may be expected to have a similar heavy-duty tractive unit chassis; namely, a 3 + 3 axle air suspended combination running at the maximum of 16.5 metres overall length with a 13.6-metre skeletal-framed semi-trailer to accommodate Category A (i.e. 13.6 metres long) swap bodies or 40-foot ISO containers, complete with twist-lock fittings at standard centers. It would have an engine producing 400/500 horse power, or even more, and similar transmission details.

12.1.3.1 Prototype 44-tonne road–train combination

In 1994, a prototype intermodal road–train combination was produced jointly by Foden trucks, King Trailers, RSG (demountable equipment), and Southfields bodybuilders. This vehicle (shown in Figure 12.2) was road tested by *Commercial Motor* in December 1994. It was capable of carrying two Category C swap bodies or 20-foot ISO containers. All-up cost of the rig at that time (1994) was in the region of £105 000. The combination comprised Foden's 4000 series 26-tonne, three-axle (6 × 4) rigid chassis equipped for intermodal drawbar operations with a design gross train weight of 45 tonnes (44-tonne pmw on UK roads). It had a Cummins L10 CELECT (with electronically-controlled engine management system), six-cylinder, direct-injection, charge-cooled 10-litre diesel engine producing 350 horse power at 1900 revolution per minute. Drive was via an Eaton 12-speed, twin-splitter, constant-mesh gearbox through Eaton tandem-drive with spiral bevel differential and lockable inter- and cross axle differentials. The rear axles had air suspension and the vehicle was equipped with RSG ISO truck and Airswop equipment to facilitate swap body mounting and demounting. A VBG air-actuated coupling was fitted to provide

Fig. 12.2 Prototype intermodal road–train combination (*Source*: Ray Smith Group plc).

close-coupled operation within maximum legal length limitations. The vehicle was used in conjunction with a King Trailers, air-suspended, turntable-steered, three-axle (i.e. tri-axle) drawbar trailer operating at 24-tonne gross weight. Like the drawing vehicle described above, this had RSG ISO-standard equipment (bolster-mounted twist-locks set at standard centres) and an air-operated demount system for raising and lowering the body. Southfields-built 7.45 metres Category C swap bodies with curtain sides were carried on both drawing vehicle and trailer. Many vehicles, broadly similar in specification to the prototype described here, are in common use these days in intermodal operations.

12.1.3.2 Rolling sub-frame semi-trailers

An interesting development in intermodal semi-trailers is the rolling sub-frame model designed and built by UK manufacturer Maxilode. These skeletal semi-trailers, designed for 44-tonne operation, are equipped to carry either one 40- or two 20-foot ISO containers. The electro-hydraulically operated rolling sub-frame allows for correct weight distribution over the axles when single but heavier (up to 30 tonnes) 20-foot containers are carried, and to allow containers to be moved to the rear of the chassis for end loading/unloading when reversed up to a loading bay. Other trailer manufacturers produce units similar in concept to the one described here. For example, Dennison Trailers produce what they call a 'sliding skeletal' unit which they describe as being a 14-lock patented lightweight design weighing, unladen, just 4740 kilograms. This trailer can carry containers up to the 45-foot maximum length, but when loaded with 20- and 30-foot containers, these can be moved to the rear of the chassis for rear discharge of the container.

12.1.3.3 Piggyback (Huckepack) semi-trailers

Semi-trailers intended for use in piggyback, or to use the continental term, huckepack, operations, while conforming largely to normal road-going specifications and legislative dimensions, are special in being designed and built to withstand the stresses of being lifted in a laden state by heavy-duty forklift truck or by overhead crane with grapple arms from road onto rail and vice versa. For this purpose they have strengthened chassis and sub-frames, and are fitted with reinforced lifting pockets in the underside to accept the forklift tines or the crane's grapple arms. Typical of such semi-trailers was the Railmaster model built by UK manufacturer M&G Tankers and Trailers for the TIP trailer rental company. The specification of this model followed that of the company's standard 13.6-metre tri-axle models. It had electronic air suspension and was fitted with a curtain-sided body, running within the European 4-metre height limit, and providing capacity to load 34 Euro-pallets within a 25-tonne payload. Overall weight of the semi-trailer was about 7000 kilograms, which is said to be some 500 kilograms above that of a standard road semi-trailer. The unit was fitted with hinged under-run rear bumper and equipped with lashing points.

12.2 Bimodal semi-trailer systems

Two brand-named road–railer semi-trailer or bimodal systems are currently in use in Europe as mentioned previously: namely, the RoadRailer and Kombirail systems which use a dual-purpose reinforced semi-trailer complying with road-going legislation, and capable of mounting onto special rail bogies for the rail haul. Transfer is simply effected, the whole road-going outfit being backed up to a rail bogie (with a specially-fitted kingpin mounted on the underside rear of the semi-trailer locating in a locking arrangement on the rail bogie), the semi-trailer landing legs are lowered and its road wheels raised by air pressure from the tractive unit. An intermediate bogie is backed under the front of the semi-trailer (raising it to running height) locating and locking on the front trailer kingpin. An articulated coupling that prevents stress being passed down the line from trailer to trailer links each trailer unit. Additional rail bogies and semi-trailers are added to form a complete train. Safety locks built into the system prevent any risk of the train moving if semi-trailers are not properly secured on the rail bogies and the whole system is connected to the train's air brakes.

Such bimodal systems have the advantages of simplicity, by employing only a single load-carrying module (i.e. the special semi-trailer, which can be a box van, curtain-sider, reefer, tanker, tipper, or flat-platform model) compatible with both road and rail operations, by eliminating the need for high-cost transfer/lifting equipment in terminals, and by allowing rapid, horizontal, transfer from road to rail and vice versa. They are also claimed to have greater payload within fixed train weights (for example, a 980-tonne payload within a train grossing 1500 tonnes) compared with swap-body and piggyback trains, and a greater number of units (49) within an overall train length of 700 metres compared with other systems, in both cases resulting from the close coupling arrangement. Potential payloads with these trailers would be in the region of 27 tonnes within the 44-tonne maximum gross weight limit for road vehicles.

The RoadRailer system developed by Wabash National in the USA has been operating in Europe for some years, as well as in many other countries worldwide, and claims to have covered many thousands of kilometres in European trials operating from rail terminals as far apart as Glasgow and Budapest, Stockholm, and Verona. The RoadRailer system has achieved approval for 120 kilometres per hour running on the UK rail system and full UIR approval for high-speed freight running on European railways. It also achieved (in 1996) the notoriously difficult-to-obtain approval for Channel Tunnel operation. UK operator, the Transport Development Group (TDG) has operated a regular Glasgow, London service using this type of equipment.

12.2.1 The Kombirail system

Kombirail's bimodal system is a joint venture between the German companies Ackermann-Fruehauf and Talbot together with French companies Fruehauf France and Remafer who design and build the special Kombitrailers and the rail bogies. This system, which is virtually identical in the principles of its operation to the RoadRailer described above, is now in widespread use in Europe, having been extensively tested in the most extreme of operating conditions and met the approval of European railway companies. Major Kombirail user, Hupac SA, the Swiss road–rail (piggyback) operator, claims that this system effectively utilizes some 68 per cent of train load capacity against only 51 per cent for container and swap body traffic, 50 per cent for semi-trailers carried piggyback and 45 per cent for rolling motorway (complete vehicle) systems.

The Kombirail trailer (i.e. Kombitrailer) requires no complicated infrastructure to change its transport mode. The economic efficiency of any bimodal system depends on the time and costs involved in making the transfer from road to rail. In contrast to conventional intermodal transport systems, the Kombirail road–rail interface does not require huge investments in infrastructure (such as expensive siding and crane systems or special approaches to the road network) that can only be used for one specific purpose. All the Kombirail system needs are modifications to the mainly existing infrastructures. The special bimodal trailers used in the system, appear very similar to normal road-going semi-trailers. They comply fully with all national and EU legislative requirements and satisfy user requirements for road vehicles providing some 26/27 tonnes of payload capacity (i.e. 33 Euro-pallets), depending on body type, within the European 44-tonne gross weight limit for intermodal vehicles. On the road, the Kombitrailer travels on its conventional road running gear with air suspension. The air suspension system is also responsible for adjusting the height of the trailer when it is shunted onto the rail running gear. On rail, the road running gear is pneumatically raised, automatically locked in place for travel and it is automatically monitored for safety.

To keep the unladen weight on the road and the resulting costs for the vehicle owner to an absolute minimum, all components required for rail travel and for safety purposes are accommodated in the rail running gear, with the exception of the main pneumatic pipe which links the air-braking system from the train via the trailers to the rail bogies. The floor assembly of self-supporting Kombitrailers is reinforced to withstand the forces generated in railway operation, consequently, the trailer is also more robust on the road, its weight being approximately 900 kilogrames greater than that of conventional road haulage trailers. On rail too, the deadweight is far lower than with other forms of intermodal traffic, as the trailer forms the actual rail wagon. All conventional trailer body types may be used in the Kombirail system; for example, Kombitrailers can be supplied as flatbeds, tankers, silos, skeletal chassis, or box trailers.

Rail running gear turns the trailer into a rail wagon. An adapter forms the connection between bogie and trailer, and at the same time acts as a coupling between the trailers. The rear of one trailer and the front of the next one are supported by a middle adapter on a joint bogie, the adapter conducting the bearing forces into the bogie while the tractive forces are transmitted through the adapter and the trailer chassis. The actual coupling is a component (i.e. a 'fifth-wheel' coupling) that has been used successfully in millions of road vehicles and reinforced in this type of operation for rail operations. The adapter consists of two couplings in a laterally reversed arrangement, one to take the rail kingpin under the front part of one trailer, and the other for the kingpin under the rear of the trailer in front of it. Once the kingpin reaches its final position in the coupling, a closure hook automatically closes around the kingpin. The rear of the trailer is rigidly coupled to the adapter. The rail kingpin in the coupling and the two horizontal pins form a continuous positive joint, with the taper adapters arranged accordingly under the rear of the trailer. At the front end, the trailer is supported on the coupling plate and is held positively in its transverse and longitudinal position by the intermediate adapter. An end adapter with standard push–pull device is fitted at the front and the tail end of the Kombirail train to provide a fully compatible interface to conventional railway vehicles. The adapter coupling also allows distances between trailers to be kept to less than 400 millimetres with considerable advantages, such as shorter trains, less air resistance, and reduced noise emission, as well as providing the added security that trailer doors cannot be opened once connected into a train.

Changing equipment from road to rail and vice versa is a simple operation carried out without the use of any additional equipment and taking a minimum amount of time. According to Kombirail, when changing vehicles, the truck driver:

> Simply does what he has done a thousand times before: he manoeuvres with the truck. Anyone who can drive an articulated truck can also manoeuvre the Kombirail trailer onto the track. The driver folds up the under-ride guard (rear under-run bumper) and bolts it in position. On a level track, he simply reverses the trailer onto a braked rail running gear. The next rail running gear is shunted into position and automatically coupled on. The driver can now retract the support feet and connect the main pneumatic pipe. It takes about 3 minutes to shunt each unit on or off the track, a procedure every driver can handle. Depending on the length of the track, several groups of Kombirail trailers can be formed at once and coupled together to form a train.

12.3 Rail wagons

Combined road–rail operations are as heavily dependent on the availability of purpose-built rail wagons as on the specialist road vehicles discussed above. In this part of the chapter we look at some of the various forms of rail wagon in use for swap-body, container and piggyback trailer loading, and for rolling motorway operations. Broadly, in the context of this book, rail wagons may be categorized into three distinctive types.

- Mainly flat, skeletal-framed wagons equipped with twist-lock securing devices for swap body and container traffic.
- Special low-height wagons for piggyback (huckepack) carriage of unaccompanied semi-trailers with recessed pockets to accommodate the trailer road-wheels, or spine wagons.
- Purpose-built drive-on/drive-off wagons for rolling motorway (complete, driver-accompanied, road vehicle) traffic – as, for example, Eurotunnel's freight shuttle wagons used on Channel Tunnel services for carrying heavy goods vehicles.

12.3.1 Container and swap-body wagons

Typical of this type of wagon is the French, Multifret-design, low-platform intermodal rail rolling stock, built by Arbel Fauvet Rail (AFR) in Douai, Northern France, a large number of which are in service with English, Welsh and Scottish Railway (EWS), and numerous others of similar design have been supplied to SNCF (French railways) and Intercontainer (the European railway marketing organization). These wagons, designed basically for carrying ISO containers and swap bodies, operate in sets of two with a maximum load capacity of 38 tonnes for each wagon and 38 wagons to a train. They are UIC (Union Internationale des Chemins de Fer) approved and are fully compatible with both British and continental railway loading gauges and capable of running at speeds up to 140 kilometres per hour (90 miles per hour, mph). The wagons have 15.64-metre load platforms to accommodate either a single 13.6-metre swap body or two 7.15-metre units, or a single 40/45- or two 20-foot ISO containers. With a deck height of only 945 millimetres, containers up to 9 feet 6 inches and swap bodies up to 2.77-metre high can be carried within the existing loading gauge on Network Rail's upgraded principal routes to the Channel Tunnel.

German rail wagon manufacturer Graaff has a two-axle flat rail car primarily intended for loading containers and swap bodies. This design has a shock-protected loading platform equipped with 16-foldable locking pins to carry loading units in different combinations. The wagons can accommodate either one 30 foot or one 40/45-foot ISO container or two 20-foot units, and has a load capacity of 32 tonnes. These wagons are approved for operation at up to 120 kilometres per hour.

12.3.2 Spine wagons for piggyback

Prototypes of a 'revolutionary' new rail wagon for piggyback road–rail transport were first trialled in the UK in the late 1990s. Designed by American company Thrall Car Manufacturing Company and operated in the UK through subsidiary Thrall Europa of Glasgow, the EuroSpine is an articulated, four-segment wagon (see Figure 12.3) that can carry road-going semi-trailers (and, if required, containers) within the width and height restrictions imposed by the UK loading gauge. Designed to run as an integral unit to

Fig. 12.3 Schematic illustration of the Thrall Eurospine wagon concept (*Source*: Thrall Car Manufacturing Co).

maximize the number of trailers that can be carried per train, this wagon carries four typical 4-metre high, 13.6-metre long intermodal trailers. The EuroSpine's box beam design and minimalized bodywork is claimed to provide considerable weight savings over full-body types, including pocket wagons, and the lowest cost per trailer or container of any wagon designed for such loads. Similar models, operating as five segment or stand-alone wagons, have become the standard method of trailer transport in North America.

EuroSpine characteristics include:

- Four distinct segments weighing 14.5 tonnes each joined in an articulated manner through a proprietary connector that eliminates the need for specialized bogies. Each segment can carry a fully loaded semi-trailer weighing up to 36 tonnes, including those with nose-mounted refrigeration units. Alternatively, 40- or 45-foot ISO containers can be loaded on all segments.
- Total wagon length over the buffers of 59.6 metres with a total unladen weight of 54.8 tonnes, and a minimum curve radius (uncoupled) of 60 metres.
- Depressed wheel support platforms, cantilevered out from the main spine, allow the trailer wheels to travel at just 330 millimetres above the rail. The wagon occupies just slightly more space than a trailer alone while meeting most applicable UIC/W6 gauge clearance requirements.
- Wagon structure stiffness keeps the wheel support platforms from moving out of the clearance envelope, even under extreme conditions.
- Five two-axle bogies; one under each far end and one under each of the three articulated connections. These have a maximum capacity per axle of 22.5 tonnes.
- Disc braking with tread-assist on all bogies, which allows wagon to run at 120 kilometres per hour in high-speed slots.
- Wheels of 920 millimetres diametre, standard in all positions.
- Fewer buffers and coupling points to improve train longitudinal forces and acceleration. The longitudinally rigid articulated connectors reduce a train's tendency to bunch and stretch over uneven terrain and improve ride quality.

12.3.3 Tiphook-type piggyback wagon

Claiming to make terminal turnaround faster for savings in time, money, and manpower, the piggyback system of rail wagon rental company Tiphook Rail enables semi-trailers complying with European weights and dimensions legislation to be driven on to a specially built rail wagon via a swinging centre load-platform (see diagram in Figure 12.4). To carry out the simple (10 minutes) loading/unloading procedure, the wagon-mounted platform is swung out to the loading/unloading position, powered by an auxiliary power pack or by the tractive unit's power system, and the semi-trailer is reversed on to the platform via independent loading ramps. When the semi-trailer is in position, the tractive unit is unhitched and the platform is swung back to its closed (travelling) position. The semi-trailer is secured by its fifth-wheel kingpin locating in a locking plate on the forward superstructure of the wagon. Since no specialized handling equipment is necessary, this operation can take place anywhere adjacent to a rail siding with hard standing. Besides the loading/unloading system described for this wagon, there are other operating possibilities including loading of semi-trailers by means of traditional overhead lifting equipment at existing terminals, or more versatility can be added by incorporating locking points for containers and swap bodies on the wagon platform. These piggyback wagons can form part of a mixed freight train, or a sufficient number of wagons can form a dedicated trainload.

12.3.4 The Modalohr system

French company MODALOHR has developed a combined rail–road, rolling motorway-type, transport concept, which, it claims, meets both road hauliers' and rail operators' current needs by providing simple

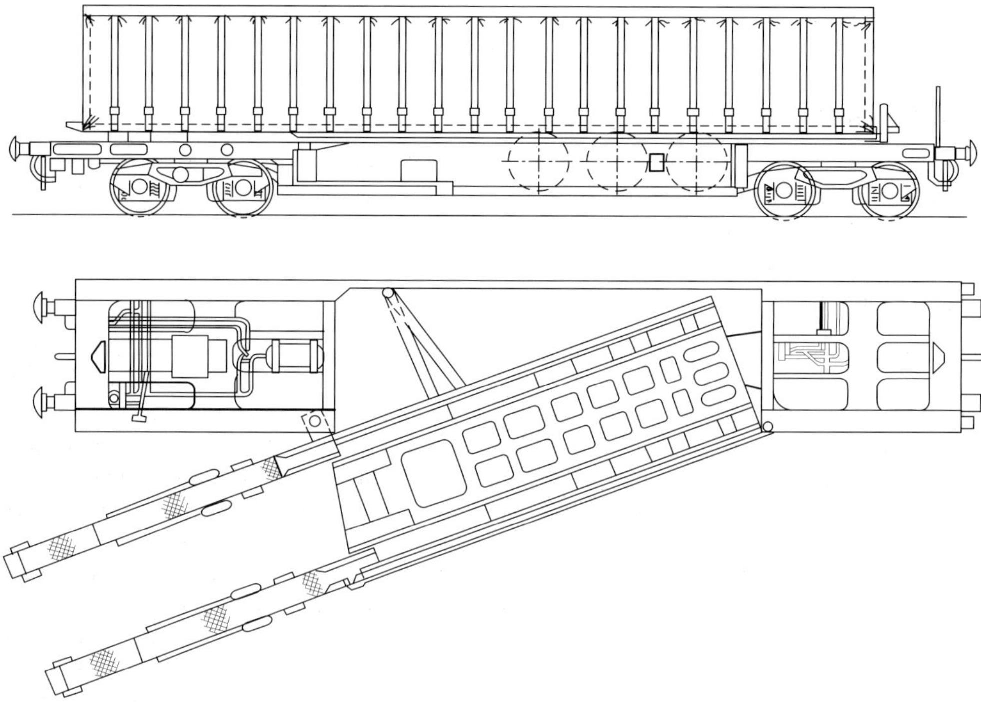

Fig. 12.4 Detail of Tiphook Rail's piggyback loading system showing side view with a 40-feet container/semi-trailer in position and a plan view with hydraulic ramp in loading position.

technical solutions which are compatible with existing railway infrastructures and impose no restrictions for users. The concept is based on the company's design of a low-frame, articulated, rail wagon that can transport standard road semi-trailers. The advantages for the carrier, apart from the possibility of using his own standard road equipment, are that:

- trucks are loaded horizontally directly with the road tractor, without the need for handling equipment,
- the lateral loading of trucks in a 'herringbone' pattern allows simultaneous and fast trans-shipment taking less than 30 minutes in all for the entire train,
- a very low floor means that trucks up to 4-metre high can be loaded within existing UIC and GB1 railway gauges,
- the system is reliable and trucks are strongly secured on the wagons.

The advantages for the rail operator are that:

- the wagons have standard bogies and wheels that can be used in the same way as a conventional wagon and with identical maintenance costs,
- the system of articulation and opening the wagons to allow vehicle access is simple and entirely mechanical, providing reliability, and independent loading and unloading of one or more vehicles at each intermediate stop,
- higher average speeds can be achieved with MODALOHR shuttles travelling at 120, or even 140 kilometres per hour,
- only simple trans-shipment terminals are needed, consisting merely of an asphalt surface on either side of the tracks (i.e. no platforms are necessary).

More details of the Modalohr system can be found on the company's web site, in French, but with moving diagrammatic images showing how the system works, at: www.modalohr.com.

12.3.5 Eurotunnel shuttle wagons

The special open lattice-sided rail wagons used on Eurotunnel's freight shuttle service are specially designed and built by Breda Fiat, Italy, to accommodate lorries up to 44-tonne laden weight, 18.5 metres in length, 2.6 metres width, and 4 metres height.

12.4 Maritime vessels

Consigning freight by waterway transport has numerous advantages, not least the fact that it is a very 'green' mode, being particularly quiet in operation, not subject to congestion, an economical user of fossil fuels and relatively pollution free. It also has the bonus of using a largely natural resource in the form of navigable rivers and coastal waters, and the ability, depending on the type of craft used, to carry the equivalent of many lorry loads in a single journey. Most inland waterway craft are defined as 'barges' which are described as large, flat-bottomed, open-deck type vessels normally towed or pushed by a tug, albeit some types have their own motive power. Their size, for inland waterway use is usually governed by two key factors: the size of locks along the route and the wake made by the vessel. As an example of the type of craft used, the smallest vessel is generally a Class 1 'Spitz' type with a load capacity of some 350 tonnes, which equates to 14, 25-tonne lorry loads. At the other extreme a push convoy of four barges, as typically seen on the River Rhine, for example, would remove up to 440 maximum capacity vehicles from the road network, while a Jowi-class container ship can carry up to 470 teu containers on navigable river, canal or coastal waters. A full list of the waterway vessel classifications and an illustration of the various vessel types and their road-vehicle equivalent capacities are shown in Figure 12.5.

12.4.1 The LASH system

LASH is an acronym for Lighter Aboard Ship. This type of vessel is operated by an American ship operator, Waterman Steamship Company, on its trans-Atlantic services between east coast US ports, such as Baton Rouge, Birmingham, Catoosa, Cincinnati, Little Rock, Louisville, Memphis, Pittsburgh and both Rotterdam on mainland Europe and Immingham in the UK. It is designed to carry up to 82 LASH barges, all of a standard size and with a cargo capacity of 385 tonnes each (giving a total ship payload of some 31 500 tonnes). The barges are towed within ports and on inland waterways by tug (see Figure 8.2 and 8.3 in Chapter 8) to various points where they are loaded and then returned to the ocean-going vessel (i.e. the mother ship) where they are hoisted aboard by a special shipboard gantry crane ready for trans-ocean shipment. LASH ships do not need special docks or terminals and the cargo rarely requires trans-shipment, invariably moving from origin to destination on a single bill of lading. A wide variety of commodities and manufactured goods may be handled by LASH including pipes, vehicles, pallets, bagged cargo, forest products (i.e. timber), and bulk loads. They can also accommodate abnormal loads and heavy cargo such as machinery and construction equipment, which can often be stowed in one piece, eliminating costly disassembly, boxing, and reassembly. Once loaded, LASH barges are water tight with secured hatch covers for each barge and as the barges are legally defined as 'vessels' this means that once cargo is loaded an onboard bill of lading may be issued.

Cargo can be loaded into LASH barges at inland waterway points, ocean ports, and at shallow draft terminals. In the USA, for example, inland waterway points include the Alabama, Arkansas, Mississippi, and Ohio River estuaries from which barges are towed to the port to await the arrival of the LASH mother vessel which does not require special docks or terminals and is often worked at anchor in rivers, roadsteads, and light traffic port areas. This is another advantage of the system; namely, the ability of LASH ships to load and discharge cargo from anchorage so shipments are never delayed by a lack of dock facilities or

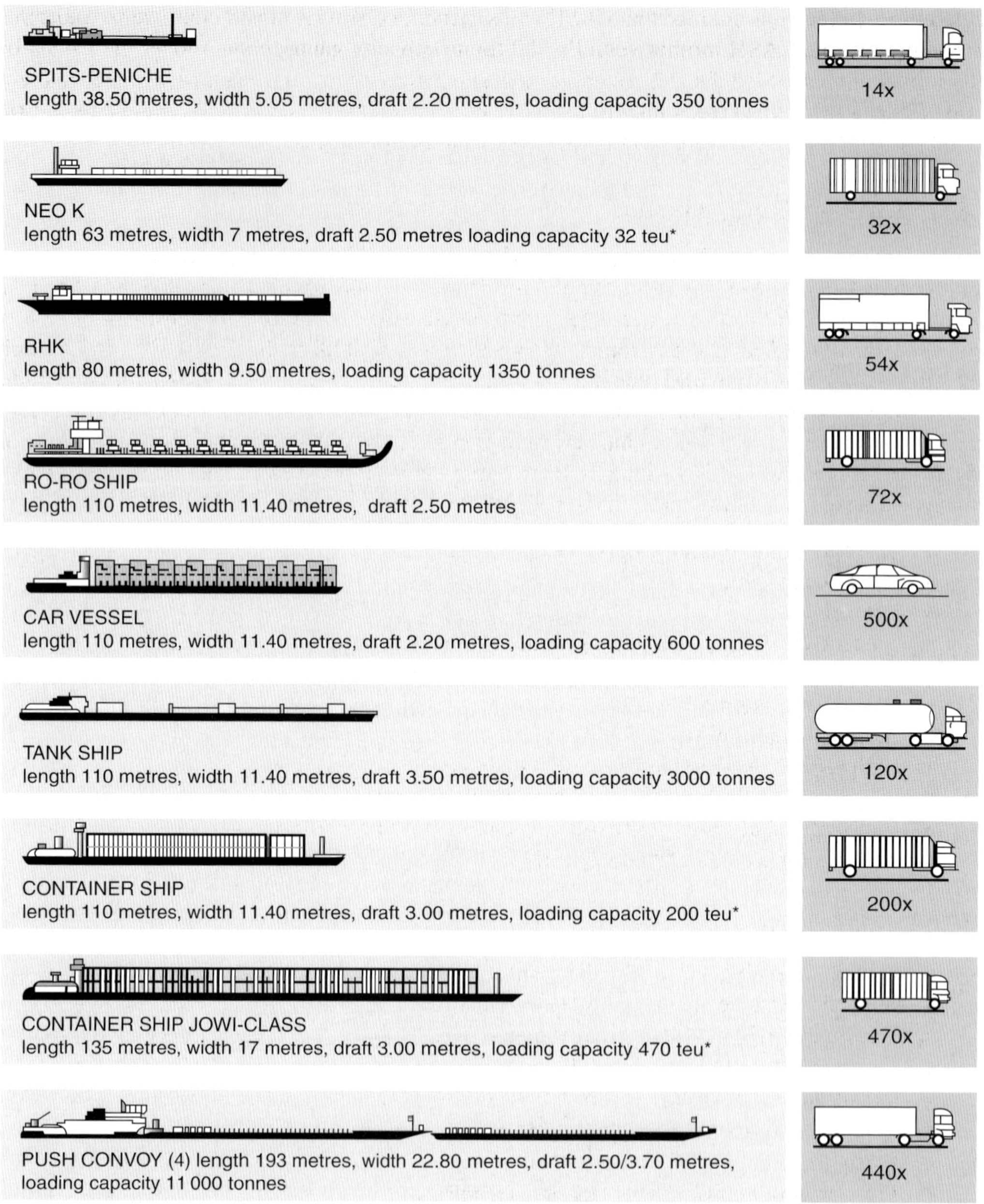

*1 teu = 20-foot equivalent unit
Source: INE

Fig. 12.5 European inland waterway vessels shown in profile and with payload comparison against road vehicles (*Source*: European Commission, Energy and Transport DG, Brussels, *Inland Waterway Transport*, 2003).

port congestion. In fact, it is claimed that LASH ships spend far less time in port compared to other vessels. On arrival in port, the LASH mother vessel's 500-tonne capacity gantry crane moves the LASH barges from their positions on board the ship, over the stem sponsons and into the water to be towed to their destinations. In succession, the same onboard gantry crane lifts the outbound loaded LASH barges from the water and places them in the holds, and on the deck of the mother vessel.

12.4.2 The 'Water Truck' concept

A new-concept vessel for inland waterway use has been developed for the Shipbuilders and Shiprepairers Association (SSA) by Armstrong Technology. The idea for the new vessel, designated the 'Water Truck' was first announced in 2004. The design is for small and flexible vessels for transporting freight on inland waterways in an economical and environmentally friendly way. It offers flexibility for carrying unitized or bulk cargoes, a self-loading and unloading arrangement, and has been developed with consideration to the intermodal transportation of goods. The basic hulls can be fitted out in various ways and can carry alternative cargo containment units. One variation is suitable for small inland waterways and is aimed specifically at economical and environmentally friendly freight movement in the UK; for example, along the Thames between the estuary and Windsor. Key features of the vessel include:

- sized for small inland waterways,
- shallow draught,
- low profile,
- intermodal freight transport capability,
- maximum hold capacity for unitized cargoes, bulk or combinations,
- low-environmental impact,
- high manoeuvrability,
- high reliability and safety through dual system design,
- propulsion and manoeuvring control from either end,
- optional and alternative self-loading and unloading arrangements,
- suitable for series build and operation.

12.4.2.1 Principal dimensions of the 'Water Truck'

- Length, overall 50.0 metres
- Breadth, extreme 6.0 metres
- Depth, amidships 3.0 metres
- Draught, operating 2.0 metres
- Air draught, operating 5.0 metres
- Deadweight, approximate 375 tonnes
- Crew 2 persons
- Crew accommodation up to 3 persons

12.4.2.2 Performance

- Propulsion units 2 × azimuthing
- Propulsion motors 2 × 200 kilo watt electric
- Speed, maximum 8 knots

The propulsion installation is designed to provide a high level of control and manoeuvrability. The power generation and propulsion systems are dualled to accommodate component failure and maintain propulsion at reduced speed should any part of one system be lost.

12.4.2.3 Main machinery

- Generators 2 × 300 kilo watt
- Fuel LPG (optional) or diesel

The generators are designed for use in two principal modes for propulsion, manoeuvring and ship systems, or for loading, unloading, and ballast operations. Machinery space arrangements maximize the utilization of available space. LPG is proposed for good economic and environmental performance.

12.4.2.4 Loading and unloading

The concept as shown includes self-loading and unloading equipment. The equipment comprises a self-contained elevating gantry crane with spreader for 20-tonne containers, a ballast system for draught adjustment and heel control and optional telescopic stabilizer outriggers to facilitate rapid load transfers. Alternative load transfer equipment can be specified or the equipment can be removed if land-based facilities are available. For bulk cargoes the holds can accommodate compatible hopper-sided removable liners.

The concept for this vessel was been developed for the SSA as part of the Integrated Technology for Marine Construction (ITMC) LINK project supported by a grant from the Department of Trade and Industry and with contributions from the following industry partners: A&P Tyne Limited; Lloyd's Register of Shipping; McTay Marine Limited; Richards Dry Dock and Engineering Limited; and Wynns Transportation Consultants. Both the Environment Agency and the Maritime and Coastguard Agency provided information to the project.

12.4.3 Cargo catamaran (the PACSCAT)

A major new European research project was launched at the University of Southampton in December 2003 to develop a novel cargo vessel (i.e. the Partial Air Cushion Support Catamaran, PACSCAT) able to attract freight from roads to major inland waterways. MarinetechSouth was co-ordinating the project, working closely with Independent Maritime Assessment Associates (IMAA) who are providing the principal technical expertise and the University of Southampton who were to carry out the hydrodynamic research and performance assessment. According to IMAA:

> Development of a PACSCAT for Freight Transportation on inland waterways will address the problem of speed and wash which limits conventional freighters on inland waterways … at the same time, such a craft presents significant environmental advantages.

With European freight transportation expected to rise by around 50 per cent over the next 10 years, inland waterways are a genuinely viable alternative to road and rail given the efficiencies of the PACSCAT vessel concept. Carrying some 2000 tonnes (equivalent to some 80 heavy truckloads) at speed of up to 45 kilometres an hour along the Rhine and Danube, PACSCAT is said to be a major step forward in waterborne logistics. The vessel is designed to utilize existing berthing and loading facilities, and its draught height can be optimized to cope with shallow water conditions and with bridge height limitations.

12.4.4 Geest North Sea Line

Geest North Sea Line is a typical example of a successful short-sea operator that leads in its technology developments. For example, it was an industry leader in introducing a new-patented design for the corner castings of 45-foot ISO containers (see Figure 12.6) which enables these units to be carried legally within the maximum permitted turning circle dimension for 16.5 metre-long articulated vehicles, and is in the process of replacing its entire fleet of 40-foot containers with 45 feet long, 9 feet 6 in high units. Similarly, Geest has recently (i.e. in 2004/05) introduced two new 804 teu (i.e. 20 feet equivalent units) container

Fig. 12.6 Geest patented 45-foot container corner casting, which allows vehicles to operate within the limits set by EU legislation (*Source*: Geest North Sea Line nv).

vessels designed to carry the maximum number of 45-foot containers, with a good deadweight able to accommodate heavy 20-and 30-foot containers and a high level of stability because it needs fast turn round times.

The company describes itself as being:

> Totally committed to the expansion of European short-sea shipping and, in particular, to greater use of intermodal options including rail and inland waterway transport in Europe. To this end, we are in the process of replacing all of the 40-foot boxes in our fleet with 45-foot high-cube containers that enable us to compete effectively with 13.6-metres road trailers. Unfortunately, no one yet, to our knowledge at least, has built a containership specifically designed around the 45-foot box and so those vessels available on the charter market are inevitably not ideally suited to the new profile of our equipment fleet ... The result is a ship which has holds designed around 45-foot modules rather than 20- and 40-foot modules while the stackweight specifications are such that we will be able to maximize stowage of both 45-foot containers and heavy 30-foot boxes.

It is interesting to note that a typical short-sea container vessel of 340 teu operating for Geest is required to discharge and reload a full cargo in around 8–10 hours. Because of the size of the new ship, more cranes will be in use to discharge and re-load the ship simultaneously. Therefore, they have had to plan very carefully so that a safe level of stability is maintained at all times. With an 804-teu ship, the company will be looking to achieve around 520 lifts during a port call, this figure being based on a typical mix of 45/40/30/20-foot containers. These new 18-knot vessels will have more than twice the container capacity of those currently operated by Geest and will be far bigger than anything that has been used in the North Sea shipping trades before. Once in service, they will operate between Rotterdam and the UK East Coast.

Geest considers its main competitor to be the unaccompanied 13.6-metre semi-trailer market. This has led to it championing the 45-foot container. According to Mr Wout Pronk, Geest North Sea Line's managing director:

> We must offer our customers the same pallet capacity as a 13.6-metre trailer but for total intermodality, we need to be able to top lift and stack. Swapbodies were not an answer. Initially, we thought we would be unable to use 45-foot containers because it seemed that they would be outlawed in Europe but we have found a way around this problem with the patented Geest Euro corner casting.

13

Intermodal Loading Units, Transfer Equipment and Satellite Communications

It could be said, as we have done so earlier in the book, that if the Channel Tunnel was the catalyst, from a UK perspective at least, for developing intermodal road–rail transport across international boundaries, then undoubtedly the standard swap body, along with the ubiquitous International Standards Organisation (ISO) shipping container, of which many millions are in use worldwide, have been crucial to the concepts of load unitization and containerization on a world-wide scale. In fact, achieving efficiency in intermodal transport operations is highly dependent upon, among other key factors, the use of standardized loading units built to conform to a regime of commonality in dimensional and constructional standards; in other words, in accordance with provisions laid down by the ISO.

The concept of standardizing types, dimensions, and technical standards for loading units is absolutely vital to ensure compatibility between the many types of rail wagon, road vehicle, waterway barge, or coastal ship, particularly in regard to the manner in which the units are lifted, positioned, and secured on the transport mode. In an effort to achieve harmony in this regard (as we have already seen in Chapter 2, but it is worth reiterating the point again here) that in 2003 the European Commission published a proposal for a Directive on *intermodal loading units*, the purpose of which it said:

> is to overcome the diversity of intermodal loading units (ILUs) and their handling and securing devices, a factor that hampers the efficiency of trans-shipment operations.

13.1 Swap bodies

Swap bodies (or, in French, *caisse mobiles*) are specially designed loading units capable of being transferred, while under load, from road vehicle to rail wagon and vice versa, and are used mainly, but not exclusively, in international transport movements. They are built of steel, although some models have an aluminium superstructure or ply-panelled bodywork on a steel sub-frame, to conform to international standards of structural strength (albeit not to the stringent strength requirements laid down for ISO freight containers), and to standard dimensions compatible with road vehicle and rail wagon width and length limitations. (It is important not to confuse these units with demountable body systems used in closed-network, domestic transport operations.)

Such bodies are secured on road vehicles and rail wagons by means of quick-release, standard, ISO-type twistlock devices. One part of the fitting (the corner casting) is built into the swap body base frame, positioned at standard centres, to locate with retractable twistlock cones mounted on the vehicle/wagon load platform or skeletal sub-frame. Special pockets are incorporated in the base frame to enable loaded bodies to be lifted by crane, using bottom lift grapple arms, or by heavy-duty forklift truck. Most Category C (see below) swap bodies have folding legs to allow free standing for loading and unloading purposes. This is not possible with the longer Category A units. Unlike freight containers, current generations of swap bodies cannot be stacked when loaded, although some fold-flat models may be so when unladen, generally the open, flat, or short-sided versions. Work is in progress on developing versions capable of being stacked when loaded. Load capacities for swap bodies vary depending on size and type: generally, 7-metre long versions have a maximum gross weight of 16 tonnes, while 13.6-metre versions usually have gross weight of 34 000 tonnes, providing actual payloads of some 13 tonnes and around 28 000-tonnes respectively. In practice, however, swap body payload weights are limited by regulations that restrict the maximum gross weight at which road vehicles may operate in European Union (EU) member states (see the preceding Chapter).

In many ways swap bodies can be compared with the ISO freight containers in terms of their usability for combined road–rail freight operations. However, they do have certain specific advantages over the container, which give rise to their growing popularity. Most importantly, the shorter units being fitted with folding legs are free standing, which allows loading and unloading at loading bay height in warehouses and factories. An additional benefit in this respect is that the driver, single-handedly, can drop off or pick up a swap body using the vehicle air suspension system to raise or lower the chassis for clearance. Most general cargo swap bodies also have side-loading doors, which is another advantage in warehouse or factory loading/unloading situations, whereas ISO containers are invariably restricted to end loading. Furthermore, swap bodies offer greater load capacity, certainly in the most popular size range, with the smallest, 7.15 metres, unit being over 1 metre longer than a standard 20-foot ISO container.

13.1.1 Swap body standards

A number of standards are applied to swap body construction, in particular the international standard for swap bodies in Europe established by the Comite Europeén de Normalisation (CEN). Many swap bodies are built and tested to conform to the German DIN standard and to the German Railway, Deutsche Bahn (DB), standard for swap bodies carried on the European rail network. Some swap bodies also conform to the Convention for Safe Containers (CSC) standard described below. Manufacturers marketing this equipment strongly feature the standards approvals achieved by their products.

13.1.1.1 CEN standard

CEN represents the individual national standards organizations of the EU member states plus those of non-member states such as Iceland, Norway, and Switzerland. Its purpose is to combine individual national standards to form a single-unified standard that is accepted and adopted, without alteration, in each national state to ensure the unimpeded circulation and transfer of swap bodies between road and rail systems throughout Europe. In the UK the relevant national body is the British Standards Institution. The European (CEN TC119) standard for swap bodies is designated EN 284 (testing of such bodies is carried out to EN 283 standard).

Three categories of swap body are identified under the CEN standard as follows:

- Category A: units of 12.2, 12.5, and 13.6 metres length.
- Category B: units of 9.125 and 10.216 metres length.
- Category C: units of either 7.15, 7.45, or 7.82 metres length.

Within the three categories mentioned above, four body types are identified as follows:

- box type (i.e. rigid, closed body with doors),
- opensided (with folding side walls),

- curtainsider,
- drop sides.

Closed-body types may be insulated and fitted with refrigeration equipment.

13.1.2 Maximum width for swap bodies

Most swap bodies were formerly built to a maximum width of 2.5 metres to conform to the legal limit that hitherto applied to road vehicles throughout Europe. However, from 1 January 1996 this dimension was increased to 2.55 metres: by comparison refrigerated road vehicles may be up to 2.6-metres wide.

13.1.3 Overall height

Overall heights for closed-type swap bodies vary according to type and loading requirement from the standard 2.45-metre high unit, which meets rail wagon loading gauge up to C45, the latter being equivalent to a 9-foot 6-inch-high ISO container (i.e. 2.45 metres plus 0.45 metres clearance). For rail carriage, swap bodies must display a UIC (i.e. International Union of Railways) codification plate indicating the loading-gauge standard, for example, ‘C30’. It is also important to bear in mind the limitation within most European countries (except the UK) of a maximum overall height of 4 metres for road vehicles. In the UK, while there is no specific legal height restriction for road vehicles (apart from the normal considerations of bridge clearances), nevertheless regulations require the overall travelling height of those carrying containers (including swap bodies), where this exceeds 3.66 metres, to be indicated to the driver in the vehicle cab by means of a notice with letters and figures at least 40-millimetre high. This is a safety measure to guard against the danger of vehicles colliding with and damaging bridges (especially those carrying main line railway trains) and overhead structures. For the purpose of these legal requirements, containers are defined as equipment, other than the vehicle itself, which has a volume of at least 8 cubic metres and which is made wholly or mainly of metal and is intended for repeated use in carrying goods, while overall travelling height is defined as the distance measured from the ground to the highest point of the equipment on the vehicle or trailer, plus 25 millimetres.

13.1.4 Gross and payload weights

While gross weights for swap bodies tend to be standardized, tare weights vary considerably depending on the type of body construction (particularly whether solid or curtain sided), loading door arrangements and the construction materials used. Typical gross and tare weights for standard swap bodies are as follows:

Body length (metres)	7.15	7.45	7.82	13.6
Gross weight (kilograms)	16 000	16 000	16 000	34 000
Tare weight (kilograms)	2850	2900	2950	5100
Potential payload (kilograms)	13 150	13 100	13 050	28 900

13.1.5 Swap bodies in use

Most swap bodies in European, and more recently UK, combined transport operations are either 7.15 or 13.6 metres units, these being the most popular sizes. In fact, the 7.15-metre standard swap body is becoming the norm in international combined road–rail operations. Some 90 per cent of European output of new units is reported to be of this dimension with numbers increasing, while industry estimates indicate that as much as 97 per cent of the Continental intermodal market is served by 7.15-metre units. Two of these units can be carried on an 18.35-metre long road–train combination, one on the rigid drawing vehicle and one on the trailer, located on the combination's standard 20-foot ISO twistlock positions. In the case of 13.6-metre swap bodies, these are carried on maximum length articulated road vehicles. However, these have the disadvantage that they do not offer the same load space or weight capacity as two standard 7.15 metre units, and they cannot be left on their legs as free-standing units at warehouses or in terminals like the shorter units, and are not readily interchangeable between vehicles. Less popular are the 7.45-, 7.82-, 12.2-, and 12.5-metre units. Although two of the former 7.45 or 7.82-metre units can be carried on a standard, maximum length EU-specification road–train, it is unlikely that many operators will change to this dimension (only some 2–4 per cent of current in-use units are of this size) due to their lack of interchangeability and flexibility, and other logistical problems.

In countries where articulated vehicle operation is more popular than road–train operation (such as in the UK where, until March 1994, articulated vehicles held a significant weight advantage over road–train combinations – that is 38 tonnes against 32.52 tonnes), the 12.2-metres (40-foot) swap body had been more commonly used. Such units would readily load on to a 40-foot skeletal-framed semi-trailer and look no different from a normal maximum-length articulated vehicle. A number of major continental multimodal operators use units of this size for road–rail operations across Europe, notably Danzas, Transfesa and Merzario.

13.2 Freight containers

Freight containers in this context are those conforming to the ISO standards of construction and dimensions. They are specially designed to withstand usage in maritime (deep sea) shipping operations (although many are never actually used in sea transport) and are fitted with universal corner, and in some cases intermediate, castings set at standard (i.e. 20-, 30-, 40-, and 45-foot) centres, top, and bottom, which allow them to be secured with matching twistlock equipment fitted to road vehicles, rail wagons, cellular container ships, and to lifting and transfer equipment. Containers are designed and built to have sufficient inherent strength and rigidity for top overhead lifting when fully laden, and for safe stacking eight or more high inside a container ship, on the dockside or in a freight terminal.

Containers are generally of all-steel (mild or stainless) construction, although some have aluminium, plywood (sandwich construction) or GRP panels. Where lightweight (and rust free, corrosion resistant) construction is essential, some containers are built wholly of aluminium, but these are normally limited to three-high stacking. Most ISO containers are of the simple box construction with end-loading doors for dry freight, but a variety of other types are constructed to suit the requirements of particular traffics. For example:

- tankers (for liquids, granules, or powders) incorporated within a standard dimension steel frame (some tankers are insulated and steam heated),
- insulated and refrigerated (reefer) containers (the latter with attached, recessed refrigeration, and controlled atmosphere (CA) units),
- open top (e.g. for top loading/unloading of heavy machinery),
- half-height (for low-density high-weight products: e.g. steel),
- side loading,
- fold flat: flatrack (for ease of stowage when returning empty).

The bulk of the Freightliner domestic road–rail container operation is conducted with 20-foot ISO boxes carried for the road journey on short-length semi-trailer articulated vehicles.

13.2.1 Standard dimensions

Most containers ('boxes' in industry jargon) are built, despite metrication, to standard, imperial, 20-, 30-, 40-, and 45-foot lengths, by 8-foot wide (although Europe has its own domestic containers which are

2.5-metre (8-foot 2-inch) wide to accommodate standard Euro-pallets) with heights varying between 8 feet and 9 feet 6 inches. Standard 20- and 40/45-foot, and 8-foot × 8-foot 6-inch boxes are predominant in European trade. However, increasing use is being made of 45-foot long 9-foot 6-inch-high (maxi-cube) containers (see page 164) which can load 33 Euro-pallets, but these need to be fitted with special tapered Euro-corner castings; for others, they marginally infringe the maximum legal swing clearance dimension by a few centimetres when being carried on road vehicles.

Capacity in container ships and flows of container traffic is measured in teu, representing 20-foot equivalent units. Thus a 40-foot container is counted as two teu. All ISO containers are assigned an ISO type code comprising four alphanumeric characters; the first character indicates the container length (Code 2 being 20 feet; Code 3 being 30 feet; Code 4 being 40 feet), the second character indicates the height and width (i.e. a 9-foot 6-inch-high, 2.5-metre-wide unit is designated Code letter E). The third and fourth characters indicate the type of unit and can be a group designation or a detailed type code designation, of which there are a great number. A full list can be freely downloaded from the SMDG (User Group for Shipping Lines and Container Terminals) website at: www.smdg.org/documents.

13.2.2 Weights and capacities

Gross weights, payloads, and cubic capacity for ISO containers vary quite considerably according to size, type, and method of construction, but for standard dry-freight boxes they may be generally summarized as follows:

	20 feet	30 feet	40–45 feet
Gross weight (kilograms)	24 000	26 000	30 000
Tare weight (kilograms)	2300	3250	4200
Average payload (kilograms)	21 700	22 000	25 800
Cubic capacity (cubic metre)	32.8	49.8	67.0

13.2.3 Other standards for containers

Besides the principal ISO standards of construction and dimensions for freight containers mentioned above, other standards and legal requirements apply to their construction, testing, and use as follows.

13.2.3.1 Convention for safe containers

Owners, lessees, and others in control of freight containers must ensure that they comply with the *International Convention for Safe Containers*, Geneva 1972. In the UK, this international convention is applied by domestic legislation namely, *The Freight Containers (Safety Convention) Regulations 1984* which apply to containers designed to facilitate the transport of goods by one or more modes of transport without intermediate reloading, designed to be secured or readily handled or both, having corner fittings for these purposes and which have top corner fittings and a bottom area of at least 7 square metre or if they do not have top corner fittings, a bottom area of at least 14 square metre. Containers must have a valid approval issued by the Health and Safety Executive (HSE) or a body appointed by the HSE (or under the authority of a foreign government which has acceded to the convention) for the purpose of confirming that they meet specified standards of design and construction, and should be fitted with a safety approval plate to this effect. If they are marked with their gross weight, such marking must be consistent with the maximum operating gross weight shown on the safety approval plate. Containers must be maintained in an efficient state, in efficient working order, and in good repair. Details of the arrangements for the approval of containers in Great Britain (GB) are set out in a document *Arrangements in GB for the*

Approval of Containers available from the HSE. The safety approval plate (issued by the HSE) as described in the regulations must be permanently fitted to the container where it is clearly visible and not capable of being easily damaged (it must be non-corroding and fireproof), and it must show information described in Chapter 18 (see page 246).

13.3 Lifting equipment

Perhaps more than in any other equipment sector for intermodal transport, the variety of different makes, models, and types of lifting and loading equipment available is quite outstanding. The range extends from small mobile units that may be found in a road–haulier's premises to the huge rail-mounted gantry cranes seen towering over docksides and freight terminals. Of course, choice depends on operational requirements; lifting unladen containers from the ground onto a vehicle load platform is one thing, but loading or unloading from ship to shore, or stacking 30-tonne containers, eight or ten units high (as is often the practice) is a completely different matter. For the purposes of this chapter, equipment has been divided into four categories and typical specifications have been outlined by way of example in each category. The four categories are as follows:

- Forklift trucks and reach stackers:
  - for empty container handling,
  - for loaded container handling.
- Vehicle/trailer-mounted loading equipment.
- Rubber-tyred straddle carriers.
- Rail-mounted gantry cranes.

To add to the complications of equipment choice, but not for discussion here due to its complexity, there are the ranges of lifting equipment attachments, most notably spreaders, which by their special or particular design features facilitate the actual handling of containers, swap bodies, or road vehicle semi-trailers. Many crane builders include these attachments as part of their equipment, but there is a whole industry of specialist manufacturers concentrating on nothing else but, for example, designing and constructing spreader attachments (both Bromma and IMA of Germany being well-known manufacturers in this field).

13.3.1 Forklift trucks and reach stackers

There are many forklift truck manufacturers and most would claim to build equipment suitable for handling, at least, Category C swap bodies and 20-foot ISO containers in an unladen condition, that is empty container handlers. Such equipment has obvious limitations of load capacity and reach. Hyster, a well-known name in forklift trucks in the UK and Europe has just such a range of diesel-powered trucks. Depending on individual specification, these trucks can stack empty containers three or four units high. They can handle a wide variety of container types, having 20–40-foot telescopic side-lift spreaders and they have a two-speed hoist system to allow them to operate fast turnround cycles. Load capacity is typically 7000 kilogram and they lift to a height of 13.8 metres. Interestingly, Fantuzzi of Italy advertizes a lift truck capable of stacking empty containers 10-unit high. In a higher weight range for handling loaded containers and swap bodies, Hyster has trucks which can handle loads weighing from 20 to 32 tonnes, or from 36 to 48 tonnes either with fork attachments, or at the higher end of the weight scale, with top-lift spreaders or a telescopic container attachment. Also typical of container handling forklift trucks in the 8–50-tonne range are those of Swedish manufacturers, Svetruck and Kalmar LMV, and also Linde AG of Germany (see Figure 13.1). Container stacking with these and other similar trucks is generally limited to two or three units high.

Reach stackers are a different type of lifting machine having an overhead two- or three-stage hydraulic lifting arm or boom to which a variety of lifting attachments can be added: typically spreaders for container

Fig. 13.1 Linde heavy-duty lift truck with container stacker capability. (*Source*: Linde AG, Germany.)

handling and grapple arms for lifting swap bodies and semi-trailers in piggyback operations. They are highly manoeuvrable machines, being equipped with rear wheel steering giving them a capability for very precise positioning especially when stacking. Many of them have a 'slewing' facility that allows loading units to be turned or to be lifted and carried end-on. They are very stable under load and can operate fast turnround cycles.

Leading manufacturers in this field are Kalmar LMV (mentioned above), French manufacturer Terex-PPM, Sisu of Finland (Figure 13.2), and again, Fantuzzi of Italy. Most of these machines can lift in the region of 45 tonnes and can stack loaded 9-foot 6-inch containers five high at short reach. As the boom is extended lifting capacity diminishes, but generally such machines can reach into a third row of a block stack. These machines are normally fitted with either spreaders to handle 20-, 35-, and 40/45-foot ISO containers or with grapple arms for other intermodal units. However, in some cases both the spreader and grapple arms are combined into a single unit, the grapple arms folding out of the way when not required. This arrangement gives added flexibility in terminal operations, being able to handle a variety of intermodal units, 'as they come', and saves considerable time by not having to make the changeover from one form of lifting attachment to another. Swedish manufacturer Elme's 850 series of top/bottom lift attachments (see Figure 13.3) accomplish this task, taking the operator no more than 20 seconds to fold each pair of grapple arms into or out of the lifting position.

13.3.2 Vehicle-mounted loading equipment

It is not always possible for container and swap body lifting operations to be carried out in purpose-built intermodal terminals or in transport depots with lifting equipment. Often loads need to be assembled or

Fig. 13.2 PPM reach stacker. (*Source*: Terex-PPM, France.)

unloaded at ground level in factory premises where no proper lifting equipment is available. For this reason, a number of manufacturers have devised on-board equipment for road vehicles to enable them to lift and load or lower load units to and from ground level or to carry out a lateral transfer from vehicle to vehicle or vehicle to rail wagon. Generally such equipment can handle 20- and 40-foot ISO containers (loading two 9-foot 6-inch-unit high) and standard swap bodies loaded to a maximum of 30-tonnes gross weight. The equipment is mounted on the semi-trailer, keeping within the European 13.6 metres overall length limit.

One manufacturer of such equipment is Hammar, another Swedish company (i.e. Hammar Maskin AG). Its semi-trailer-mounted model, which can be loaded or unloaded within just 4 minutes, has hydraulically adjustable arms that can be moved along the trailer load platform to accommodate short (i.e. 20-foot containers or 7.15-metre swap bodies) or longer 40-foot and 13.6-metre units (see Figure 13.4). Another manufacturer of such equipment is Containerlift Ltd of Essex, UK (see Figure 13.5). The semi-trailer with the equipment has a maximum payload capacity of 31 tonnes but the equipment itself has a maximum lifting capacity of 33 tonnes. An 8-foot 6-inch-high ISO container can be carried within the European 4 metre overall height limit. The equipment consists of two cranes with two separate hydraulic circuits, one for each crane with sealed pressure control valves to protect against overloading and to protect the equipment. Power for the hydraulic pump is taken from the power-take-off (PTO) on the tractive unit gearbox. Load security valves make the speed of the cranes independent of the weight of the cargo handled, that is the same lifting speed is attained whether 33 tonnes are handled or just 1 tonne. These valves also prevent the load from being dropped in case of a hose breakage. Each crane consists of an upper arm, a lower arm, and two hydraulic cylinders mounted on a heavy-duty frame of hollow beams. The cranes are built of high-grade steel to minimize unladen weight and give maximum

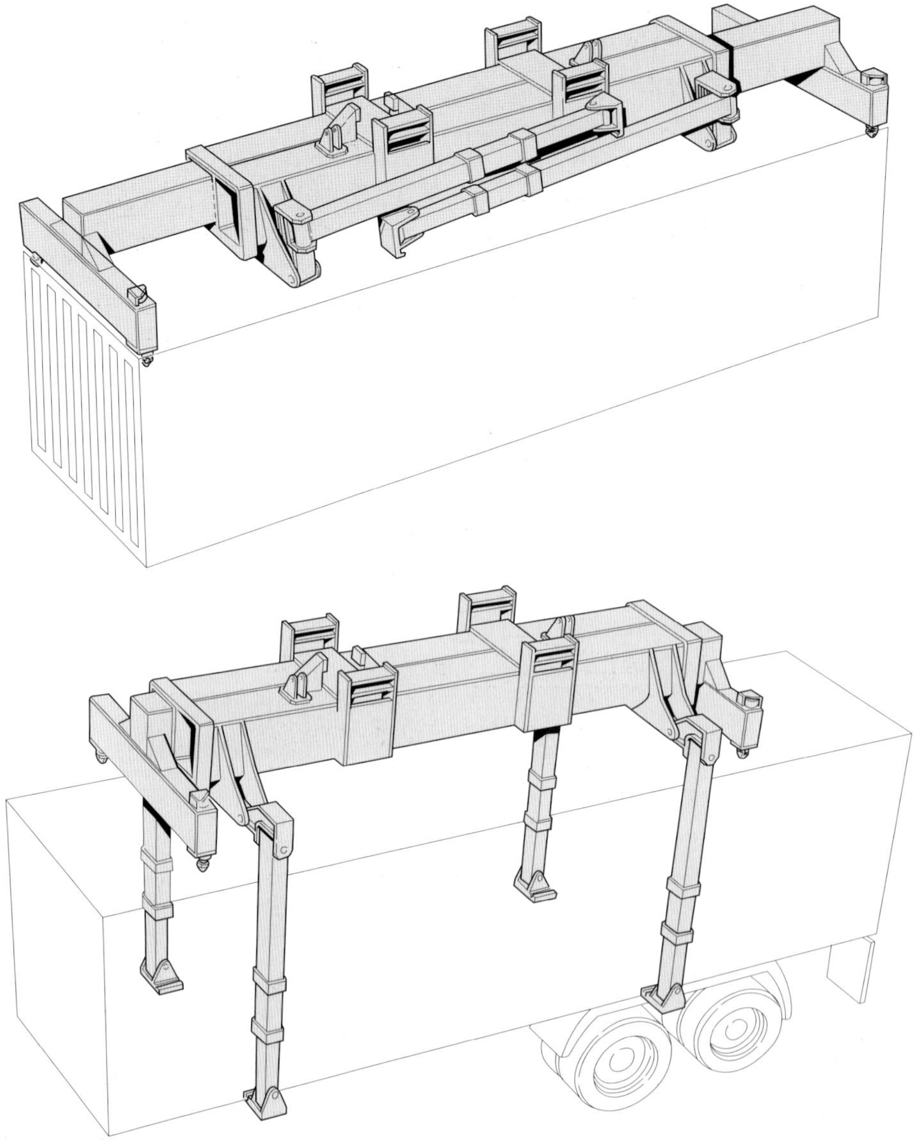

Fig. 13.3 Elme spreader for container loading with twistlock attachments and grapple arms for swap body and piggyback loading. (*Source*: Elme Swedish Spreader Systems.)

payload. Stabilizers provide support against the ground, rail wagon, trailer, or loading bay. Due to the large support area, the surface pressure is very low: a maximum of 13 kilogram/square centimetre when handling 33 tonnes. Adjustment of the lifting units for different container lengths is by means of two hydraulic cylinders that are well protected between the frame members of the semi-trailer. The lifting units are freely adjustable between 20 and 40 feet, and are carried in slide bearings of stainless steel on the semi-trailer frame. The container is secured on the Hammar semi-trailer during transport by four manual twistlocks with drop-forged tops. The Hammar equipment can be operated by one person – the driver – using

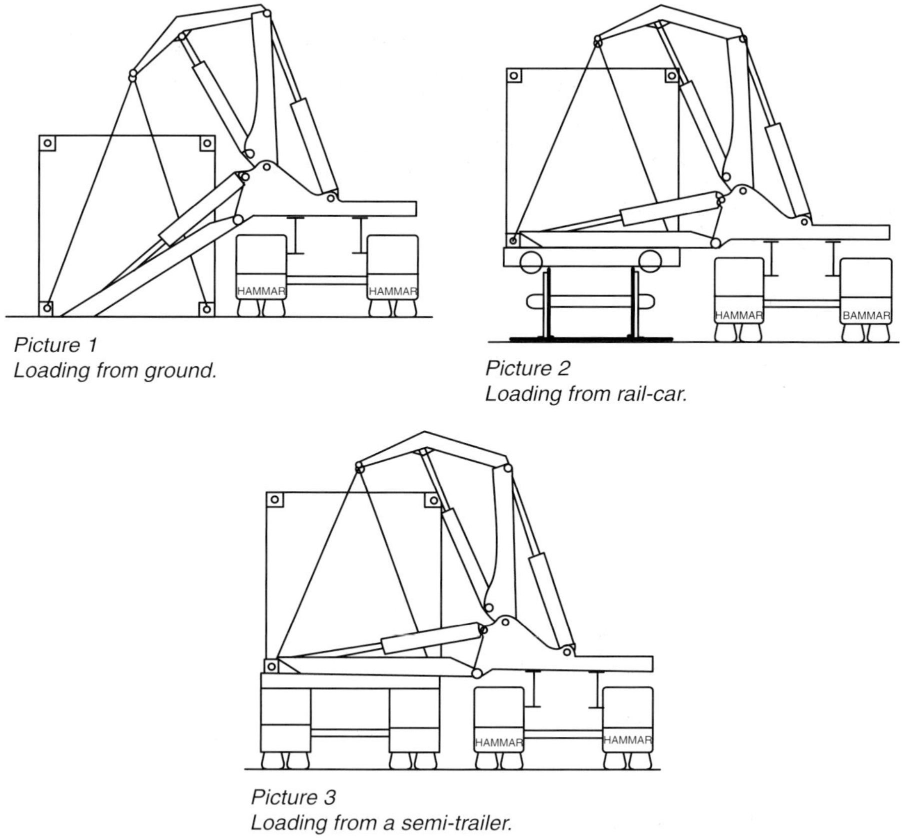

Fig. 13.4 Hammar vehicle-mounted container/swap body lifting system.
(*Source*: Hammar Maskin AB, Sweden.)

the portable control box which allows him to move around during loading and unloading, to get the best possible view of the operation. All the hydraulic cylinders can be manoeuvred separately or in parallel and two-speed operation is provided for all functions. A built-in safety system makes incorrect manoeuvres impossible. In case of emergency, for instance, if there is a malfunction in the electrical system, all hydraulic valves can be operated manually.

A UK company, Containerlift based at Stansted Airport, operates a number of semi-trailers fitted with New Zealand-built Sidelifter equipment. This is very similar in concept to the Hammar equipment described above. It can lift and load or lower both 20- and 40-foot containers and Categories C and A swap bodies at gross weights up to 30 tonnes from the ground. According to the company, many firms are catching on to the idea of safe ground loading of heavy or expensive goods, and this is where the Sidelifter scores with its ability to be taken into industrial premises where space is limited and no lifting equipment is available.

13.3.3 Rubber-tyred straddle carriers

Straddle carriers are specially designed mobile units that can be driven astride containers or swap bodies to lift, carry, or stack as required. They are highly manoeuvrable and can traverse rows of stacked containers with, in the case of the larger models, the operator sitting in a high-mounted control cabin to give a good all-round view of the working area and to allow precise positioning of containers when loading to road

Fig. 13.5 Containerlift vehicle-mounted container transfer device. (*Source*: Containerlift Ltd, UK.)

vehicle or when stacking (generally, up to a maximum of four high). Their ability to hoist while travelling adds to operating speed and provides fast turnround cycles. Telescopic spreaders adjust automatically for precise positioning and fast operation, while safety locks ensure that lifting cannot commence until all four corner twistlocks are secured. Leading straddle carrier manufacturers Noell of Germany offer, among a wide range of models, a small, four-wheel, low-cost version which they call the Easy-Lift, designed for handling 20- and 40-foot containers up to 40 tonnes, which it can carry (at up to 16 kilometre/hour) and stack two high. It has a choice of two- or four-wheel steering controlled by the operator providing exceptional manoeuvrability for negotiating small areas and positioning containers into awkward corners. Either an open or closed driver's control cabin is available and the machine can also be operated by remote control (using infrared technology) allowing the operator to select the best viewing position. In the large straddle carrier range, over 500 large Noell straddle carriers are in use worldwide. These feature three-high stacking capability, eight-wheel steering, and Noell's patented maintenance-free suspension system to provide good handling over poor surfaces. They have the ability to handle full- and half-height containers in all lengths (including 45- and 48-foot-long models) at maximum weights.

Claiming to be the world market leader in straddle carriers, Valmet of Finland (a product name of Sisu Terminal Systems) has over 40 years' experience in building this type of equipment. Specifications are similar to those described above with the emphasis on good traction from four-wheel drive and smooth running, high levels of manoeuvrability, speed of lifting and travel, exceptional visibility, and precise control for the operator from high-level cabins. Above all, this type of equipment is claimed to enable terminals to be set up with a minimum of infrastructure investment, to operate efficiently with fast cycle times (pickup, lift, carry, position, and put-down) and with an economy of labour.

13.3.4 Rail-mounted gantry cranes

The very largest of container terminals, particularly those in ports, invariably feature rail-mounted gantry, or stacker, cranes which can lift and load all types of unit-load equipment at maximum weights and can traverse as many as seven rows of stacked containers and a truck-loading lane. This type of equipment generally features electric drive (DC) with thyristor control to provide infinitely variable speed and AC-drive for hydraulic pumps, spreader operation and lighting. Cable hoists, mounted on trolleys that traverse the gantry (and on which the operator cabin is mounted), are used for lifting and lowering the spreader equipment or grapple arms for lifting swap bodies. Leading names in this field are again Noell and Valmet, while German firm Kransysteme Rheinberg GmbH (KSR) build both rail-mounted (RMG) and rubber-tyred (RTG) versions, the latter having the advantage of flexibility of operation since it is not confined to parallel tracks.

13.4 Other handling equipment

While heavy-duty forklift trucks, reach stackers, and straddle carriers are predominant in the field of container and swap body handling, there are many other forms of handling equipment from giant-sized container cranes that can perform ship to shore container loading and unloading, in some cases straddling 12 or more rows of stacked containers (e.g. by Aumund, Kunz and Hilgers of Germany) to simple roll trailers for moving containers within terminals, dockside tractors, and other forms of container lift-and-carry equipment.

13.5 Satellite tracking of vehicles and loading units

Worldwide tracking of containers, swap bodies, semi-trailers, road vehicles, and rail wagons is possible with satellite-based tracking systems. Being able to know where swap bodies, containers, unaccompanied trailers, or vehicles are at any particular point in time, and their status, is a great help in the management control of transport operations and in logistics planning.

13.5.1 ITF intertraffic – global tracker service

An example of this type of system is the Global Tracker Service of Daimler-Benz subsidiary ITF Intertraffic, in this particular case with the added advantage that it allows retrieved data to be integrated into existing company information systems (a schematic representation of the system is shown in Figure 13.6). This simple system locates intermodal loading units using a small transmitter, often called a Data Terminal or Asset Tracker Unit, unobtrusively fitted to the load unit (in a small rugged housing) which transmits signals back to a confidential electronic mailbox at the ITF Intertraffic Operations Centre identifying the unit's latest position, and if required, other information. Other data, including temperature, shock, weight, humidity, door open, and load status, as well as locally input data from hand-held communication systems, can be provided via modular hardware units and sensors attached to the load unit which feed information to the transmitters. Data terminal interfaces allow the addition of sensors or Radio Frequency Identification (RFID) tags. Data from loading units is transmitted via Low-Earth Orbit (LEO) satellites first to the satellite network owner's ground station, then onto ITF Intertraffic's operations centre and finally the system user's terminal. Operation is simple: the user makes just one mouse click on the 'connect' box on his PC terminal screen, which instructs the programme to dial the ITF Intertraffic database centre. This automatically connects the terminal to the user's personal mailbox and sends the necessary commands to retrieve the positions of loading units. The user can make a number of selections for data retrieval, choosing to collect data relating all units or just to selected containers or trailers, the latest information only or all information covering a number of days.

Information is supplied via a standard modem and in standard data format (e.g. dBase) for easy integration into logistics management software. This means that standard database management systems can access the files and the data can be used to generate charts showing routes taken, daily logs of positions, and other information. MS-Windows-compatible user software displays loading unit positions on vectored maps or in database format. The service relates all the positions provided by the satellite system to

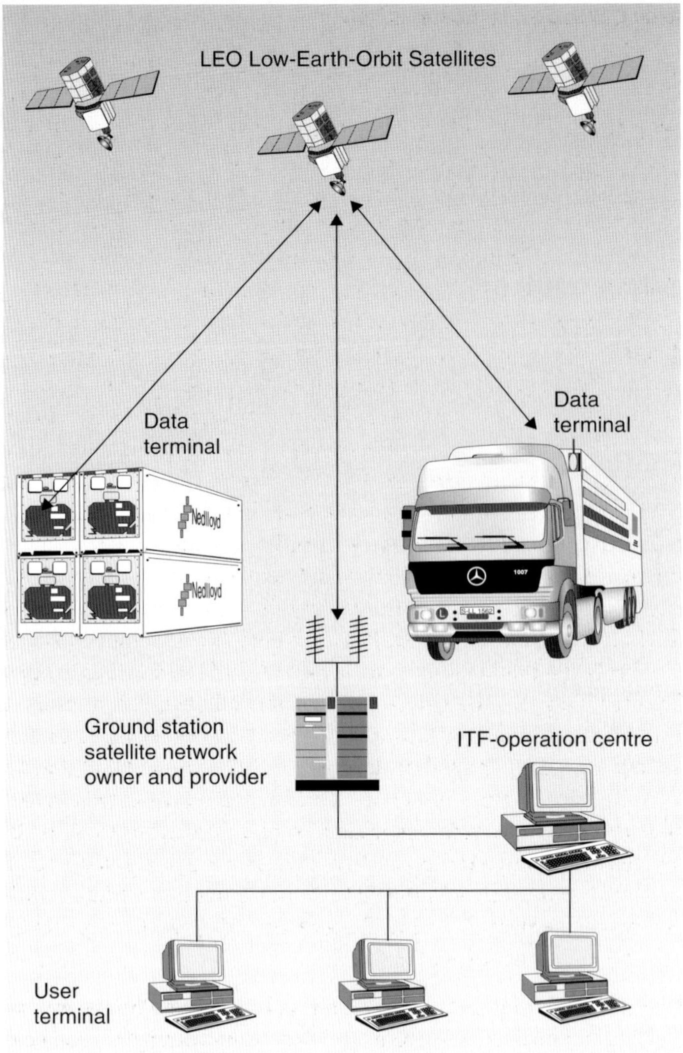

Fig. 13.6 System structure for the ITF Intertraffic Global Tracker Service. (*Source*: ITF Intertraffic.)

reference positions: the 'proximity calculation'. Each position is given as a line of screen text, such as '10 kilometre north of Munich'. The reference positions are those chosen by the user. An ordinary PC and a few simple commands are used to connect the user to ITF Intertraffic's operations centre (the software supplied to users of the service can do this). The user programme requires the following PC system:

- an 80486-compatible computer with colour screen,
- a mouse,
- at least 40 Megabyte hard disk,
- at least 4 Megabyte of random-access memory (RAM),
- an Hayes-compatible modem,
- WINDOWS 3.0 software (or above).

13.5.2 The 'Galileo' global satellite system

The EU's global navigation and positioning satellite system 'Galileo' is beginning to take off in a big way with support from the Community's Trans-European Transport Networks (TEN-Ts) funding programme: in fact, it appears as project number 15 in the TEN-T list of projects in page 148. Galileo is Europe's initiative to create a global satellite navigation system offering precise position and timing services for commercial and personal uses anywhere in the world, using a small and inexpensive receiver. When fully deployed, the system will consist of an array of 30 satellites, together with associated infrastructure on the ground and newly developed applications and services. Importantly, Galileo will make Europe independent in a technological field of strategic importance. Cost benefit analyses carried out for the European Commission estimate that Europe's share of the global market for satellite navigation products and services may be worth as much as €9 billion each year from 2015 as a result of using Galileo, and that up to 140 000 new jobs could be created. A wealth of promising applications is already emerging especially in the fields of transport. The Galileo system is designed to respond to the specific needs of every transport domain, including aviation, maritime transport, and road and tall transport even pedestrians will benefit. But Galileo will also benefit other professional and personal activities, from civil engineering, social and emergency services to agriculture and fisheries, banking and finance, environmental protection, and civil protection. From the user's perspective, Galileo will offer the advantages of complete reliability and unprecedented accuracy. It will allow goods, vehicles, and people to be located with approximately 10 times greater accuracy than global positioning satellite (GPS), to within a few metres. And unlike GPS, the continuity of its signal will be guaranteed. Galileo has been designed specifically for civilian use worldwide, and will provide both a freely available signal and ones restricted to specific groups such as commercial service providers, safety-of-life applications such as aviation, and government users. Galileo will comprise a constellation of 30 satellites orbiting at an altitude of nearly 24 000 kilometres. Ground stations will be responsible for the management and control of the system. It is claimed that thanks to the compatibility and interoperability of Galileo, users throughout the world will gain easier access to signals emitted by the navigation satellites and will benefit from much greater efficiency: the system will also offer greater accuracy than GPS. The services offered by Galileo will be covered by a guarantee of continuity, which can be laid down in a contract; this is seen as very important when risk to human life is involved, as with air transport for example. In a Press statement issued in December 2004, the European Commission stated that Galileo would definitely become operational in 2008:

> A decisive stage had been reached,' it said, and 'it is without a doubt the most wonderful European technological project. We are now on the home straight: next year will see the, launch of the first satellites'; said Jacques Barrot, Vice-President of the Commission, 'Galileo will be as much of a technological revolution as mobile telephony.

14

Carrier Liability in Intermodal Transport

Intermodal and combined road–rail transport offers many practical benefits to freight shippers, to road haulage contractors, and to the public at large through reductions in environmental pollution of one form or another, as this book shows. While such operations are running smoothly there is a tendency to conclude that a more efficient and effective means of transportation would be hard to conceive. However, this is not to say that it is a perfect solution to all long-haul freighting requirements, far from it. When things go wrong and the problems arise of having to resolve claims for loss of, or damage to, the goods in transit (GIT), the incompatibility of the respective legal regimes relating to carrier liability is quickly highlighted. In other words, it is at this point that otherwise efficiently conjoined systems appear to leak significantly at the seams. Confusion reigns, not only among the lay freight shippers and carriers themselves, but also among the legal fraternity (both lawyers and courts) whose job it is to determine whether there was negligence, by whom and to what extent, and to assess liability and award compensation.

One undoubted benefit that can be cited for the through-carriage, door-to-door, of international freight from its point of origin to its final destination by road vehicle is that liability for any loss, damage, or delay to the goods can clearly be laid at the door of the road haulier whose vehicle undertook the whole operation. There can be little dispute in such circumstances as to where loss or damage claims should be directed. A road haulier who has made proper provision by way of adequate GIT insurance cover to protect against international carrier liabilities (Convention Marchandises Routiers (CMR), see later), should have little to worry about if such claims arise. Similarly, the original shipper of the goods need to have little worry either since, if he suffers losses, he can readily make a claim on the haulier. Even if more than one road haulier is involved in the international freight movement – for example, by the first, collecting, haulier handing over the loaded semi-trailer to another haulage firm for delivery – CMR liability is clearly established. The same liability regime applies to both carriers, and claims can be resolved accordingly.

Straightforward though this may sound in principle, in practice matters are not so clear, especially where freight loads are sub-contracted between hauliers, one or more of whom may have been under the impression that normal UK domestic carrier liability under his own, or Road Haulage Association (RHA), Conditions of Carriage would apply. As this chapter clearly shows, where a contract for an international haulage journey exists, domestic conditions of carriage (and relevant levels of liability) are superseded by CMR. When multimodal operations are introduced into international freighting, an even more complex and confused situation obtains. For a start, completely different liability regimes apply to the various modal stages of the journey: for example, when road and rail stages are introduced, and possibly even carriage by sea. Added to this are the not inconsiderable problems arising from variations in liability law between the different countries through which the international consignment passes.

14.1 International agreements

A clutch of international agreements is currently in existence with each mode having its own agreement as follows:

Mode	*Liability regime*
Carriage by sea and inland waterway	The Hague Rules (1921) as superseded by the Hague-Visby Rules (1968) and the, as yet unratified, Hamburg Rules of 1978
Carriage by air	The Warsaw Convention 1929 as amended by the Hague Protocol 1955 and the Guadalajara Convention 1961
Carriage by rail	The COTIF Convention 1980 (covering all international carriage by rail), and within this Convention the CIM Rules which relate specifically to the carriage of goods by rail. Applied in the UK by the International Transport Conventions Act 1983
Carriage by road	The CMR Convention 1956 (amended by protocol in 1979). Applied in the UK by the Carriage of Goods by Road Act 1965

For our purposes here it is useful to examine in detail the liability regimes that apply individually to domestic road and rail transport (since it must not be assumed that *all* combined transport operations relate to international freighting albeit, currently, most do so), international road and rail freighting, and then the United Nations Conference on Trade and Development/International Chamber of Commerce (UNCTAD/ICC) Rules for Multimodal Transport Documents.

14.2 Liability in domestic road and rail operations

Many combined transport movements (both in the UK and Europe) are conducted on a purely 'domestic' basis between inland or internal destinations which do not involve the crossing of national frontiers. (*Note*: In the UK, domestic transport excludes freight journeys to the Republic of Ireland which remains an international destination whether approached directly by sea from the British mainland or overland via Northern Ireland.) For such operations, although not so complex and therefore not so legally fraught as those involving the crossing of national boundaries, the road haulier still needs to ensure that he is adequately safeguarded as to his legal liabilities. This can be done, firstly, by limiting liability for the loss of or damage to goods entrusted to his care by applying appropriately worded conditions of carriage (e.g. those of the RHA but only if he is a member of that organization, it is illegal to do so otherwise) to all contracts he undertakes for the carriage of goods; and secondly, by taking out appropriate GIT insurance cover. It is useful to mention here that protecting one's liability in this way is of little use without also insuring against the risk of claims arising unless adequate funds are held in reserve from which to meet any liabilities which may arise, which could result in very substantial damages.

Without venturing into the extreme complexities of the law of carriage as it applies to domestic or inland transport by land (Glass and Cashmore do an excellent job in this respect in their *Introduction To The Law of Carriage of Goods*, Sweet & Maxwell, London 1989), it is suffice to say that both the privately-owned railways and most professional road hauliers (certainly those who are members of the RHA) are, in law, 'private', as opposed to 'common' carriers. The latter, by definition, hold themselves out as being prepared to carry goods for reward for all and sundry without reserving the right to refuse to carry any particular goods or for any particular person.

Private carriers on the other hand do reserve such rights and thus achieve this status. Furthermore, they do so by virtue of restricting their liability for loss of or damage to goods entrusted to their care (i.e. as 'bailees for reward') only to instances where they are held to have been negligent or guilty of conversion. Such restriction of their liability is set out in their respective conditions of carriage as published and brought to the attention of all contracting customers of their businesses.

For example, liability under [the former] British Rail's (BR's) August 1986 conditions (now obviously superseded by the relevant conditions of the new individual private owners of the former BR's freight businesses), the RHA's current conditions, and those of the British International Freight Association's (BIFA), which many combined transport operators use, is restricted to:

1. BR: £2000 per tonne.
2. RHA: £1300 per tonne gross weight or per 2.25 cubic metres, whichever is the greater.
3. BIFA: approximately £2100 per tonne (based on current value of Special Drawing Right (SDR), see below).

By contractual arrangement, the carrier may undertake to shoulder a greater level of liability for high value goods, in which case their value should be declared by the consignor/customer and notified by the carrier to the GIT insurer. If this is not done, subsequent claims for loss or damage may be refused by the insurer or met only on the basis of the normal (basic) level of cover which may be far from adequate recompense to the customer who has suffered the loss.

In the event of loss or damage to goods in domestic transport, the claimant will need to pursue the relevant carrier (i.e. road, or rail) to whom the consignment was entrusted, bearing in mind that the original contract may have been with a road haulier, who subsequently ships the load by swap body or container on rail, or with a private rail-freight company which carries the goods by rail but uses its own road vehicles to effect collection and delivery.

14.3 International carriage of goods by road: CMR

Transport operators, whether just a one-man haulage business (i.e. owner driver) or a large firm, carrying goods for reward on international road haulage journeys must comply (*Note*: Since, in effect CMR applies automatically to an international road haulage journey, the haulier has no choice in the matter, but he may be carrying out the operation in ignorance of this fact.) with the Convention on the Contract for the International Carriage of Goods by Road, 1956 (*Convention Relative au Contrat de Transport International de Marchandises par Route*, commonly referred to as the CMR Convention). This Convention is applied in the UK under English law by the provisions of the Carriage of Goods by Road Act 1965. The Convention defines the carriers' liability and the documents to be carried on vehicles engaged in the international movement of goods between different countries of which at least one is a party to the CMR Convention.

The following countries (in addition to all European Union (EU) Member States except Cyprus and Malta) are party to the CMR convention:

- Belarus
- Bosnia–Herzegovina
- Bulgaria
- Croatia
- Kazakhstan
- Macedonia
- Moldova
- Morocco
- Norway
- Romania
- Russian Federation

- Switzerland
- Tajikistan
- Tunisia
- Turkey
- Turkmenistan
- Uzbekistan
- The former Yugoslavia

14.3.1 Applicable law

International Conventions such as CMR override the relevant provisions of national law but only insofar as the Convention covers the point at issue. However, not all points of issue that may arise by way of dispute or claim out of a contract for international carriage are covered by the Convention, in which case relevant national law would apply (basically this means the law of the country in which the dispute or claim arose and to whose courts the matter is referred). In deciding the outcome of any dispute or claim, the courts may refer to either the French or English language texts of the Convention, although the French text is not reproduced in the schedule to the Carriage of Goods by Road Act 1965. When referring to this Act, a court is still free to consider the French language version of the Convention for clarification should difficulty be found in determining the precise meaning of any aspect of the English text.

14.3.2 Non-CMR operations and journeys

For the CMR Convention to apply there must be clear evidence of a contract for the international carriage of goods for reward. In other words, the carriage of goods on an international journey at no charge, and therefore, outside a contract for carriage as specified in the Convention, would not be covered by CMR. Furthermore, the carriage must be of *goods* to allow the Convention to apply. For this purpose, it may be necessary to examine closely a translation of the French term *marchandises* to which the Convention specifically applies in its original form. This word primarily translates to mean 'merchandise' which may not be held by a court to mean precisely the same as 'goods' as we understand the term in normal English usage.

Besides these exclusions arising from defining the precise terms of applicability, the Convention has a number of more specific exemptions. Namely, its terms do not apply to:

- own-account operations which involve international journeys,
- furniture removals,
- funeral consignments which are transported abroad,
- carriage under an International Postal Convention.

The Convention is also *not* applicable in respect of international haulage operations between the UK (including Northern Ireland) and the Republic of Ireland, or to contracts for the carriage of goods between the UK mainland and the Channel Islands (i.e. Guernsey, Jersey, Alderney, Sark and Herm), such operations not being legally classed as international journeys. Cabotage journeys (internal journeys within a country by a road haulier from another country) are also outside the provisions of the CMR.

14.3.3 Basic requirements of CMR

The CMR Convention automatically applies to every contract for the international carriage of goods by road in vehicles for reward, even when the vehicle containing the goods is carried over part of its journey by sea, rail or inland waterway, although other conventions may also apply and take precedence over CMR. A CMR-type consignment note must be completed for the journey. There is no escape from the

Convention's provisions and road hauliers may not opt out (by agreement with consignors or otherwise) from its legal (liability) requirements. Thus, for example, if a road haulier were to be induced to carry goods for reward on an international journey knowing nothing of the CMR and its provisions and his consequential liability responsibilities, and even without a CMR consignment note in force, the provisions of the Convention would still apply. Should a legal dispute or claim subsequently arise, he could be liable to pay substantial compensation not covered by his domestic GIT insurance policy.

Even in circumstances where a road haulier may not be aware that a load being carried on one of the firm's vehicles is destined to continue on, or has previously been moved on, an international journey, and where he is not aware of the conditions and implications of CMR, the Convention's provisions still apply. This is particularly relevant where loaded articulated semi-trailers are collected or delivered. It is the load and that part of the vehicle which are together at the time of crossing national boundaries which determines that a particular journey is legally an international journey to which CMR applies. The Convention is *not* relevant where containers are carried domestically after having been transferred from the vehicle or rail wagon which crossed national boundaries; or where loads are trans-shipped between the vehicle which undertook the international part of the journey (to which CMR did apply), and local delivery vehicles (to which CMR does not apply).

14.3.3.1 CMR conditions for international road haulage journeys

An outline of the principal conditions of the Convention is given here.

1. The Convention applies to every contract for carriage of goods, whether wholly by road or partly by road and partly by rail, sea or inland waterway, as long as the goods remain in the original vehicle, on a journey from one country to another, one of which is a contracting party to the Convention (with the exception of UK–Eire and UK mainland–Channel Islands journeys which are ruled not to be international journeys for this purpose). Exemptions to CMR apply to carriage under international postal conventions, funeral consignments and furniture removals.
2. The carrier (road haulier) is responsible under the Convention for the actions and omissions of his agents and any other persons whose services are used in carrying out the movement. Even if the original road haulier contracted to undertake the movement sub-contracts the whole of the operation to another road haulier (whose name appears on the CMR consignment note) the first (original) haulier remains fully liable under CMR should a dispute or claim arise. This can present problems in a case where, for example, the first (original) haulier operates only in domestic transport and thus is covered only by, say, RHA conditions of carriage and is insured accordingly, not being insured to the much higher CMR level of liability.
3. A contract for the international carriage of goods for reward is confirmed by making out a CMR consignment note in three original copies, which should be signed by the sender and carrier. Each keeps a copy and the third copy travels on the vehicle with the goods. While a CMR consignment note confirms that a contract exists, the absence of, or failure to raise such a note does not invalidate the contract or dis-apply the terms of the Convention.
4. If the goods are carried under a single contract in different vehicles or are divided owing to their different nature, the carrier or the sender can specify that a separate consignment note should be made out for each vehicle or each load of goods.
5. The consignment note must contain certain specified details (see p. 194) and may also contain additional information of use to the parties to the contract. It must state that the carriage is subject to CMR. Although not a 'title' to the goods, the consignment note is evidence of the facts it contains (the details shown are presumed to be correct), such as the number of packages etc., and any claim which disputes such facts would have to be backed by substantial independent evidence to the

contrary. Normally a standard note such as that available from the International Road Transport Union (IRU) is used (or in the UK from the RHA, or the Freight Transport Association, both IRU members).

6. The sender is responsible for all expenses, loss and damage sustained by the carrier as a result of inaccuracies in completion of the consignment note in relation to information supplied by him, even if the road haulier completes the note from information given to him by the sender.
7. On receipt of the goods, the carrier must check the accuracy of the details shown in the consignment note particularly, for example, as to the number of packages, the apparent condition of the goods, their packaging and how they are marked. Any discrepancies or comments about other relevant matters such as the condition of the goods or packages should be noted by a 'reservation' on the note. Should the sender request the carrier to check the contents of packages or to have the consignment weighed he must reimburse the carrier any costs incurred in doing so.
8. The sender is liable to the carrier for damage and expenses due to defective packing of the goods unless the defects were known to the carrier, when taking over the goods and he indicated this fact by way of a 'reservation' on the note. The absence of such a reservation means that the carrier, if he was aware of the damage, accepted any likely risks of subsequent claims.
9. The sender must attach to the consignment note or make available to the carrier the necessary documents to complete customs formalities. The sender is liable to the carrier for any damage caused by the absence, inadequacy or irregularity of such documents.
10. The sender has the right of disposal of the goods and may stop transit of the goods or change the delivery address up to the time of delivery to the consignee unless he has stated on the consignment note that the consignee has this right. Once the goods are delivered to the address on the consignment note, the consignee has the right of disposal.
11. A carrier who fails to follow the instructions on the consignment note or who has followed them without requesting the first copy of the consignment note to be produced is liable for loss or damage caused by such failure.
12. The carrier must provide the consignee with a second copy of the consignment note at the time of delivering the goods.
13. If the carrier cannot follow the instructions on the consignment note for any reason, he must ask the sender or the consignee, depending on who has the right of disposal (see item 10 above), for further instructions.
14. The carrier is liable for the total or partial loss of the goods and for any damage to them occurring between the time when he takes over the goods and the time of their delivery unless the loss, damage or delay was caused by a wrongful act or neglect of the claimant. The burden of proof in this case rests with the carrier.
15. Failure to deliver goods within 30 days of a specified time limit, or within 60 days from the time when the first carrier took them over if there is no time limit for delivery, results in the goods being considered to be lost.
16. When goods of a dangerous nature are consigned, the carrier must be informed of the nature of the danger and the precautions to be taken.
17. Calculation of compensation in the event of loss or damage is related to the value of the goods at the place and time they were accepted for carriage but will not exceed a set value (related to SDR, see p. 186–7 for explanation).
18. Carriage charges, Customs duties and other charges in respect of the carriage are refunded in the case of total loss of the goods and proportionately in the case of partial loss.
19. Higher levels of compensation may be claimed where the value or a special interest in delivery has been declared or where a surcharge has been paid in respect of a declared value exceeding the limit mentioned in item 17 above.

20. In the case of damage the carrier is liable for the amount by which the value of the goods has diminished.
21. The claimant may demand interest in respect of the amount of any claim at 5 per cent per annum from the date on which the claim was sent to the carrier.
22. A carrier cannot avail himself of exclusions or limiting clauses if damage to goods was caused by his wilful misconduct or default, which constitutes wilful misconduct.
23. The consignee is considered to have accepted the goods in a satisfactory condition if he does not indicate his reservations at the time of delivery or within 7 days (excluding Sundays and public holidays).
24. In legal proceedings, the plaintiff may bring an action in any court or tribunal of a contracting (i.e. CMR contracting) country, or of a country in which the defendant is normally resident or has his principal place of business, or of a country where the goods were taken over by the carrier or where they were designated for delivery, and in no other courts or tribunals.
25. The period of limitation for an action under the Convention is 1 year, or 3 years in the case of wilful misconduct.
26. Where successive road carriers are involved in a contract under the Convention, each one of them is responsible for the whole operation as a party to the contract. Each successive carrier must give the previous carrier a dated receipt and must enter his name and address on the second copy of the consignment note.
27. A carrier who has paid compensation arising from a claim may recover the compensation plus interest, costs and expenses from other carriers who were parties to the contract subject to:
    (a) the carrier responsible for the loss or damage paying the compensation
    (b) each carrier responsible for loss or damage jointly caused shall be liable to pay proportionate compensation or compensation proportionate to its share of the carriage charges if responsibility cannot be apportioned.
28. If a carrier who is due to pay compensation is insolvent, his share must be paid by other carriers who are parties to the contract.

In particular, road hauliers should note that the terms of the carriage contract require carriers taking over the goods to check the accuracy of the statements in the consignment note as to the number of packages, their marks and numbers, the apparent condition of the goods and their packaging, and they should obviously do so for their own protection in the event of later disputes or claims. Under the Convention the carrier is responsible for loss, damage or delay from the time of taking over the goods until the time of their delivery.

Furthermore, where goods are handled by a number of carriers on an international journey (e.g. by transferring a loaded articulated semi-trailer from one to another), provisions are contained in CMR to apportion the liability for loss or damage between all the carriers (based on the relative proportions of the total carriage cost charged by each one of them). This is because of the difficulties which may arise in pinpointing the exact time and place when the damage occurred (unless specific responsibility can be determined), but should one or more of the carriers in this situation default (through insolvency) in meeting its share of any claim for damage to or loss of the goods then the remaining carrier or carriers will have to meet the share of those defaulting.

14.3.4 CMR liability

As stated above, in international road haulage operations (but not own-account road transport operations), the carriage automatically comes within the terms of the CMR Convention under which the carrier's liability for claims resulting from loss of or damage to the goods carried is determined by comparison with a measure known as Special Drawing Rights (SDRs), whereby compensation must not exceed 8.33 units of account per kilogram of gross weight short (gws). SDRs are defined by the International Monetary Fund

(IMF) as being a unit for converting currency values based on a 'basket' of the currencies of the key Member States of the IMF and are converted to the national currency of the country in which any claim is dealt with in court, and is assessed as to value on the date of the judgement, or on a date agreed to by the parties. A treasury certificate stating the value for that day is taken to be conclusive proof of that fact.

Note: The value or exchange rate of SDRs on the date of judgement or agreement referred to above must not be confused with the date of calculation of the value of the goods which are subject to the claim as referred to in item 17 on p. 185.

14.3.4.1 SDR conversion rate

The daily conversion rate for SDR to national currencies can be found in the financial newspapers. With changing values it is essential that the current value should be established at any particular point in time, and adequate insurance cover to at least this level of liability should be carried. To give an approximate idea of what this value represents, on 23 February 2005 (bearing in mind the rates change daily), a conversion of SDR to sterling at a rate of £1.24915 per kilogram represented a value of £10 405.42 per tonne (i.e. calculated as 8.33 units × £1.24915 × 1000 kilogram).

14.4 International carriage of goods by rail: CIM

The Uniform Rules for the International Carriage of Goods by Rail, commonly known and referred to as the CIM rules, are an extensive and complex body of legislation, similar in style and structure to CMR for international road haulage as described in detail above, but with certain obvious differences relating to rail carriage. These Rules are too extensive to reproduce here, but it is useful to outline the important liability provisions in the context of this chapter.

First, however, a number of basic facts should be clarified. For instance:

1. For these rules to apply, the carriage by rail must cross over from the national state in which the originating rail station is located into another different state, both of which states must have adopted this Convention.
2. The railway company to which freight is offered, is bound under the rules to accept such goods for carriage if:
   - it is in complete wagon loads,
   - the sender complies with the Uniform Rules,
   - the carriage can be undertaken by the railway's normal staff and transport resources,
   - the carriage is not prevented by circumstances which the railway company cannot avoid and which it is not in a position to remedy.
3. The railway company is not bound to accept goods for carriage if special handling facilities are needed, or if the carriage cannot take place without delay (i.e. if temporary storage is required).
4. Certain goods are not accepted for carriage, particularly articles which:
   - are prohibited in the territories over which carriage would take place,
   - form the monopoly of the postal authorities in one of the territories through which carriage would take place,
   - which due to their mass (i.e. weight) or packaging are not suitable for carriage having regard to the rolling stock of the railway company, or successive railway companies,
   - are dangerous goods not acceptable for carriage under the regulations for dangerous goods carriage (International Carriage of Dangerous Goods by Rail (RID), see Chapter 16).
5. Certain other goods are accepted for carriage subject to conditions. These comprise:
   - substances which are permitted under the RID regulations,
   - funeral consignments,

- live animals (which, generally, must be accompanied by an attendant provided by the consignor),
- consignments of a difficult size, weight or packaging.

14.4.1 Liability provisions

The Rules make clear the fact that the railway company which accepts goods for carriage, along with a consignment note (as described in detail in Chapter 15) relating to those goods (a separate note being required for each consignment or wagon load) is responsible for the carriage over the entire route up to the point of delivery. Succeeding railways, by the very act of taking over the goods along with the consignment note(s), become parties to the contract of carriage and must assume the same obligations as the accepting railway company.

Normally, the railway company's liability is for loss or damage resulting from the total or partial loss of, or damage to, the goods between the time of their acceptance for carriage and the time of delivery, and for any loss or damage resulting from the transit period being exceeded. However, it is relieved of liability if the loss or damage, or exceeding of the transit time, was caused by the following:

1. A fault on the part of the person entitled to the goods.
2. An order given by the person entitled to the goods.
3. Other than as a result of a fault on the part of the railway.
4. Inherent vice of the goods (decay, wastage, etc.).
5. Circumstances which the railway could not avoid and the consequences of which it was unable to prevent.
6. The special risks inherent in one or more of the following circumstances:
   (a) carriage in open wagons;
   (b) absence or inadequacy of packing for goods which are liable to loss or damage when not packed or when not properly packed;
   (c) loading operations carried out by the consignor or unloading operations carried out by the consignee;
   (d) defective loading by the consignor;
   (e) completion by the consignor, the consignee or an agent of either, of the formalities required by customs or other administrative authorities;
   (f) the nature of certain goods which renders them inherently liable to total or partial loss or damage, especially through breakage, rust, interior and spontaneous decay, desiccation or wastage;
   (g) irregular, incorrect or incomplete description of articles not acceptable for carriage or acceptable subject to conditions, or failure by the consignor to observe the prescribed precautions in respect of articles acceptable subject to conditions;
   (h) carriage of live animals;
   (i) carriage which (under provisions applicable or by agreement) must be accompanied by an attendant, if the loss or damage results from risks which the attendant was intended to avert.

The burden of proving that the loss, damage or exceeding of the transit period was due to one of the causes listed above rests with the railway company. If it establishes that the loss or damage could have arisen from one or more of the special risks referred to in item 6 above, 'it shall be presumed that it did so arise' but the person entitled to the goods has a right to prove that the loss or damage was not attributable either wholly or partly to one of those risks. Goods may be presumed lost, without the claimant being required to furnish further proof, when they have not been delivered to the consignee or are not being held at his disposal within 30 days after the expiry of the transit periods. After having received compensation for lost goods, claimants may request in writing to be notified should the goods be recovered within 1 year.

14.5 Compensation for loss

In the event of total or partial loss of the goods the railway must pay compensation calculated according to the commodity exchange quotation or, if there is no such quotation, according to the current market price, or if there is neither such quotation nor such price, according to the normal value of goods of the same kind and quality at the time and place at which the goods were accepted for carriage. However, the amount of compensation payable is limited to a maximum of 17 *units of account* per kilogram of gross mass short (see below), and it may be further limited if the carriage took place under special conditions of carriage through special or exceptional tariffs; in other words, if reduced carriage charges were negotiated then compensation may also be reduced in the event of a loss or damage claim arising. When the calculation of compensation necessitates the conversion of amounts expressed in foreign currencies, the rules require that conversion is made at the rate of exchange applicable at the time and place of payment of the compensation. In addition to any compensation paid, the railway company is required to refund carriage charges, Customs duties, and other amounts incurred in connection with carriage of the lost goods.

Limits are placed on the amount of compensation payable where a claim is for loss that arises through natural wastage of certain goods. Generally, compensation is only payable when the loss exceeds a stipulated allowance. This is 2 per cent of the mass (i.e. weight) in the case of liquid goods, goods consigned in a moist condition and certain other listed products (see list in the full rules at Appendix I, Article 41(1(a)). For all other dry goods the allowance is 1 per cent.

The rules also place limits on the amount of compensation payable for damage to goods and when transit periods are exceeded. In the former case, compensation is payable by an amount equivalent to the loss in value of the goods but not exceeding:

1. If the whole consignment has lost value through damage, the amount which would have been payable in the case of total loss.
2. If only part of the consignment has lost value through damage, the amount which would have been payable had that part been lost.

Proportionate refunds are made of the carriage and other charges.

In the case of loss or damage resulting from the transit period being exceeded, the railway company must pay compensation not exceeding three times the carriage charges. However, if the delay results in total loss of the goods, the amount of compensation payable will be for the total loss only and will not include an additional amount in compensation for exceeding the transit period. In fact, in no case will the total amount of compensation payable under these rules exceed that which would be payable in the event of total loss of the goods. When the loss or damage to goods, exceeding the transit period, the failure to perform, or failure to perform properly, the railway company's additional services provided for in the Uniform Rules, has been caused by wilful misconduct or gross negligence on the part of the railway company, full compensation for the loss or damage proved is to be paid to the person entitled by the railway. In case of gross negligence, liability is limited to twice the normal maximum amounts specified under the rules.

Where there has been a declaration of interest in delivery (i.e. an amount above the normal value of the goods to represent other losses which may result from loss or damage to the goods), for which the railway company may levy additional charges, further compensation for any loss or damage proved may be claimed, above the amounts normally payable under these rules, up to the amount declared. A claim for interest on compensation payable may be made, calculated at 5 per cent per annum, from the date of the claim or if no claim has been made, from the day on which legal proceedings are instituted. Interest is only be payable if the compensation exceeds four *units of account* per consignment note.

14.5.1 Unit of account

The unit of account referred to above is the SDR as defined by the IMF. In the case of a state which is not a member of the IMF and whose legislation does not permit the application SDRs, the unit of account referred to in the Uniform Rules is deemed to be equal to 3 gold francs (the gold franc is defined as 10/31 of a gram of gold of millesimal fineness 900).

SDRs for these purposes are exactly as described above in the context of the CMR Convention and have the same daily changing value. The calculation of maximum compensation value on a per tonne basis is carried out by taking the daily SDR rate (see financial newspapers or visit the IMF web site at www.imf.org) and multiplying this by 17 (see above), and then by 1000 kilogram. For example, based on the SDR rate for 23 February 2005, the figure would be £1.24915 $\times$ 17 $\times$ 1000 = £21 235.55 per tonne.

Note: It may be noted that this is, in fact, almost double the rate for maximum compensation payable under the CMR Convention for claims arising in international road transport operations.

14.6 Liability rules for multimodal transport

No single agreement currently made under international law fully covers the diverse liabilities applicable to a transport journey involving a combination of modes. The problem with these international conventions, certainly so far as multimodal transport operations are concerned, is that of proving the mode of carriage, and thus the individual liabilities of intermodal operators, being used at the precise time when loss of or damage to the goods occurred. The Commission of the European Communities, in fact, identified this very matter (in its report *The Future Development of the Common Transport Policy*, Brussels, December 1992) as a problem that needed further investigation. It suggested the necessity of determining the extent to which a more uniform approach to the different regimes of liability of intermodal operators between member states and between the different modal stages of an intermodal journey, would increase the attractiveness of multimodal services.

Past attempts to produce a single, all-encompassing solution have failed; take, for example, the non-adopted TCM Draft Convention of 1971 on the Combined Transport of Goods (*Transport Combines des Marchandises*), and the United Nations (UN) Convention on International Multimodal Transport of Goods (the so-called MT Convention) which was adopted as long ago as 24 May 1980, but which to this day has never been brought into force. In the absence of an accepted and ratified international convention, applicable solely to multimodal transport, such as the one that the UN had in mind (see above), UNCTAD and the ICC in more recent times, jointly devised a set of rules for Multimodal Transport Documents, based on the TCM draft convention mentioned above. These rules, which include provisions relating to the liability of the multimodal transport operator, became effective from 1 January 1992. They provide such operators with a private transport contract having a uniform legal regime, so as to avoid the application of a multiplicity of other different regimes, as is currently the case.

However, even these rules are not widely adopted. For a start, as the ICC advises in its booklet (ICC publication no. 481), they are not comprehensive. The shipper in devising a multimodal freighting contract would still have to devise and add other suitably worded clauses dealing with such matters as routeing, freight and charges, liens, jurisdiction and arbitration, applicable law, and so on to meet his own particular requirements. And, secondly, because so many different commercial arrangements are applied throughout the industry, no single, uniform set of rules or standard contract terms would appear to be capable of providing satisfactory solutions to all the liability problems which might arise in the event of an international freight shipment being conveyed by multimodal transport going awry.

International multimodal transportation insurers Through Transport Mutual Services (UK) Limited (the TT Club), which has been providing this type of insurance for over 30 years, commenting on the confused situation, made the following points:

> Multimodal operations are, of course, organized by somebody. This 'somebody' is often a freight forwarder who is using his expertise to bolt together various sectors of a journey and offering a through transport document (e.g. a multimodal bill of lading) for the entire carriage. Alternatively, other carriers such as shipping lines or railway companies can offer a through transport option, including collection from the supplier's premises and delivery from the port/terminus to the consignee's premises.
>
> No matter how the transport is organized, generally a through document is issued which incorporates contractual conditions of carriage applying between the multimodal operator and the shipper/consignee. Normally these conditions include provisions dealing with the carrier's responsibility and liability in the event of loss or damage to the cargo. Typically the provisions are worded so that if the sector of the journey where the loss or damage occurred can be identified, and liability for that particular method of transport is governed by an international convention or national law, then the carrier will be liable according to those legal conditions. If the exact sector of the journey cannot be identified or if there is no compulsory convention applicable, then the carrier can limit his liability to an amount of (usually) 'SDR 2' or 2 US dollars per kilogram.
>
> In today's world, although some containers may move direct to consignee's premises by rail, almost all multimodal movements will involve some road transport. The road haulier will, of course, be responsible for any loss or damage that occurs while the goods are in his care. His primary liability is, however, to the multimodal transport operator who retains overall responsibility for the safe completion of the whole journey.
>
> Road haulier's liability will be determined by the nature of the journey which he undertakes. In Europe this will depend on whether the actual road sector (as distinct from the total journey) is international or not. Movement from Madrid to Barcelona will be subject to the Spanish law on road transport, LOTT, whereas the same consignment moved from Madrid to Marseilles by road will be subject to CMR. As both these are compulsorily applicable to the road journey, the liability of the multimodal carrier will be the same as that of the haulier to whom he has sub-contracted the road carriage.
>
> It must be stressed that the road haulier will be liable only if it can be proved that he was responsible for the loss or damage. In many multimodal operations the haulier collects a container which is already sealed and delivers the same to the consignee. Provided the container remains sealed and intact the road haulier will normally be able to avoid liability for any damage. The multimodal carrier (i.e. the company which issued the through transport document) has no such luxury and will remain responsible for any loss or damage which is established at destination, unless the cause comes within one of a limited number of exclusions and exceptions. Multimodal operators dearly require insurance cover for these liabilities.

14.7 GIT insurance protection

14.7.1 Domestic liabilities

In domestic road haulage operations, insurance cover against the risks of damage to or loss of goods carried by vehicles is essential in addition to the normal requirement for vehicle cover under national legislation (in the UK, e.g. under the Road Traffic Act 1988). Motor vehicle insurance cover (whether covering only minimum legal requirements, or more extensive cover such as comprehensive insurance against damage to the vehicle) *does not* provide for compensation in the case of loss or damage to the goods carried.

GIT insurance is normally arranged to cover such liabilities as are accepted by professional haulage operators in accordance with the published conditions under which they accept goods for carriage. Invariably such conditions of carriage stipulate a maximum value per tonne of gross weight or by cubic dimension (where capacity is relatively greater than weight) above which liability is not accepted without prior notification and agreement; the RHA sets a limit of £1300 per tonne gross weight or 2.25 cubic metres, whichever is the greater. When goods carried are in excess of this level of value, it is incumbent on the operator to notify the insurance company accordingly so that additional cover for an appropriate amount can be arranged at an extra cost, which is normally passed on to the customer.

Own-account vehicle fleet operators, while not carrying their own goods under conditions of carriage in the same way as professional hauliers, nevertheless need to arrange GIT cover for an amount appropriate to the value of the goods normally carried. Again, additional cover can be arranged, where goods of higher value are carried on a one-off or occasional basis.

14.7.2 International liabilities under CMR

For professional road hauliers operating internationally under the terms of the CMR Convention, insurance cover for the goods carried is a more complex matter. They need to obtain additional cover sufficient to meet the liabilities which they are bound to carry under the terms of the Convention (based on the current value of SDR) as described above. By comparison, current UK domestic GIT liabilities are normally rated at £1300 per tonne (i.e. RHA conditions of carriage level) while the current SDR conversion shows a CMR liability (which may vary on a daily basis) at around £10 400 per tonne (see p. 187 for calculation).

15

Intermodal Documentation and Authorizations

It is a widely recognized fact that in transport operations, and especially in the case of operations that involve cross-border journeys; in other words, international transport journeys, it is essential to get the documentation absolutely right. Failure to do so can result in a whole range of problems that may arise in the way of delays to consignments, penalties being imposed, additional costs incurred, even loads being rejected at the point of delivery. In this context, the term 'documentation' can be taken to include a wide range of commercial documents (such as consignment notes) and legal authorizations (such as community authorizations), and road haulage bilateral- and multilateral-journey permits depending, of course, on the journey destination.

In domestic transport operations, and particularly in domestic road haulage, a whole mishmash of consignment notes of one form or another are used, ranging from printed, multipart, self-copying document sets to hand-scribbled collection and delivery instructions on scraps of paper. In general terms, these are all satisfactory in their own particular way, and are fine when systems are operating without hiccup, but when it comes to tracing consignments and when handling loss or damage claims, scraps of paper are invariably found to be wanting. Such documents, if they can be called that, rarely make consignor instructions clear, delivery instructions are imprecise, other crucial details are omitted or incorrectly recorded, and then trying to determine who, or whose employee, scrawled a signature (if indeed anybody at all signed) for safe receipt of the goods in good condition can be a nightmare.

In international transport there is no room for such laxity, particularly since such operations – whether conducted by road transport, by rail transport in isolation, or by a combination of modes – are invariably undertaken in accordance with, and under the contract terms of, international conventions on road, rail or combined/multimodal transport, as the previous chapter in this book shows. In fact, as mentioned in the previous chapter, international road haulage operations are subject, automatically, to the provisions of the Convention Marchandises Routiers (CMR) Convention. This clearly defines the form of consignment note which must be used, and which, in fact, constitutes the contractual basis upon which the carriage is undertaken. Similarly, for through-rail movements which commence in one state and travel to another state (both of which are party to the convention) a consignment note conforming to the requirements of the CIM Rules of the COTIF Convention must be raised, confirming both the making of the carriage contract and its content. With the complexity introduced when both road and rail modes are combined in a single transport movement, and when the liability issues and documentation requirements are taken account of, there is an increasing movement towards the use of combined transport (CT) documents where the issuing carrier accepts responsibility for the whole operation as the combined or multimodal transport operator (i.e. the CTO or MTO).

This chapter examines the three main forms of consignment note in use, namely those of the CMR and the CIM Conventions (respectively; The Conventions on the Contract for the International Carriage of

Goods by Road and by Rail) and the CT document which forms the basis of the United Nations Conference on Trade And Development (UNCTAD)/International Chamber of Commerce (ICC) Rules on Multimodal Transport Documents. Mention is also made of the form of consignment note needed when undertaking international journeys by road on an own-account basis. This chapter also examines the legal provisions relating to road haulage issues such as community authorizations, cabotage, road haulage-journey permits, and various other transit documents and road toll Eurovignettes, for example.

15.1 CMR consignment notes for international haulage journeys

Road hauliers carrying goods for hire and reward on international journeys under the provisions of the Convention on the Contract for the International Carriage of Goods by Road (CMR) as described in the previous chapter must complete special CMR consignment notes. A copy of the consignment note (see Figure 15.1) must be carried on the vehicle to confirm that the carriage is being conducted under a contract subject to the terms of the CMR Convention. Even in the absence of a CMR note, the carriage will still be subject, under international law, to the terms of the Convention although the carrier may not have been aware of this fact.

The consignment note is made out in four original copies, all of which should be signed by both the carrier and the consignor of the goods. One copy of the note (printed with red lines) is retained by the consignor; the second copy (printed with blue lines) is for the consignee; and the third copy (printed with green lines) is for the carrier and must travel forward with the vehicle and remain with it while ever the goods are on board. A fourth copy (printed with black lines) may be retained on file by the originator of the document. Where a consignment is divided to travel by different vehicles or by separate means, separate CMR consignment notes should be made out for each individual part of the consignment. The following details must be entered on CMR consignment notes:

Box 1	Sender (name, address, country)
Box 2	Customs reference/status
Box 3	Sender's/agent's reference
Box 4	Consignee (name, address, country)
Box 5	Carrier (name, address, country)
Box 6	Place and date of taking over the goods
Box 7	Successive carriers
Box 8	Place designated for delivery of goods
Box 9	Marks and numbers, number and kind of packages, description of goods*
Box 10	Gross weight (kilogram)
Box 11	Volume (cubic metre)
Box 12	Carriage charges
Box 13	Sender's instructions for customs
Box 14	Reservations
Box 15	Documents attached
Box 16	Special agreements
Box 17	Goods received
Box 18	Signature of carrier
Box 19	Company completing the note
Box 20	Place, date, signature
Box 21	Copies to: (i) Sender (ii) Consignee (iii) Carrier

LETTRE DE VOITURE INTERNATIONALE (CMR) INTERNATIONAL CONSIGNMENT NOTE

193922

Approved by FTA/RHA/SITPRO UK 1981:1987 COPY 1 SENDER
COPY 2 CONSIGNEE
COPY 3 CARRIER

Sender (Name, Address, Country Expéditeur (Nom, Addresse, Pays)) 1	Senders/Agents References Référence de lèxpéditeur/de làgent 2/3
Consignee (name, Address, Country) Destinataire (Nom, Addresse, Pays) 4	Carrier (Name, Address, Country) Transporteur (Nom, Addresse, Pays) 5
Place & date of taking over the goods (place, country, date) Lieu et date de la prise en charge des marchandises (lieu, pays, date) 6	Successive Carriers Transporteurs successifs 7
Place designated for delivery of goods (place, country) Lieu prévu pour la liveraison des marchandises (lieu, pays) 8	This carriage is subject, notwithstanding any clause to the contrary, to the conversion on the Contract for the international Carriage of Goods by Road (CMR) Ce transport est soumis nonobstant toute clause contraire à la Convention Relative au Contract de Transport International de Marchandises par Route (CMR)

*NB FOR DANGEROUS GOODS
INDICATE
1. CORRECT TECHNICAL NAME (PROPER SHIPPING NAME)
2. HAZARD CLASS
3. UN NUMBER
4. FLASHPOINT (IF ANY) IN°C.

Marks & Nos: No & Kind of Packages: Description of Goods* Marques et Nos: No et nature des colis; Désignation des marchandises* 9	Gross weigh (kg) 10 Poids Brut (kg)	Volume (m^3) 11 Cabage (m^3)

Carriage Charges Prix de transport 12	Sender's Instructions for Customs, etc... Instructions de l´Expéditeur (optional) 13

Reservations Réserves 14	Documents attached Documents Annexés (optional) 15 Special agreements Conventions particuliéres (optional) 16	
Goods Received/Marchandises Rescues 17	Signature of Carrier/Signature du transporteur 18	Company completing this note Société émettrice 19 Place and Date; Signature Lieu et date; Signature 20

Fig. 15.1 CMR consignment note used in international road haulage.

*For dangerous goods indicate:

(i) Correct technical name (i.e. proper shipping name);
(ii) Hazchem class;
(iii) UN number;
(iv) Flashpoint (°C), if applicable.

Where applicable, the consignment note must also contain the following particulars:

1. A statement that trans-shipment to another vehicle is not allowed.
2. The charges which the sender undertakes to pay.
3. The amount of 'cash on delivery' charges.
4. A declaration of the value of the goods and the amount representing special interest in delivery.
5. The sender's instructions to the carrier regarding insurance of the goods.
6. The agreed time limit within which the carriage is to be carried out.
7. A list of documents handed to the carrier.

The consignor or consignee can also add to the consignment note any other particulars that may be useful to the road haulier.

15.2 Consignment notes for own-account carriage by road

Own-account operators are not required to use the CMR consignment note described above for international journeys. For most journeys a simple consignment note is all that is necessary to prove that the journey is on own account (i.e. that it is not for hire or reward). It should contain details of the following:

1. The vehicle operator.
2. His trade or business.
3. The goods being carried.
4. Their loading and delivery points.
5. The vehicle being used (by registration number, etc.).
6. The route to be followed.

In the case of own-account traffic to Germany a more detailed document is required containing the following particulars:

1. The place at which the document was made out and the date it was made out.
2. The name and address of the own-account operator and an accurate description of the nature of his business.
3. If the goods are to be accepted from, or delivered to, any other person, the name and address of that person, and an accurate description of the nature of his business.
4. Details of the loading point or points.
5. Details of the place or places at which the vehicle is to deliver.
6. Details of the nature of the load (i.e. a description of the goods carried).
7. The gross weight or other indication of the quantity of the load.
8. The carrying capacity of the vehicle by weight.
9. The index mark and registration number of the vehicle or, if these do not exist, the chassis number.
10. The distance of the loaded journey in Germany in kilometres.

11. The point or points at which the frontier is to be crossed.
12. A signature of the operator or his authorized representative.

15.3 CIM consignment notes for international rail journeys

Under the CIM Uniform Rules for international carriage by rail, a contract of carriage comes into existence as soon as the forwarding railway has accepted the goods for carriage together with the consignment note on which has placed its stamp or made an accounting machine entry, showing the date of acceptance. This procedure must be carried out immediately after all the goods to which the consignment note relates have been handed over for carriage and, where applicable, any charges due have been paid or a security deposited. Once the stamp has been affixed or the accounting machine entry has been made, the consignment note becomes evidence of the making and content of the contract.

The International Union of Combined Road–Rail Transport Companies (UIRR) consignment notes (see Figure 15.2) are in duplicate, the second copy of which is stamped by the railway company indicating receipt of the goods and the date of their acceptance for carriage. This copy does not have effect as the original note accompanying the goods which is handed over to the consignee on delivery, nor is it a bill of lading (indicating legal title to the goods). The duplicate copy of the consignment note must be produced by the consignor when making any loss or damage claim, if it was he who paid the carriage charges. If the charges were paid by the consignee then he must produce the actual consignment note when making a claim, if the note has been handed over to him. Without the duplicate copy or the original consignment note, or authenticated copies of these documents, making a claim and establishing a right to compensation for loss or damage can be extremely difficult, if not impossible.

It is the duty of the consignor to present to the railway company with a properly completed consignment note or, where there is more than one consignment, a separate note for each consignment. A single consignment note may not normally relate to more than one wagon load. He must attach to the note any documents necessary for the completion of customs formalities which may be necessary prior to delivery, it being the consignor's duty (not that of the railway company) to ensure that these documents are sufficient for the purpose and correct. The consignor may be held liable by the railway company for loss or damage resulting from the absence, insufficiency, irregularity, or inaccuracy of such documents. The railway company's duty is to prescribe, for both *petite vitesse* and *grande vitesse* traffic, a standard form of consignment note, which must include a duplicate for the consignor. The consignor is required to choose whichever of these notes is relevant to the service he requires. Only by agreement with the railway company (or companies) concerned may the consignor select *grande vitesse* over one part of the route and *petite vitesse* over the remainder. A simplified form of consignment note may be used in the case of certain traffic, notably between adjacent countries.

Consignment notes must be printed in two or, where necessary, in three languages, at least one of which is one of the working languages of the Intergovernmental Organisation for International Carriage by Rail (OTIF), headquartered at Berne in Switzerland and whose working languages are French and German. The consignment note must contain the following information:

1. The name of the destination station.
2. The name and address of the consignee; only one individual or legal person may be shown as consignee.
3. A description of the goods.
4. The mass (i.e. weight), or comparable information.
5. The number of packages and a description of the packing in the case of consignments in less than wagon loads.
6. The number of the wagon and, for privately owned wagons where the loading is the duty of the consignor, the tare weight.

UIRR

Contrat
Vertrag
Contratto
Contract

TEL.(1) 43 87 41 79 - TELEX 285 625 F

(10) Gare de départ / Versandbahnhof / Stazione di partenza / Forwarding station

(11)

(A) No :

(13) Référence / Referenz / Riferimento

(12) Déclaration / Erklärungen / Dichiarazioni / Declaration

(C) Expéditeur / Absender / Mittente / Consignor

(B) Date / Datum / Data

CENTRE D'EXPLOITATION DE

(14) Destinataire / Empfänger / Destinatario / Consignee

(15)

(19) Wagon / Wagen / Vagone / Wagon No.

(E) Client à facturer / Rechnungsempfänger / Da fatturare a / Firm paying charges

Le transport est soumis à la CIM et aux conditions générales de Ferroutage de l'UIRR et des sociétés de Ferroutage

(F) Douane de départ / Versandzoll in
Dogana di partenza / Export customs
A dédouaner à / Zu verzollen in
Da sdoganare a / To be cleared through customs at

(23) Tarifs et itinéraires demandés
Verlangte Tarife and Wegs
Tariffe e itinerari richiesti
Tariffs and routes required

Es gelten die CIM und die allgemeinen Geschäftsbedingungen für den Internationalen Huckepackehr (UIRR - Bedingungen) und der beteiligten Huckepackgesellschaffen
Sono valide le norme della CIM e la condizioni generali dell'UIRR e delle società di ferroutage
Transport is subject to the terms of business of the UIRR the piggy-back companies involveld and to CIM conditions of carriage

(24) Gare destinataire
Empfangsbahnhof
Stazione destinataria
Destination station

(48)

(D)

Exemplaire pour / Ausfertigung für / Copia per / Copy for.

BLANC	: Facturation	ROSE	: Transit
JAUNE FONCÉ	: Centre destinataire	VERT	: Expéditeur
BLEU	: Destinataire	JAUNE CLAIR	: Souche

(25) Matériel / Kombisendung / Genere / Piggy-pack

N° Identification / Kennzeichen No. / No. Identificazione / Registration N°	A	B	C	D	Net Kg	Tare / Tara Kg	(33) RID oui / ja / si / yes — Frigo Kg	(26) Poids / Gewicht / Peso / Weight Kg
								+
								+
J 1								=

Timbre en Compte avec le Trésor Autorisation du 7 juillet 1969

(H) Description de la marchandise
Bazeichnung der Güter
Descrizione della merce
Description of Goods

(32) Documents douaniers
Zolldokumente
Documenti doganali
Customs documents
Plombage / Plombiert
Piombatura / Seal

Codification de la case / Erläuterung zu Feld 25
Codificazioni della casella 25 / Code to box 25

(25) A
Type / Art / Genere / Material
0 Chàssis / Sattelanhänger horizontal verladen / Telaio / Chassis
1 Semi-Remorque / Sattelanhänger
Semirimochio / Semi-trailer
Véhicule articlulé / Sattelzug
Autoarticolato / Articulated vehicle
2 ou / oder / od / or
Train routier / Lastzug
Autotreno / Road train
3 Carnion / Tracteur / 1.KW solo
Zugmaschine / Carnion / Tractor unit
Caisse Mobile / Wechselbehälter
Cassa Mobile / Swap body :
4 12.19 ≤
5 9.51 ≤ 12.18 m
6 6.05 ≤ 6.30 m
7 6.31 ≤ 7.15 m
8 7.16 ≤ 8.30 m
9 8.31 ≤ 9.50 m

(25) B
Spécialisation / Auflare-Art
Specializzazioni / Specialised
0 Plateau / Offen
Pianale / Flat
1 Savoyarde / Mit Plane
Telonato / Tilt
2 Citerne / Tankaufbau
Cisterna / Tank
3 Frigo / Thermosaufbau
4 Fourgone/ Kofferaufbau
Furgonato / Box
5 Silo

(25) C
Plein-Vide / Voll-Leer
Pieno-Vuoto / Landen-Empty
0 Vide / Leer
Vuoto / Empty
1 Plein / Voll
Pieno / Laden
2 Emballages vides / Leereladegäte
Imballagi vuoti / Empty unit

(25) D
Jumelage / Zusammengehörigkeit
Abbinamento / Mariage
1 1" / 1. / I / 1 st
2 2' / 2. / II / 2 nd
3 3' / 3. / III / 3 rd

(J) Aide aux constatations éventuelles / Hilfe fur evlt feststellungen
Eventuale constatazione di danni / Potential damages

1 Déchiré / Gerissen
Strappato / Ripped
2 Détaché / Lose
Staccato / Lose
3 Bosselé / Eingebeult
Ammaccato / Dented
4 Cassé / Gebrochen
Rotto / Broken
5 Troué / Durchlöchert
Bucato / Perforated
6 Plié / Verbogen
Piegato / Bent
7 Absent / Fehlend
Assente / Missing
8 Hors d'état / Ausser Betrieb
Fuori uso / Out of order
9 Déformé / Verformt
Deformato / Out of shape
0 Griffé / Zerkratzt
Graffiato / Scratched

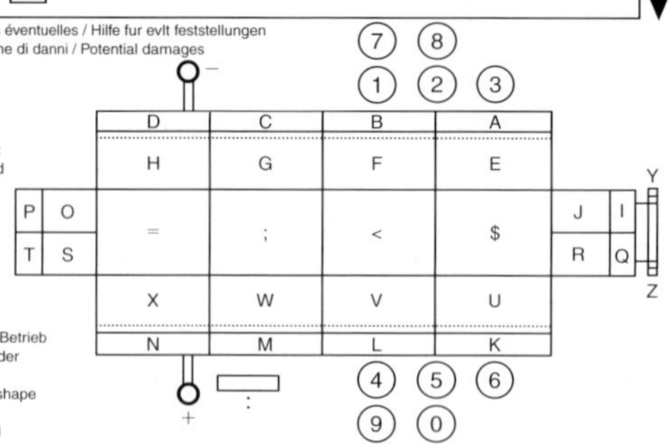

L'ABSENCE D'INDICATION NE PROUVE PAS L'ÉTAT INTACT DU VÉHICULE LORS DE SA REMISE.

(K) Observations / Vermerke / Osservazioni
Remarks

(L) Nom du chauffeur / Namen der Fahrer
Nome dell'au tista / Name of Driver

(M) Signature de l'expéditeur / Unterschrift des Absenders
Firma del mittente / Signature of Sender

Destinataire / Empfänger
Destinatario / Consignee

(O) Nom du chauffeur / Name des Faheres
Nome dell'autista / Name of Driver

(N) Signature/ Unterschrift / Firma

(P) La société de ferroutage expéditrice
Versand Huckepack gesellschaft
La società di ferroutage mittente
For warding piggy back company

(Q) La société de ferroutage destinataire
Empfange-Huckepackgesellschaft
La società di ferroutage destinataria
Receiving piggy back company

(82) Pays-Land-Paese — Gare-Bahnhof-Starione — No
Etiquetic de contrôle
Kontroll-Etikette
Etichetta di controllo
Exp.-Vers.-Sped.-Exp. — No

(83) Date d'expédition (jour-mois) Versanddatum (Tag-Monat) Date di spedizione (giorno-mese)

(84) Arrivage-Empiang Arrivo No

VOIR CONDITIONS GENERALSES AU VERSO

Fig. 15.2 UIRR contract/consignment note used combined road–rail transport.

7 A detailed list of the documents that are required by customs or other administrative authorities and are attached to the consignment note or shown as held at the disposal of the railway at a named station or at an office of the customs or of any other authority.
8. The name and address of the consignor. Only one individual or legal person may be shown as the consignor. (If required, the consignor must add to his name and address his written, printed, or stamped signature.)

The consignment note must, where appropriate, contain all the other particulars provided for in the Uniform Rules, but it must not contain any other details unless they are required or allowed by the laws and regulations of a state and are not contrary to the Uniform Rules. Nevertheless, the consignor may add as information for the consignee, in the space set apart for the purpose, remarks relating to the consignment, without involving the railway in any obligation or liability. He may also stipulate the route to be followed, indicating it by reference to frontier points or frontier stations (but only those which are open to traffic between the forwarding and destination places concerned) and, where appropriate, to transit stations between railways. The consignment note must not be replaced by other documents or supplemented by documents other than those prescribed or allowed by the Uniform Rules, the supplementary provisions or the tariffs.

15.4 Combined/multimodal transport documents

15.4.1 TCM Convention

In 1969, a Convention covering combined transport operations (CTOs) was agreed, with the title TCM Convention. A further draft Convention was produced in 1971, and although never formally adopted, various bodies concerned with CT have followed its provisions. Under the Convention, CTOs could issue a document headed 'Combined Transport Document Governed by the TCM Convention'. The Convention permitted this to be either a negotiable document, giving title to the goods, or a simple receipt for the direct delivery of the goods to a named consignee. When goods are carried in containers it is often impossible to ascertain at which point loss or damage has occurred. In such circumstances the Convention makes the CTO liable, by the issue of a CT document, for the loss or damage to the goods occurring while the goods are in his charge, subject to certain exceptions embodied in the TCM Convention. However, if it is known where the loss or damage occurred, liability is normally determined by the 'network' system of liability. Under this system, liability is determined in accordance with the appropriate convention or national law, or in certain exceptional circumstances and with prior agreement, by the subcontract between the CTO and the actual carrier.

The principal contents of the Convention are as follows:

1. Voluntary application of the Convention by the issue of a clearly identified document likely to be accepted by all parties.
2. It is the responsibility of the CTO to undertake (or arrange for others to undertake) the carriage, and to accept liability throughout the whole CT. At any particular stage the CTO may be either a basic carrier or an operator, arranging for others to provide the actual carriage. Thus, the document mentioned above may be the CT equivalent of either a shipping line or forwarders' bill of lading.
3. The information which should be contained in the CT document, leaving it to commercial practice which party should complete the document. The Convention avoids the ambiguity arising from the use of the word 'containers' by such expressions as 'coverings in which goods are packed' (i.e. traditional packings) and 'unit loads' (i.e. palletized or containerized cargo).
4. The responsibility of consignors and the CTO, respectively, for information given in the CT document. They are now satisfactorily based on commercial and transport practices.
5. Provision for the alternatives of a negotiable- or non-negotiable-type document, that is a document of title enabling a holder to claim delivery of the goods or a simple receipt for the direct delivery of

the goods to a named consignee. It also introduces a 'fail–safe' mechanism to govern a case where a document does not state which of these it is.

6. A 'dangerous goods' clause follows precisely the wording of certain existing conventions.
7. Stipulation of the period in which notice of loss or damage to the goods should be given to the CTO, including loss or damage not immediately apparent with, for example, containerized cargo.
8. The CTO's overall liability, and certain exemptions plus the limit of financial liability of the CTO, leaving the amount to be fixed at a diplomatic conference.
9. Establishment of the 'network' system of liability; it also deals with the problem of containerized cargo being carried on deck without specific reference being made to this fact in the document.

In the absence of acceptance of a formal Convention on CT documents, the ICC in Paris, in conjunction with the UNCTAD, has adopted a set of Uniform Rules based on the draft TCM Convention mentioned above and the so-called 'Tokyo Rules' of the Comité Maritime International (CMI). The UNCTAD/ICC Rules are now widely recognized and extensively used, their provisions being incorporated within such standard CT documents as the International Federation of Freight Forwarders (FIATA) combined bill of lading and the Baltic and International Maritime COnference (BIMCO) 'Combidoc'. However, as the ICC has pointed out in its document; *UNCTAD/ICC Rules for Multimodal Transport Documents*, published by the ICC in Paris in January 1992, these rules cover only part of the normal contents of a multimodal contract, leaving the CTO to add other clauses to suit his particular needs.

15.5 Legal requirements for international road haulage journeys

Despite increasing volumes of international freight being shipped in swap bodies and containers via combined road–rail systems, a substantial amount of UK–Europe traffic is still carried door-to-door by heavy lorry, crossing the Channel either on the short-sea ferries or on Eurotunnel's freight shuttle rolling motorway service. Whichever of these routes is chosen, the fact remains that a heavy lorry is carrying out an international journey and therefore is subject to a whole raft of legal provisions, particularly concerning authority to operate outside its country of registration.

For example, international road hauliers whether from the UK journeying into Europe or European hauliers entering Great Britain must, by law, comply with European Union (EU) community authorization and cabotage regulations and, if travelling outside the EU, must comply with requirements on road haulage permits issued under bilateral agreements, or with the permit requirements of Member States of the European Conference of Ministers of Transport (ECMT). Failure to comply fully with any of these requirements, as appropriate, could lead to severe delay of the vehicle by the relevant national enforcement authorities and, potentially, heavy financial penalties could be imposed on the operator who could face loss or suspension of his right to operate international haulage services.

15.6 Community authorization

One of the most important legislative steps taken in opening the Single European Market (SEM) in 1993 directly affected road transport by allowing the free movement of goods between Member States. The complex and restrictive system of quota allocations for bilateral permits formerly needed for most international road haulage journeys within the EU, and for transit traffic to and from certain non-EU member countries, was abolished. In its place, a new system of 'community authorizations' was implemented from 1 January 1993, enabling EU road hauliers to operate freely (i.e. to undertake as many journeys as they wish) *between* Member States, including three European Economic Area (EEA) states, namely Liechtenstein, Norway, and Iceland – not to be confused with the quite separate cabotage authorizations which are needed by hauliers wishing to collect and deliver goods *within* EU Member States other than their own (see below). International hauliers operating within the EU (apart from when operating domestically within their own state) must hold a community authorization issued by the transport authority in their own Member State. It is important to note the continuing requirement for bilateral and ECMT permits for road haulage journeys *outside* the EU (and the three EEA states mentioned above) and for international furniture removals, and the particular requirements relating to journeys to and through Austria which are described later in this chapter.

15.6.1 Regulation 881/92/EEC

The system of community authorizations for intra-EU road haulage operations is established under Council Regulation (EEC) 881/92 and implemented in the UK by the Goods Vehicles (Community Authorisations) Regulations 1992 (SI 1992/3077). Regulation 881/92 amended earlier legislation (Council Regulation 3164/76/EEC as amended by Council Regulations 1841/88/EEC on Access to the Market in the International Carriage of Goods by Road) by effectively introducing qualitative criteria in place of the previous system of quantitative restriction. The qualitative criteria are as specified in Council Regulation 561/74/EEC as amended by Council Regulation 438/89/EEC, namely a requirement that the road haulage operator be of good repute, of adequate financial standing and professionally competent in road haulage operations. (This latter subject is discussed in detail in *A Study Manual of Professional Competence in Road Transport Management* by David Lowe and published by Kogan Page, London, 11th edition, 2004.)

15.6.1.1 International carriage

For the purposes of this regulation, 'international carriage' is defined as:

(a) where a goods vehicle departs from one EU Member State and arrives in another, different, Member State (whether transiting other Member States or non-member countries *en route*);
(b) where a vehicle departs from a non-member country and arrives in a Member State or vice versa (with or without transit through one or more Member States or non-member countries *en route*);
(c) where a vehicle departs from one non-member country and arrives in a another, different, non-member country but travels via one or more Member States *en route*.

This definition also includes all *unladen* journeys undertaken in conjunction with the carriage of goods on the defined journeys.

15.6.2 UK issue of community authorizations

Community authorizations are issued in the UK by Traffic Commissioners (TC) located at Traffic Area Offices (TAOs) to all standard international operator ('O') licence holders. It should be noted that such licences are granted only to those applicants who fully satisfy the qualitative standards of the EU; in other words, the legal requirements for good repute, adequate financial standing, and professional competence in road haulage operations. The authorization comprises an original document to be retained safely at the licence holder's main place of business, and a number of certified true copies equalling the total number of vehicles authorized on the operator's licence. One of the certified true copies of the authorization must be carried in each vehicle undertaking international journeys within the EU. Community authorizations and the certified true copies carried on vehicles must be produced for inspection on request. Failure to do so, and to carry the certified true copy on a vehicle while on an international journey within the EU, is an offence.

15.6.3 Penalties for infringement of the law

UK-based international hauliers who jeopardize their 'O' licences by failing to meet the requirements of good repute, financial standing, or professional competence also jeopardize their community authorization.

In other words, where circumstances arise which, as a result of infringement of the law or failure to meet the qualitative requirements of good repute, financial standing, and professional competence, require the TC to suspend, curtail, or revoke an 'O' licence, the community authorization will also be automatically suspended, curtailed, or withdrawn (i.e. revoked). The precise action taken by the TC will depend on the seriousness of the offence or offences. Serious or repeated minor infringement of carriage regulations (i.e. the community authorization regulation itself) will result in temporary or partial suspension of the certified true copies of the authorization.

In Member States where 'O' licences of the type issued in the UK or its equivalent are not used, failure by international hauliers based in those states to meet (or maintain) the standards of good repute, financial standing, and professional competence required under the EU regulation will, nevertheless, result in jeopardy of the community authorization. It is a specific requirement of the regulation that where one Member State becomes aware of infringement of community authorization legislation by a haulier from another Member State, it shall inform the authorities in that Member State and may ask that state to impose sanctions on the haulier in accordance with the regulations (i.e. for temporary or partial suspension of certified copies or withdrawal of the community authorization).

15.6.4 Validity and duration of authorizations

Community authorizations are made out in the original licence holder's name and are not transferable to any third party, and remain valid while the 'O' licence is in force unless otherwise revoked. Certified copies as mentioned above must be carried on the relevant vehicle when on an international journey and must be produced by the driver for examination whenever he is required to do so by an authorized inspecting officer.

On the expiry of a community authorization after 5 years, it is a requirement that the issuing authority (in the UK, the TC on behalf of the Department for Transport) must verify whether the operator still satisfies the legal conditions for its issue. Since these conditions are identical to those on which renewal of the haulier's 'O' licence depends, namely good repute, financial standing, and professional competence, in the UK at least, operators whose 'O' licences are renewed can rest assured that their community authorization will be automatically renewed at the same time.

15.6.4.1 Community authorization documents

Annex I to EU Regulation 881/92/EEC illustrates a model for the community authorization, the front page of which contains details of the haulier (name and full address), the date from which it is valid and the name of the authority by whom it is issued and the date of issue. On the rear are printed the general provisions for the use of such authorizations, in particular that while within the territory of any Member State the holder (i.e. both the road haulage operator and the vehicle driver) must comply with the 'laws, regulations, and administrative provisions in force in that state', especially in regard to transport and traffic.

15.6.5 Exemptions from community authorization procedure

Certain transport operations are specifically exempt from the requirement for community authorizations in accordance with Annex II to the EU regulations as follows:

1. Carriage of mail as a public service.
2. Carriage of vehicles which have suffered damage or breakdown.
3. Carriage of goods in vehicles with a permissible laden weight (including that of any trailer drawn) which does not exceed 6 tonnes or the maximum permitted payload of which does not exceed 3.5 tonnes.

4. Carriage of goods in vehicles owned (including hired) by an own-account firm solely for its own purposes and where the transport is no more than ancillary to its overall activities and where the vehicle is driven only by an employee of the firm.
   Note: The goods concerned may be the property of the firm, or have been sold, bought, let out on hire or hired, produced, extracted, processed, or repaired by the firm.
5. Carriage of medicinal products, appliances, equipment, and other articles required for medicinal care in emergency relief, in particular for natural disasters.

15.7 Road haulage cabotage

The link between CTOs and road haulage cabotage may seem to be tenuous, and thus inclusion of an explanation of the subject here may seem superfluous for the intermodal CTO. However, it may be interesting to the reader to note, for background information only, that EU legislation in the form of Council Directive 92/106/EEC (of 7 December 1992) freed from cabotage the initial and final road legs of any CT journey within the community so they could be carried out by a haulier from any EU Member State, providing that haulier satisfies the requirements of community authorization (see above). Previously the activity of a 'foreign' haulier collecting container or swap body loads for delivery to a rail terminal within a Member State other than the one in which his business is established may well have been regarded as cabotage and therefore illegal. Since cabotage is such an important element of international transport operation, and liable to lead the road haulier and CTO into legal complexity if the ground rules are not properly and fully adhered to, the subject is dealt with in some detail here.

15.7.1 What is cabotage?

Cabotage operation is provided for under the Treaty of Rome in which Article 75 of the Treaty states that:

> for the purposes of implementing [a common transport policy] … the Council shall … lay down … the conditions under which non-resident carriers may operate transport services within a Member State.

A UK Government transport minister was reported as saying that:

> the liberalisation of cabotage is essential to the creation of a true single market in road haulage', and furthermore, 'it would help to reduce the wasteful costs associated with empty running, bringing both economic and environmental benefits, and would open up exciting new markets for hauliers.

Cabotage is quite simply internal haulage by foreign transport operators; namely, the collection and delivery of goods by road within a country by a road haulier whose business is established in another country. The significance of cabotage, of course, is that it protects internal haulage markets against incursion – or in this case the abstraction of domestic traffics – by outsiders (see below). Hence the reason why, hitherto, it has always been an illegal practice, but now with the liberalization policies of the SEM in force, such restrictive practices have been swept away and road freight cabotage within EU Member States is permitted by regulation. Cabotage by EU own-account road transport operators is permitted, but only on the same basis as defined above (i.e. item 4 under exemptions from community authorization requirements).

15.7.2 Distortion of domestic haulage markets

Provision is included in the EU regulation for safeguard measures to be implemented – on the authority of the Commission of the EU – where cabotage operations cause or lead to serious disturbance of the national transport market in a given geographical area. In practice, the operation of road haulage cabotage in the early days of the liberalization process (up to late 1995) appeared to have had negligible impact on domestic haulage markets – no more than 0.25 per cent, according to EU Transport Commission – and most of that probably within Germany by Dutch and Belgian hauliers.

15.7.3 The permanent cabotage regime

Council Regulation 3118/93/EEC (of 25 October 1993) – as subsequently amended to take account of newly joining Member States – initially limited access to cabotage within the EU on the basis of a quota system only with specified annually increasing numbers of cabotage permits being available for Member States. These permits could be obtained by road hauliers who wished to operate internally within other Member States so long as they held valid a community authorization permitting international road haulage operations. The total allocation of cabotage permits throughout the EU (acknowledged by the issuing authorities to be more than adequate to meet demand at that time) increased by some 30 per cent or more annually until the quota system was legally abolished and replaced by the so-called 'permanent cabotage regime' (i.e. total liberalization of the EU road haulage market) from 1 July 1998.

Since the implementation of the permanent cabotage regime from 1 July 1998, all international road hauliers holding community authorizations have been entitled to operate 'temporary' (see below for definition) road haulage services in Member States other than their own without any restriction as to quantitative limits or any requirement for a registered office, or any other establishment in that state. Since this date there has been no requirement for transport operators to obtain cabotage authorization or for goods vehicles to carry cabotage permits. They must still, however, comply with the laws, regulations, and administrative provisions in force in the 'host' country including those concerning the following:

1. Rates and conditions incorporated in haulage contracts.
2. Weights and dimensions of road vehicles – which may, in fact exceed those of the home country, but must not in any case exceed the vehicle's design standards.
3. Dangerous goods, perishable foodstuffs, and live animals.
4. Goods vehicle drivers' hours and rest periods.
5. Value-added tax (VAT) on transport services (see also below).

The word 'temporary' used in the context above means that cabotage permit holders may enter a Member State (temporarily) and carry out internal road haulage journeys as required. It does not mean that they have the right to establish a permanent haulage operation in that country. If a haulier is established on a permanent basis within a Member State, or wishes to be so, so that he can operate domestic haulage, then he must conform with the relevant national legislation of that state relating to internal haulage.

15.7.3.1 VAT on cabotage operations

Internal transport operations under cabotage authorization require operators to comply with national VAT regulations. For this purpose, operators may need to register in the Member States in which they are operating or appoint a suitable VAT agent or fiscal representative to handle these matters on their behalf.

15.7.4 Prohibited operations, offences, and penalties

Cabotage by hauliers and own-account operators in non-EU states is prohibited. Non-resident hauliers who either infringe the cabotage rules when operating in a state other than that in which their business is established, or in another EU Member State, or otherwise offend against community or national transport legislation while in such states, may be penalized by the host nation, on a non-discriminatory basis. Penalties may comprise an official warning or, in the case of more serious or repeated infringements, a temporary ban on cabotage. Where falsified cabotage documents are found these will be confiscated immediately and returned to the appropriate authority in the haulier's own country.

15.7.5 Issue of cabotage authorizations

In the UK, cabotage authorization permits are issued by the International Road Freight Office (IRFO) and in other Member States by the relevant transport authority, for journeys to countries where cabotage prohibitions are still relevant; for example, Belarus, Bulgaria, Croatia, Czech Republic, Estonia, Hungary, Kazakhstan, Latvia, Lithuania, Macedonia, Morocco, Poland, Romania, Russia, Slovakia, Slovenia, Turkey, Ukraine, and Uzbekistan – the countries listed here that have become new Member States of the EU since May 2004 will eventually concede permission for cabotage operations along with the rest of the community. Cabotage authorizations are issued in the name of the applying haulier for vehicles authorized in his name (i.e. in the UK, for those specified on his 'O' licence). They are valid for one vehicle for 2 months only, but that vehicle may undertake any number of cabotage movements within the 2-month period. Authorization permits must accompany the vehicle at all times – in the case of articulated vehicles it relates to the tractive unit.

15.8 Bilateral road haulage permits

Certain road haulage operations from the UK and other EU Member States to non-Member States require, in most cases, the authorization of a bilateral road haulage permit. At the present time road haulage journeys to or through Belarus, Estonia, Georgia, Morocco, Russia, Turkey, Tunisia, and Ukraine require such permits for specified transport operations. A 'third country permit' may be required for journeys that necessitate transit of a third (i.e. intermediate) country. Road haulage permits are not required for transport operations within the EU; in the case of Austria, which is an EU Member State, a special Eco-points system formerly applied as described below, but this has now been abolished and hauliers are free to enter the country. However, road hauliers on transit journeys across EU territory to a bilateral destination country must be in possession of a community authorization.

15.8.1 Validity of permits

Where bilateral road haulage permits are required as described above, such permits are available covering single journeys only, allowing just *one* return journey to be undertaken between the dates shown on the permit. Outside these dates the permit is invalid and it would be illegal to commence or continue the journey. In the case of Turkey, single-journey permits as described above are available as well as multiple-journey permits authorizing four journeys.

15.8.2 Issue of permits

Road haulage permits where necessary are issued by the relevant authority in each Member State, in the UK this is the IRFO based at the Eastern TAO in Cambridge. This involves completion of application forms, advance payment of the relevant fee, and submission by the applicant of a copy of his authority to operate (e.g. his community authorization). Used and expired permits must be returned to the issuing authority not later than 15 days after the relevant journey has been completed or the permit expiry date, whichever is earlier. Journey record sheets issued with period permits (e.g. the four-journey permit for Turkey) must be returned within the same time scale.

15.8.3 Lost or stolen permits

Road haulage permits are valuable transit documents and as such should be treated with care and appropriate security. They are not transferable to another operator and such misuse is illegal throughout the

Community, with harsh penalties imposed on offenders (see also below). Replacement of lost or stolen permits is not normally automatic, and in any case a full written explanation of the circumstances surrounding the loss or theft is required, together with a copy of the police report.

15.8.4 Journeys to or through non-agreement countries

If vehicles are to travel to or through a country with which an EU Member State has no bilateral agreement, permission to operate in that country must first be sought direct from its transport authority. Application should be made well before the journey is due to commence and full details of the vehicle, the load, and the route should be given.

15.8.5 ECMT permits for non-EU journeys

A number of ECMT permits are allocated to the UK each for haulage journeys between ECMT member countries (i.e. all EU Member States plus Albania, Armenia, Azerbaijan, Belarus, Bulgaria, Bosnia Herzegovina, Croatia, Czech Republic, Estonia, Georgia, Hungary, Latvia, Liechtenstein, Lithuania, Macedonia, Moldova, Norway, Poland, Romania, Russia, Slovakia, Slovenia, Switzerland, Turkey, and Ukraine). However, the validity of some permits is limited in certain countries, particularly Austria. These ECMT permits allow journeys between member countries, including laden or empty transit journeys, and third country journeys to other ECMT countries, which are prohibited by certain bilateral agreements. However, they cannot be used for transit of ECMT countries on journeys to non-ECMT states or for cabotage. They are for hire or reward journeys only and may not be used by unaccompanied trailers or semi-trailers. They are valid for one calendar year and allow an unlimited number of journeys within that period but they may be used with only one vehicle at a time. The quota for their issue is limited, so these permits are allocated before the beginning of the year in which they are issued. Usually no further supplies are available during the course of the year, but should the quota be increased an announcement is made in the trade press.

15.9 Eco-points for transit of Austria

Transit permits that were previously required for authorizing journeys through Austria were abolished and replaced by a system of Eco-points (see page 207). This scheme was intended to reduce the effects of air pollution created by exhaust emissions from heavy lorries in transit through the country, hence 'Eco-' (i.e. ecology) points and to benefit operators who used 'less polluting' vehicles. The number of Eco-points available (to the UK and other EU Member States) was controlled annually, thereby limiting the total number of transit journeys permitted through Austria unless progressively greater use was made of ecologically friendly vehicles. The broad principle of the system was that the greater the potential exhaust emission, the greater the number of Eco-point stamps the haulier needed to submit to fulfil his journey. Conversely, the lower the potential exhaust emission, the fewer the number of stamps required. Verification of vehicle exhaust emissions was by means of a Conformity of Production (COP) document issued to vehicle operators, which had to be produced at the border on entry to the country. Eco-point stamps were *not* required for journeys terminating in Austria, neither are they required for operations carried out under an ECMT permit.

15.9.1 The Eco-points system

The Eco-points system comprised Eco-point stamps and Eco-point cards (plus the issue of the COP document for relevant vehicles as described below). To undertake international road haulage journeys which involved a transit crossing of Austria, operators needed a supply of Eco-point stamps and an Eco-point cards on which to stick the stamps for each leg of the journey (i.e. one each for the outward and homeward bound transit of Austria).

Eco-point cards were available on application to national transport authorities (i.e. in the UK, the IRFO), usually free with the issue of Eco-points stamps (see below), or they could be purchased from the Austrian authorities on reaching the border. The card comprised three pages which had to be completed by the haulage operator or the driver prior to entering Austria:

1. Page one had space for affixing the Eco-points stamps, which had to be cancelled by the driver signing across their face before crossing into the country. This page was detached and retained by the Austrian authorities.
2. Page two (with carbon copies) required details to be completed of the vehicle, load and journey (including, where possible, the postcode of both loading and unloading locations – but an offence was not committed if this information was omitted). This page was stamped by the authorities at the border, confirming the number of Eco-points stamps used, and a copy given to the driver to be carried for the rest of the journey as proof that Eco-points stamps had been paid.
3. Page three listed the appropriate codes for the Austrian border controls and international distinguishing signs to be used when completing page two of the document.

Eco-points stamps (each worth one Eco-point) were issued solely by national transport authorities in connection with international road haulage journeys involving transit of Austria. For this purpose, the Austrian authorities 'charged' vehicles with Eco-points in accordance with the following rules:

1. Vehicles first registered prior to 1 October 1990 and those not carrying a COP document (see below): 16 Eco-points (i.e. 16 Eco-points stamps).
2. Vehicles carrying a COP document: the number of Eco-points equal to the rounded (up or down) COP value shown on the COP document.

15.9.2 COP documents

COP documents were issued by the relevant transport authority (in the UK, the IRFO) on application by road hauliers for vehicles first registered from 1 October 1990 whose engines had a lower NO_x emission than older vehicles. Operators were required to supply the following information in respect of each vehicle (i.e. those to be used for journeys involving transit through Austria):

1. Vehicle registration number.
2. The date of first registration.
3. The type approval number.
4. The chassis number.

The COP document, which was individual to a vehicle (and was non-transferable), showed the NO_x emission value and the COP value (i.e. the NO_x emission value plus 10 per cent) for the vehicle and indicated the corresponding number of Eco-points stamps that would be needed for each single-leg journey by that vehicle. The document had to be carried on the vehicle to verify its so-called 'greener' performance. When shown at the Austrian border, the authorities would charge fewer Eco-points stamps to permit the transit journey through the country.

15.9.3 New Austrian 'Maut' system

The former Austrian Eco-points system described on page 206 and above, was completely abolished in 2004 and replaced by a 'Maut' lorry tolling system similar in concept to the German LKV Maut described on page 77; the old Austrian 'Vignette' sticker system has also been replaced by the new system. The system requires relevent vehicles to be fitted with an easy-to-install, DSRC-TAG electronic (microwave) transponder, colloquially called a 'G0-Box'. The system applies nationwide to trucks over

3.5 tonnes gross weight and payment of the tolls is via the internet or at point-of-sale terminals, mainly located at roadside fuel filling stations.

15.10 Permit checks

Stringent regulations exist to prevent permit fraud and vehicles travelling on international journeys without valid permits (where relevant, see above), and checks are made on vehicles to ensure that these regulations are complied with. A vehicle will be prevented from continuing its journey if it does not carry a valid permit. In the UK it is an offence to forge or alter permits, to make a false statement to obtain a permit or to allow one to be used by another person.

15.11 Own-account transport operations

Own-account transport operations within EU territory are free from all bilateral permit requirements (under the provisions of EU Regulation 881/92/EEC Annex II) provided that goods are carried solely in connection with the trade or business of the vehicle user and are not carried for hire or reward, and that the following conditions are also met:

1. The goods carried must be the property of the business (of the vehicle user) or must have been sold, bought, let out or hired, produced, extracted, processed, or repaired by the business.
2. The purpose of the journey must be to carry the goods to or from the business or to move them, either within the business or outside for its own needs.
3. Motor vehicles used for the carriage of own-account goods must be driven by employees of the business.
4. The vehicles carrying the goods must be owned by the business or have been bought by it on deferred terms or hire (this does not apply where a replacement vehicle is used during a short breakdown of the vehicle normally used).
5. Road haulage must not be the major activity of the business.

In all cases, own-account vehicle operators (and their drivers) should be aware that they may be asked to provide satisfactory evidence to help the authorities to determine the ownership of the goods and that they are being carried solely for own-account purposes.

15.12 Other documents

Besides the above-mentioned documents, and the customs documents described in Chapter 16, the road haulier may need to ensure that his vehicles when on international journeys carry certain other documents, such as the following.

15.12.1 'De Suivi' document

Drivers of UK haulage and own-account vehicles entering France must be able to produce a 'de suivi' document in addition to the normal CMR or own-account consignment note. There is no official or formal layout for this additional document, it being left to operators to determine how they present the relevant information, which can be added to the existing CMR document if space permits. The relevant information required is as follows:

- Date the consignment was drawn up.
- Transport operator's name and address.

- Transport operator's VAT number.
- Date when goods were taken into charge.
- Nature, weight, and volume of goods.
- Name of the consignor.
- Address of loading point.
- Name of consignee.
- Address of unloading point.

Space must be provided on the note for entering of the following information:

- Time of vehicle arrival at loading point.
- Time of departure of loaded vehicle.
- Required time of arrival at destination.

A separate document must also be made out to show:

- Date and time of arrival at destination.
- Date and time of departure of unloaded vehicle.

On international journeys the driver must also carry a further document showing the relationship between the transport operator and the driver (i.e. whether he is a direct employee, an owner-driver or an employee of a sub-contractor).

16
Customs Procedures

Intermodal road hauliers undertaking cross-border journeys both within and outside the European Union (EU) are faced with the important need to comply with complex Customs procedures and documentation requirements. It is a certain fact that failure to comply with these procedures as necessary, or to carry the right documents, or indeed, if errors are found in the manner in which documents have been completed, will inevitably lead to delays and even vehicles being turned back from the port of exit or entry, or from a border crossing. This can result in a frustrating and costly chain of events.

In this chapter the various Customs systems are described. These are as follows: Community Transit (CT) for movements within EU territory; the Common Transit system for movements through European Free Trade Association (EFTA) countries; the transport international routier (TIR) Carnet system for international journeys which commence or terminate outside EU territory; the system of Admission–Temporary Admission (ATA) Carnets that applies to temporary imports of goods used, for example, for international exhibitions; and the Carnets de Passage system which relates to temporarily imported goods vehicles and trailers. There are also Customs procedures for short-sea shipping. It is also appropriate to mention the risks of and penalties for smuggling.

In dealing with international carriage, which effectively means the import and export of goods, the various parties (e.g. the shipper, the forwarding agent, the intermodal operator, etc.) become involved with HM Revenue and Customs whose role, on behalf of the government, is to control such movements. In particular their duties are to ensure that movements are legal and that prohibited goods are not moved; that appropriate declarations are made; that documentation is fully and correctly completed; and, where appropriate, that relevant duties are paid.

16.1 Community Transit

There have been many changes in Customs' procedures in recent years as the European Community (EC) has gone through various transitions, most notably the removal of barriers to intra-Community trade with the opening of the Single European Market (SEM) from 1 January 1993, and many former complicated document systems have been replaced with simplified paperwork systems, and the term 'export' no longer applies to shipments within the territories of the EU. The EC's keystone procedures for dealing with this trade, known as CT, and Common Transit in their current (2004) manifestation (as described in the *HM Customs and Excise Guide to Exporting and Importing*,version 5 of December 2004), are Customs procedures that allow, respectively:

- Goods, that are not in free circulation, and in certain cases, EC goods to move between two points in the EC, including two points in a single Member State, with the duties and other charges suspended. The procedure is also used to control the movement of goods to and from Andorra, San Marino, and the 'special territories' of the Community (such as the Channel Islands), (Community Transit).

- The movement of both Community and 'non-free circulation' goods to, from and through EFTA countries such as Iceland, Norway, Liechtenstein (Common Transit).

For CT purposes goods are said to have a particular Customs 'status' which is based on their liability to Customs duty and other charges (e.g. excise duty and value added tax, VAT) or claims to Common Agricultural Policy (CAP) refunds. Goods are divided into two distinct categories as follows:

1. Goods which have Community status (i.e. Community goods)
   These are goods which:
   - originate in the Community;
   - have been imported from a non-EC country and have been put into free circulation (see below) in the Community;
   - have been manufactured in the Community from materials or parts imported from a non-Community country provided the imported materials or parts are in free circulation;
   - The term 'free circulation' is used to describe imported goods on which all import formalities have been complied with and any Customs duties or equivalent charges which are payable have been paid and not repaid in whole or in part. Goods originating in the Community are also in free circulation unless a CAP export refund or other refund has been claimed on them. Consequently, as a general rule, Community goods can move within the EC without any Customs controls.
2. Goods without Community status (i.e. 'non-Community goods').

16.1.1 The tariff

All the information and advice needed to help with importing and exporting procedures is to be found in 'The Tariff'; that is, 'The Integrated Tariff of the UK', to give it its full name. This document, published in three volumes and to the same format in all EU Member States, contains all the relevant commodity codes (more than 65 000 of them); duty rates; procedures, including a box-by-box completion guide for the C88 documents; as well as references to the legal provisions.

16.1.1.1 Customs documentation

An eight-part Customs document (C88) known as the 'Single Administrative Document' (SAD) is used throughout the Community and EFTA countries for the purposes of import, export, and transit controls (see Figure 16.1).

A CT declaration comprises completed copies of pp. 1, 4, and 5 of the SAD. In certain circumstances a Community status document (T2L) may be used where required to provide evidence to Customs of Community status. A T2L can be either a copy 4 of the SAD, or a commercial document such as an invoice or a transport document or a manifest. These documents have no transit function.

16.1.2 CT movements

There are two types of CT procedure:

1. *The external CT procedure*: This is mainly used to control the movement of non-Community goods and to control the movement of Community goods which are subject to a Community measure involving their export to a third country; for example, CAP goods which are liable to an export refund. This procedure is also known as the 'T1' procedure.

A OFFICE OF DESPATCH/EXPORT

EUROPEAN COMMUNITY 1 2 3 4 5 6 7

1 DECLARATION

1

2 Consignor/Exporter No.

3 Forms

4 Loading lists

5 Items

6 Total packages

7 Reference No.

Copy for the country of despatch/export

8 Consignee No.

9 Person responsible for financial settlement No.

10 Country first destin.

11 Trading country

13 CAP

14 Declarant/Representative No.

15 Country of despatch/export

15 C disp./exp . Code a b

17 Country destin. Code a b

16 Country of origin

17 Country of destination

18 Identity and nationality of means of transport at departure

19 Ctr.

20 Delivery terms

21 Identity and nationality of active means of transport crossing the border

22 Currency and total amount invoiced

23 Exchange rate

24 Nature of transaction

25 Mode of transport at the border

26 Inland mode of transport

27 Place of loading

28 Financial and banking data

1

29 Office of exit

30 Location of goods

31 Packages and description of goods

Marks and numbers – Container No(s) – Number and kind

32 Item No.

33 Commodity Code

34 Country origin Code a b

35 Gross mass (kg)

37 PROCEDURE

38 Net mass (kg)

39 Quota

40 Summary declaration/Previous document

41 Supplementary units

44 Additional information/ Documents produced/ Certificates and authori-sations

A.I.Code

46 Statistical value

47 Calculation of taxes

Type	Tax base	Rate	Amount	MP
		Total:		

48 Deferred payment

49 Identification of warehouse

B ACCOUNTING DETAILS

50 Principal No. Signature:

C OFFICE OF DEPARTURE

51 Intended offices of transit (and country)

Represented by:

Place and date:

52 Guarantee not valid for

Code

53 Office of destination (and country)

D CONTROL BY OFFICE OF DEPARTURE

Stamp:

Result:

Seals affixed:Number:

Identity:

Time limit (date):

Signature:

54 Place and date:

Signature and name of declarant/representative:

C88 (1–8)

PT(July 2003)

Fig. 16.1 Form: EU-style SAD (C88): eight-part set of documents used in intra-Community trade.

2. *The internal CT procedure*: This procedure controls the movement of free circulation goods to or from the special territories, to or via the EFTA countries and between the EFTA countries. This procedure is also known as the 'T2' procedure (or the 'T2F' procedure for goods moving to or from the special territories).

16.1.3 Control procedures

All transit movements are the responsibility of a 'principal' who is the person or company who undertakes to ensure that the goods are delivered to the office of destination within the prescribed time limit. The principal must put up a guarantee to secure the relevant duties and other charges in case the goods do not arrive intact at the office of destination. The guarantee can be cash (lodged at the office of departure), vouchers (purchased from a guaranteeing association) or a guarantee undertaking (provided by a guarantor).

To start a CT movement the completed declaration must be presented to Customs at an 'office of departure' where it will be authenticated and a time limit for completing the movement will be set. The goods may also be sealed and an itinerary set. Copies 4 and 5 of the CT declaration (i.e. the SAD) must travel together with the goods until they reach the EU Member State or EFTA country where the movement is to end. Here the declaration and goods must be presented to the appropriate Customs at the 'office of destination'. This office will then stamp the copy 5 of the SAD and return it to the office of departure so that the movement can be discharged. If the copy 5 is not received within 2 months the principal is to be contacted and asked to provide proof that the procedure has ended correctly. If proof is not provided within 4 months, the enquiry procedure is initiated to recover the potential debt.

There is a range of transit simplifications available for use by compliant and reliable traders. These include the use of comprehensive and reduced guarantees or waivers, becoming Authorized Consignors and Consignees, using trader seals on vehicles, special loading lists, having an exemption from prescribed itineraries, goods moved by pipeline, use of other simplified procedures, and being able to use commercial documentation instead of the SAD in the case of air, rail, and sea environments. All of these simplifications are subject to authorization by the Customs authorities.

16.1.4 Computerization of CT

At the time of writing this text in early 2005, the Community and Common Transit systems are undergoing a planned computerization programme. The aim of this is to replace the paper CT declaration with electronic messages. This will enable Customs offices throughout the Community and Common Transit countries to exchange information electronically about goods despatched and goods received, thus speeding and tightening the whole process of control. The new system will be able to carry out certain validation and verification checks, thus reducing the workload for Customs staff. Additionally, authorized CT operators will be able to access the system directly enabling them to reduce administrative and other costs. All UK Customs transit offices are now connected to what is known as the New Computerized Transit System (i.e. the NCTS) and traders can submit electronic transit declarations to Customs and take advantage of the facilities offered by NCTS. In fact, all countries that joined the EU with effect from 1 May 2004 are now required to make electronic transit declarations.

16.1.4.1 NCTS

The NCTS is a Euro-wide system, based on the concept of electronic declaration and processing, and is designed to replace, in due course, the existing, paper-based CT system, and to provide better management and control of CT. It involves all EU Member States and the EFTA countries, 29 countries in all. Each country develops its own NCTS processing system, according to centrally defined architecture, and these systems are connected, through a central domain in Brussels, to all of the other countries, some 3000 European Customs offices.

The UK, in common with many other participating countries, is using the Minimal Common Core (MCC) software developed by the European Commission (EC), which provides all of the basic data capture and messaging functionality for effective connection to the European network. It is important to note that UK NCTS is a completely separate system from the Customs Handling of Import and Export Freight (CHIEF) (see below) and is not interconnected to it in any way, and separate declarations for transit will continue to be required.

16.1.4.2 Customs Handling of Import and Export Freight

CHIEF is one of the largest and most advanced Customs declaration processing systems in the world, providing a sound technological platform for Revenue and Customs and international trade. Its sophisticated computer software controls and records the UK's international trade movements, whether by land, sea, or air and links with several thousand businesses. It uses the latest technology and is designed to cope with continuous growth in international trade. While it enhances Revenue and Customs control of imports and exports it is also claimed to improve the service provided to the international trade Community and reduce overall administrative costs.

Readers may wish to note that further details about the extensive CT procedures can be found in *The Transit Manual* on the EC web site, at: www.europa.eu.int/comm/taxation-customs/publications/info-docleustoms/tansitmanual-en

16.2 Transport International Routier (TIR)

The TIR Convention system applies to road journeys to all 52 countries (i.e. except those that are Member States of the EU) that are party to the Convention but only where the haulier elects to conform to the Convention's requirements. It is an international Convention that provides for goods in Customs-sealed vehicles or containers to move across one or more international boarders to their final destination with minimum Customs interference (i.e. with any Customs duties and other taxes under suspension and without the need for unloading/reloading at international frontiers) provided a TIR Carnet has been issued in respect of the journey. The movement must essentially be by road to, via and from European countries and some North African and Asian countries. However, it cannot be used to move goods between EU Member States unless they transit a third country. UK vehicles may operate on international haulage journeys outside the EU without the protection of TIR but in this case they will be subject to the full weight of Customs formality and bureaucracy (and delay) at each border crossing and on arrival at destination. Anyone who has travelled on European roads will recognize the thousands of heavy lorries and trailers using the TIR system by their familiar blue and white TIR plates.

There are five main principles to the TIR system:

- Access to the system is controlled by the national guarantee associations and Customs authorities. In the UK operators must apply for authorization to use TIR Carnets to one of the two national guarantee associations, either the Freight Transport Association (FTA) or the Road Haulage Association (RHA).
- The goods must be listed on, and accompanied by, an internationally recognized document, the TIR Carnet. The Carnet is taken into use in the country of departure and serves as the control document in the countries of departure, transit, and destination.
- The duties and taxes at risk are covered by an internationally valid guarantee.
- The goods must travel in approved secure vehicles and containers.
- Customs control measures taken in the country of departure should be accepted by the countries of transit and destination.

16.2.1 TIR Carnets

The TIR Carnet (see below) is an internationally recognized Customs document intended purely to simplify Customs procedures; it is not a substitute for other documents, nor is it mandatory for any operator to use it; it does not give any operator the right to run vehicles in any European country. Use of a Carnet frees the operator from the need to place a deposit of duty in respect of the load he is carrying in each country through which the vehicle is to pass. The issuing authorities for the Carnets (in this country the FTA and the RHA) act as guarantors on behalf of the International Road Transport Union (IRU: the international guarantor), and for this reason Carnets are only issued to bona-fide members of these associations. Goods may only be carried under a TIR Carnet provided the vehicle in which they are carried has been specifically approved by the relevant authority (i.e. in the UK the Vehicle Operator and Services Agency, VOSA). This means complying with constructional requirements so that the load-carrying space can be sealed by Customs, after which it must not be possible for any goods to be removed from or added to the load without the seals being broken, and there must be no concealed spaces where goods may be hidden.

Carnets are in pairs and have counterfoils in a bound cover. Each Carnet is in four parts and contains 6, 14, or 20 pages/sections (i.e. volets in French). A 6-page Carnet is valid only for a journey between the UK and one other country. Journeys to more than one other country require 14- or 20-page Carnets valid for 2 months and 3 months, respectively. A Carnet covers only one load and if a return load is to be collected, a separate Carnet is needed (each individual voucher covers one frontier crossing) and the driver should take this with him on the outward journey. At each Customs point *en route* a voucher is detached and the counterfoil is stamped. Careful attention must be paid to the completion of the Carnet if delays and difficulties are to be avoided during the journey. Carnets are valid for limited periods only (see above) and if not used they must be returned to the issuing authority for cancellation. Those which are used and which bear all the official stampings acquired *en route* must also be returned within 10 days of the vehicle's return. Drivers should never leave the Carnet with any Customs authority without first obtaining a signed, stamped, and dated declaration quoting the Carnet number and certifying that the goods on the vehicle conform with the details contained in the Carnet. Similarly, they should also ensure that the Customs officials at each departure office, transit office, and arrival office take out a voucher from the Carnet, and stamp and sign the counterfoil accordingly. Failure to do this can lead to complications later.

The four parts of the Carnet comprise the following:

1. Details of the issuing authority, the Carnet holder, the country of departure, the country of destination, the vehicle, the weight, and the value of the goods as shown in the manifest (see point 3 below).
2. A declaration that the goods specified have been loaded for the country stated, that they will be carried to their destination with the Customs seals intact and that the Customs regulations of the countries through which the goods are to be carried will be observed.
3. A goods manifest giving precise details of the goods, the way in which they are packed (the number of parcels or cartons) and their value.
4. Vouchers which Customs officials at frontier posts will remove, stamping the counterfoil section which remains in the Carnet.

Before obtaining a Carnet the applicant must sign a form of contract with the issuing authority, agreeing to abide by all the necessary legal and administrative requirements. A financial guarantee is required to ensure that any claims that may be made against the applicant will be met.

16.2.2 Structure of the TIR vehicle

Detailed requirements are laid down concerning the structure of the body, particularly regarding the manner in which it is assembled, so that there is no possibility of panels being removed by releasing nuts and bolts, and so on. The manner in which doors and roller shutters are secured must also meet stringent specifications. Sheeted vehicles or containers may be used provided conditions relating to the construction of the sheeting are observed and as long as when the closing device has been secured it becomes impossible to gain access to the load without leaving obvious traces.

In the UK, vehicle are examined at Goods Vehicle Test Stations to ensure that they meet the technical requirements for operation under the TIR Convention and issues a certificate of approval, which must be renewed every 2 years and *must* be carried on the vehicle when it is operating under a TIR Carnet. This point is particularly important as Customs authorities carry out checks on vehicles leaving the UK to ensure that this certificate is being carried where necessary.

16.2.2.1 TIR plates

When a vehicle has been approved it must display at the front and the rear a plate showing the letters 'TIR' in white on a blue background. They should be removed or covered when the vehicle is no longer operating under TIR.

16.2.2.2 Seal breakage

If a Customs seal on a TIR vehicle is broken during transit as a result of an accident or for any other reason, Customs or the police must be contacted immediately to endorse the Carnet to this effect.

16.2.3 Non-TIR journeys

Goods may be sent abroad in vehicles without TIR cover and no Carnet is required. In this case, however, it is necessary to comply with the individual Customs' requirements of each country through which the vehicle passes. A guarantee in lieu of import duty, or a deposit against such duty, will have to be paid before the vehicle is allowed to enter the country to which it is travelling, or any country through which it needs to pass, and the vehicle will be subject to stringent Customs' scrutiny not only at the port of exit from the UK and the port of entry to the Continent, but at all further frontier crossings during the journey.

16.3 ATA Carnets

Goods which are being imported only temporarily into certain countries which are party to the scheme, such as samples, professional equipment and items for display at exhibitions and fairs, and which, eventually, will be returned to their country of origin (e.g. the UK) can be moved under an international Customs clearance document known as an ATA Carnet. These documents, valid for 12 months from the date of issue, are issued by Chambers of Commerce to members without the need for payments of, or deposits against, duty although the Carnet fees must be paid. It should be noted, however, that holding an ATA Carnet does not relieve the operator from observing Customs requirements in each individual country.

16.4 Carnets de Passage

Most countries to which vehicles are likely to travel permit the temporary importation of foreign vehicles and containers (not to be confused with the loads they carry) free of duty or deposit and without guaranteed Customs documents. However, a Customs document known as a *Carnet de Passage en Douane* is required for the following:

- Vehicles and trailers entering Iran, Iraq, Jordan, Kuwait, Saudi Arabia, Syria, Turkey, and other Middle Eastern countries.

- Vehicles remaining in Italy for more than 3 months.
- Vehicles remaining in Greece for more than 10 days.
- Vehicles remaining in Pakistan for up to 3 months.

A triptyque (i.e. a Customs permit) is required for vehicles carrying spare parts into Portugal.

Where a *Carnets de Passage* is needed for travel, in the UK it can be obtained from the Automobile Association (AA), the Royal Automobile Club (RAC), and the Royal Scottish Automobile Club (RSAC).

17
International Carriage of Dangerous Goods

With such a wide variety of goods being shipped for transportation these days and so many of them falling within scope of what are being increasingly defined as 'dangerous' products, it is essential for the intermodal operator to be aware of the risks that such carriage portends and the extent and complexities of the legislation that controls it, bearing in mind that ISO containers or intermodal swap bodies loaded with such goods may come into his jurisdiction – even unwittingly. It is, of course, widely recognized that transporting dangerous goods poses a variety of both safety and environmental risks, irrespective of the mode used; but unfortunately, transport by road exacerbates these risks quite considerably due, primarily, to the close proximity of other road vehicles, the general public and the built environment, especially in congested urban areas. Hence the need for a strict legislative regime to protect against such risks and to ensure compliance with relevant safety standards, emergency procedures, and training requirements for personnel engaged in handling, packing, and transporting these substances. The international legislative regime is in the form of international and European Agreements and European Union (EU) directives that are structured to cover most eventualities and to protect all parties. It is presented in a form common to and recognized by most countries, in addition to which many national governments have their own domestic legislative provisions, as with the UK, for example.

17.1 Dangerous goods legislation

The legislative framework is founded on the work of the United Nations (UN), which since 1956, through its Economic and Social Council Committee of Experts, has published a set of 'Recommendations' on the transport of dangerous goods – commonly referred to as the 'Orange Book'. These Recommendations take the form of a set of model regulations that apply throughout most of the world to road, rail, sea, air, and inland waterway transport and provide the basis for National Governments and International Organizations to regulate such transport in a commonly recognized manner. The key point about the 'Recommendations' is the establishment of the Dangerous Goods List (DGL). This assigns to most known dangerous substances a UN serial number and a suitable classification according to the risks it presents.

The international carriage of dangerous goods by road within Europe is governed by the 'European Agreement concerning the International Carriage of Dangerous Goods by Road' (Accord Dangereux Routier, 'the ADR Agreement'), which was made under the auspices of the United Nations and is consistent with the UN Recommendations mentioned above. Most European countries are contracting parties to the ADR Agreement, which is regularly updated, and their respective domestic legislation on dangerous goods carriage is based on it. Similarly, in regard to rail freighting, the Regulations Concerning the

International Carriage of Dangerous Goods by Rail (RID) Rules apply. These form Annex 1 to the International Convention on the Movement of Goods by Rail (CIM) Uniform Rules, which are contained within the Convention for International Carriage by Rail – abbreviated to COTIF. In the case of rolling motorway traffic via the Channel Tunnel, the applicable rules are those formulated by the Channel Tunnel Intergovernmental Commission (IGC), which was established under the Channel Tunnel Treaty to supervise the construction and safe operation of the Tunnel.

17.1.1 Dangerous goods and combined transport

In combined transport operations where elements of both international road–rail carriage are involved, and depending on whether the movement involves swap-body traffic or driver-accompanied freight carried via Eurotunnel's rolling motorway system, it may be necessary to consider both the provisions of ADR and RID. While the RID Rules equate basically to those of the ADR Agreement it is important to note that Annex B to ADR states that:

> If the vehicle carrying out a transport operation subject to the provisions of ADR is conveyed over a section of the journey otherwise than by road haulage, any national or international regulations which govern the carriage of dangerous goods on that section by that mode of transport used for conveying the road vehicle shall alone be applicable to that section of the journey.

Clearly, this implies that for a dangerous-goods-carrying vehicle on a journey through the Channel Tunnel, the rail section of the journey will be subject to the RID Rules while the road-haulage element at either end will be subject to the ADR Rules, added to which are Eurotunnel's own rules must be observed.

17.1.1.1 Civil liability for damage by dangerous goods carriage

No harmonized Euro-wide legislation is yet in place to cover civil liabilities for damage caused by dangerous goods in either national or international transport. Individual Member States of the EU have their own legal practices that are applied to civil liability claims for such damage arising from incidents or accidents in national transport operations. Clearly this can lead to considerable disparity in the way claims are dealt with by one Member State and another, and in the levels of compensation awarded. In consequence, in 1989 the Inland Transport Committee of the United Nations Economic Commission concluded a Convention on Civil Liability for Damage Caused During Carriage of Dangerous Goods by Road, Rail, and Inland Navigation Vessels (CRTD). Under this Convention, transport operators who carry dangerous substances could be held liable to pay compensation for any damage or pollution resulting from spillages or vehicle accidents. Unfortunately, the Convention has not yet been signed by a sufficient number of ECE (Economic Commission for Europe) states to allow it to enter into force, thus leaving the European scene dotted with many differing liability regimes.

17.1.2 The ADR Agreement

ADR is published in two volumes, the current edition, being applicable from 1 January 2005 and referred to as the *Restructured ADR 2005* (reference: UN ECE document number: ECE/TRANS/175). It is consistent with the United Nation's *Model Regulations on the Transport of Dangerous Goods (13th revised edition, 2003)*, the *International Maritime Dangerous Goods Code*, the *International Civil Aviation's Technical Instructions for the Safe Carriage of Dangerous Goods by Air* and is fully harmonized with the RID.

Volume 1 of this restructured edition contains the text of the ADR Agreement and the Protocol of Signature, and Parts 1 to 3 of Annex A covering the general provisions and provisions concerning dangerous substances and articles. Part 1 deals with the scope of the Agreement, definitions and units of measure, training of persons involved in the carriage of dangerous goods, safety obligations of the participants in the carriage of such goods, various derogations and transitional measures, the general requirements for the carriage of Class 7 goods (i.e. radioactive material), checks, and other measures to ensure compliance with safety requirements, the rights of Contracting Parties to the Agreement to restrict or prohibit the carriage of certain dangerous goods for reasons other than safety, and it lists in an Appendix to Part 1 the 'Competent Authorities' in each of the signatory states. In the UK this is the Department for Transport (DfT).

Part 2 deals with the general provisions concerning the classification of dangerous goods and lists, the nine main classes and sub-classes. It also describes the test methods to be used for the classification of dangerous goods and cross-references, the procedures to the UN's *Manual of Tests and Criteria*. The remainder of the volume (i.e. Part 3) is taken up with the actual DGL wherein all relevant substances are classified across 20 column headings under their UN number and also by their names in alphabetical order.

Volume 2 contains continuations of Annex A and Part 3 from volume 1, Part 4 dealing with packing and tank provisions; Part 5 deals with consignment procedures; Part 6 with the construction and testing of packagings, and intermediate bulk containers (IBCs); Part 7 with carriage, loading, and handling; Part 8 with vehicle crews and equipment; and Part 9 with the construction and approval of vehicles. International road hauliers must apply the so-called consolidated 'restructured' version of ADR 2005, which has been applicable since 1 January 2005. Further information may be found on the UN/ECE web site: (www.unece.org/trans/danger/publi/adr).

17.1.2.1 Great Britain legislation

A completely new, consolidated, set of regulations relating to dangerous goods carriage in Great Britain (GB) was introduced from May 2004 which completely replaced the previous conglomeration of more than a dozen sets of individual and separately published regulations. The new regulations, *The Carriage of Dangerous Goods and Use of Transportable Pressure Equipment Regulations 2004 (SI 2004/568)*, implement in GB three European Commission (EC) Directives, namely *Commission Directives 2003/28/EC* dealing with carriage by road, Directive *2003/29/EC* dealing with carriage by rail, and Directive *1999/36/EC* dealing with transportable pressure equipment, and align UK domestic legislation directly with the ADR Agreement (concerning the transport of dangerous goods by road) and the RID Agreement (concerning the transport of dangerous goods by rail).

17.1.3 Definitions for dangerous goods

Dangerous goods are defined as being those substances and articles the carriage of which is prohibited by ADR (or RID in the case of carriage by rail), or are authorized only under the conditions prescribed in these Agreements. In general terms the following types of articles and substances would be defined as dangerous goods:

- goods named in the ADR classification list as set out in ADR Part 2 (see also *SI 2004/568*, Regulation 15);
- other goods which have one or more hazardous properties;
- explosives; or
- radioactive material.

Dangerous articles or substances falling within the scope of the legal requirements may range from:

- quite small quantities of individually packaged goods such as domestic cleaning products; to
- tanker loads of acids and corrosive substances; plus

- explosives (both substances and explosive articles – e.g. detonators, etc.); and
- radioactive substances.

17.1.3.1 Identification of dangerous goods

The purpose of, and legal necessity for, identifying dangerous goods is to determine their classification and most hazardous properties and the dangers which may arise or be created when they are loaded, transported, and unloaded. In many cases such goods have a number of hazards, not all of which may be immediately apparent. It is for the manufacturer or consignor initially to determine the classification for dangerous goods intended to be consigned for carriage by deciding whether they are:

- explosives,
- radioactive material,
- named individually in the ADR,
- not named individually in the ADR.

ADR provides the names, individually, by their proper name, in alphabetical order and with their UN designated number, for most of the commonly transported dangerous goods (e.g. ethanol, i.e. ethyl alcohol UN No. 1170). Where the goods comprise a mixture or preparation of substances these too may be listed, as may dangerous goods that have a generic name. The DGL is to be found in ADR Chapter 3.2.

17.1.3.2 Classification systems

The carriage of dangerous goods is prohibited unless their classification, packing group, and any subsidiary hazards have been determined in accordance with the established classification system set out in the UN '*Orange Book*' (see below) and in ADR Part 2. This system classifies dangerous goods into:

- classes (of which there are nine)
  - divisions
- packing groups.

The nine *classes* cover the main types of dangerous goods (e.g. explosives, gases, flammable, and oxidizing substances, etc.). Each of these *classes* is further subdivided into a number of *divisions*, which more specifically identify dangerous goods falling within each of the classes. A further degree of classification is given by assigning goods to a *packing group* according to the degree of danger they present.

Classes and divisions The following list (Figure 17.1) shows the classes and divisions for dangerous goods:

Class 1: Explosive substances and articles
  Division 1.1: Substances and articles which have a mass explosion hazard.
  Division 1.2: Substances and articles which have a projection hazard, but not a mass explosion hazard.
  Division 1.3: Substances and articles which have a fire hazard and either a minor blast hazard or a minor projection hazard, or both, but not a mass explosion hazard
  Division 1.4: Substances and articles which present no significant hazard
  Division 1.5: Very insensitive substances that have a mass explosion hazard
  Division 1.6: Extremely insensitive articles that do not have a mass explosion hazard
Class 2: Gases
  Division 2.1: Flammable gases

Classification	Packing group	Class no.	Optional lettering
Non-flammable, non-toxic gas	–	2.2	Compressed gas
Toxic gas	–	2.3	Toxic gas
Flammable gas	–	2.1	Flammable gas
Flammable liquid	I, II, or III*	3	Flammable liquid
Flammable solid	I, II, or III*	4.1	Flammable solid
Spontaneously combustible substance	I, II, or III*	4.2	Spontaneously combustible
Substance which in contact with water emits flammable gas	I, II or III*	4.3	Dangerous when wet
Oxidizing substance	I, II, or III*	5.1	Oxidizing agent
Organic peroxide	II	5.2	Organic peroxide
Toxic substance	I, II, or III*	6.1	Toxic
Infectious substance	–	6.2	Infectious substance
Corrosive substance	I, II, or III*	8	Corrosive
Miscellaneous dangerous goods	–	9	–

**Note*: Depending on its relevant properties.

Fig. 17.1 UN dangerous goods classifications, packing groups, class numbers and optional lettering for labels.

Division 2.2: Non-flammable, non-toxic gases
Division 2.3: Toxic gases
Class 3: Flammable liquids
Class 4: Flammable solids, substances liable to spontaneous combustion and substances which, on contact with water, emit flammable gases
Division 4.1: Flammable solids, self-reactive and related substances and desensitized explosives
Division 4.2: Substances liable to spontaneous combustion
Division 4.3: Substances which in contact with water emit flammable gases
Class 5: Oxidising substances and organic peroxides
Division 5.1: Oxidizing substances
Division 5.2: Organic peroxides
Class 6: Toxic and infectious substances
Division 6.1: Toxic substances
Division 6.2: Infectious substances
Class 7: Radioactive material
Class 8: Corrosive substances
Class 9: Miscellaneous dangerous substances and articles

Besides the system of classification described above, dangerous goods are assigned:

- a UN number (comprising four digits which are used in conjunction with the letters 'UN', e.g. UN 2033 which represents the substance potassium monoxide);
- a proper shipping name (according to their hazard classification and depending on their composition).

UN packing groups Dangerous substances are allocated UN packing groups as follows:

- Packing Group I: Goods presenting high danger
- Packing Group II: Goods presenting medium danger
- Packing Group III: Goods presenting low danger.

17.1.3.3 The Orange Book

The *Orange Book* (see above) sets out a table of precedence to take account of situations where dangerous goods not specifically identified in the DGL have more than one risk, or where dangerous goods have multiple risks. In such cases the most stringent packing group (i.e. Packing Group I) takes precedence over other packing groups. Figure 17.1 indicates the relevant classifications along with their packing group number (i.e. I, II, or III), their class number and any optional lettering to be shown on packages and Figure 17.2 shows the relevant danger signs for dangerous goods indicating the symbol to be used and the colouring of the lettering and background.

17.1.3.4 Safety obligations

ADR places responsibility for general safety measures on the participants in any carriage operation (i.e. the consignor, the carrier, and the consignee), particularly to take appropriate measures according to the nature and extent of foreseeable dangers, so as to avoid damage or injury and, if necessary, to minimize their effects. Where there is an immediate risk that public safety may be jeopardized, the participants must immediately notify the emergency services and make available to them the information they need to take appropriate action. ADR specifies the individual responsibilities of the three main participants as outlined below.

The consignor The consignor must ensure, when handing over dangerous goods for carriage, that consignments conform to the requirements of ADR, in particular by:

- ensuring that the goods are classified and authorized for carriage in accordance with ADR;
- giving the carrier relevant information, data, and the required transport documents;
- using only packagings, IBCs and tanks (i.e. tank-vehicles, demountable tanks, battery vehicles, MEM, portable tanks, and tank containers) approved for and suited to the carriage of the substances concerned and displaying the markings prescribed by ADR;
- complying with the requirements on the means of despatch and on forwarding restrictions;
- where he uses the services of other participants (packers, loaders, fillers, etc.), he must still ensure that the consignment meets the requirements of ADR.

The carrier The carrier must ascertain that:

- the dangerous goods to be carried are authorized for carriage in accordance with ADR;
- the 'next test date' for tank-vehicles, portable tanks, and tank-containers, etc. has not expired;
- the vehicles are not overloaded;

Description of sign	Symbol	Lettering	Background
Non flammable, non-toxic gas	Black gas cylinder	Black or white	Green
Toxic gas	Black skull and crossbones	Black	White
Flammable gas	Black flame	Black or white	Red
Flammable liquid	Black flame	Black or white	Red
Flammable solid	Black flame	Black	Vertical white/red stripes
Spontaneously combustible substance	Black flame	Black or white	White top/red bottom
Substance which in contact with water emits flammable gas	Black flame	Black or white	Blue
Oxidizing substance	Black 'O' and flame	Black (5.1)	Yellow
Organic peroxide	Black 'O' and flame	Black (5.2)	Yellow
Toxic substance	Black skull and crossbones	Black	White
Infectious substance	Black symbol	Black	White
Corrosive substance	Black symbol	White	White top/black bottom
Miscellaneous	–	–	Vertical white/black stripes at top

Fig. 17.2 Description and colours for UN dangerous goods signs.

- visually, the vehicles and loads have no obvious defects such as leakages, cracks, or missing equipment, etc.;
- the danger labels and prescribed markings have been affixed to the vehicle;
- the prescribed documentation is on board the transport unit (i.e. the vehicle);
- the equipment prescribed in the driver's written instructions is on board the vehicle.

If, in carrying out the above checks, the carrier observes any infringement of the ADR requirements, he must not allow the consignment to be shipped until the matter has been rectified. Similarly, if during the journey, an infringement is detected which could jeopardize the safety of the operation, the consignment must be stopped as soon as possible bearing in mind the requirements of traffic safety, the safe immobilization of the consignment, and of public safety.

The consignee Consignees may not defer acceptance of a consignment of dangerous goods without compelling reasons. They must also, after unloading, verify that the requirements of ADR have been complied with.

17.1.3.5 Exemptions

There are a number of exemptions to ADR as follows:

- The carriage of dangerous goods by private individuals where the goods are packaged for retail sale and are intended for their personal or domestic use, or for their leisure or sporting activities.
- The carriage of machinery or equipment not specifically listed in the DGL and which contain dangerous goods in their internal or operational equipment.
- Carriage undertaken by enterprises which is ancillary to their main activity.
- Carriage by, or under the supervision of, the emergency services, in particular by breakdown vehicles carrying vehicles which have been involved in accidents or have broken down and contain dangerous goods.
- Emergency transport intended to save human lives or protect the environment.

Limited quantities exemptions Exemptions also relate to dangerous goods packed in limited quantities where a 'limited quantity' code is given in the appropriate entry of the DGL, in which case that article or substance is exempted from the requirements of ADR, but only if:

- it meets packaging and marking requirements;
- the prescribed maximum quantities per inner packaging, and per package, are not exceeded.

Transport unit exemptions This class of exemptions is sometimes also referred to as being a 'limited quantities exemption', but it should not be confused with the limited quantities exemption described above. In this case, the exemption applies where:

- a specified maximum total quantity per transport unit, depending on its transport category is not exceeded (see Figure 17.3);
- the dangerous goods carried in the transport unit are in more than one transport category and the total quantity carried does not exceed the value calculated in accordance with ADR.

Transport category	Maximum total quantity per transport unit*
0	0
1	20
IA	50
2	333
2A	500
3	1000
4	unlimited

**Note*: For articles, the total figure represents gross mass in kilogram (for articles of Class 1, net mass in kg of the explosive substance); for solids, liquefied gases, refrigerated liquefied gases and dissolved gases, the total figure represents net mass in kg; for liquids and compressed gases, the total figure represents nominal capacity of receptacles in litres.

Fig. 17.3 ADR transport unit exemptions.

ADR also includes exemptions relating to:

- the carriage of gases,
- the carriage of liquid fuels,
- empty uncleaned packagings,
- carriage in a transport chain including maritime or air carriage,
- the use of portable tanks approved for maritime transport,
- carriage other than by road.

17.1.3.6 Carriage in bulk

ADR states that goods may not be carried in bulk in either vehicles or containers unless a special provision, explicitly authorizing this mode of carriage is indicated in the ADR DGL.

17.1.3.7 Loading and unloading

Requirements concerning the loading, unloading, and handling of dangerous goods are specified in ADR. Among the general provisions that apply are requirements that are as follows:

- On arrival at a loading or unloading site, the driver must ensure that he and his vehicle comply with the relevant legal provisions especially those concerning safety, cleanliness, and satisfactory operation of the vehicle equipment used in loading and unloading.
- Loading must not be commenced if examination of the documents and a visual inspection of the vehicle and its equipment show that the vehicle or driver do not comply with legal requirements.
- Unloading must not be commenced if the above-mentioned inspections reveal deficiencies that might affect the safety of the unloading operation.
- In the case of loading mixed packages bearing different danger labels into the same vehicle, this is permissible only in accordance with the tables shown in ADR.
- Dangerous goods loads must be properly stowed on the vehicle or in the container and adequately secured to prevent them from being displaced.
- Drivers or other crew members do not open any package containing dangerous goods.
- Vehicles and containers must be properly cleaned before reloading.
- There shall be no smoking during handling operations in or in the vicinity of vehicles and containers.
- To avoid electrostatic charges a good earth connection between the vehicle and the ground must be made.
- The additional provisions shown in the DGL should also be observed.

17.1.3.8 Supervision and parking of vehicles

Vehicles carrying dangerous goods, when parked, must be supervised or alternatively they may be parked in an isolated position unsupervised in a secure depot or factory premises. If such facilities are not available the vehicle, if properly secured, may be parked in any one of the following:

- Vehicle park supervised by an attendant who knows the nature of the load and the whereabouts of the driver.
- Public or private vehicle park where it is not likely to suffer damage from other vehicles.
- Suitable open space separated from the public highway and from dwellings, where the public does not normally pass or assemble.

17.1.4 Responsibilities of parties

Responsibility for the classification of dangerous goods rests with the following:

- The appropriate competent authority (e.g. the UN) in publishing the classification system.
- The manufacturer of the goods and the consignor (i.e. the sender) – these may be one and the same or separate entities who must:
  - identify their hazards,
  - determine the most appropriate description for them;
  - establish their correct UN identification number (i.e. a four-digit number) by reference to the ADR DGL and their packing group.

17.1.4.1 In respect of ADR

For the purposes of applying ADR requirements, which relate to the carriage of dangerous goods by road on international journeys, in the UK the competent authority is the DfT.

17.1.4.2 For ensuring legal requirements are met

Ensuring that relevant legal requirements are met (including the task of correctly identifying and classifying dangerous goods) is the responsibility of:

- the competent national authority (i.e. to enforce the law), which in the UK is the DfT;
- consignors of dangerous goods;
- those who load, unload, and transport such goods, namely:
  - employers,
  - dangerous goods safety advisers (DGSAs),
  - lorry drivers,
  - employees and other people.

17.1.4.3 Employer responsibilities

Undertakings or individuals involved in the loading, unloading, or transport of dangerous goods by road must comply with statutory requirements on the prevention of risks to the health and safety of persons, to damage of property, or the environment, including the need to appoint a qualified safety adviser. An 'undertaking' is any legal body (e.g. limited liability companies), association or group (e.g. voluntary and charitable organizations), or official body that transports, loads, or unloads dangerous goods.

Employers are responsible for ensuring that:

- a qualified DGSA is appointed, where relevant;
- dangerous goods vehicle drivers:
  - hold vocational training certificates (VTCs);
  - are trained in emergency action procedures.

17.1.4.4 Rules for drivers

Drivers must not:

- carry unauthorized passengers on vehicles transporting dangerous goods;
- open any package containing dangerous goods unless authorized to do so;
- carry matches or lighters (or anything else capable of producing a flame or sparks) on vehicles carrying dangerous goods (except where the only goods on the vehicle are infectious

substances), nor must the vehicle crew smoke during handling operations inside or in the vicinity of the vehicle.

17.1.5 Packaging and labelling

ADR specifies that:

- dangerous goods must be packed in good quality packaging, strong enough to withstand the shocks and loadings normally encountered in transit and during trans-shipment between vehicles and between vehicles and warehouses, as well as during any removal from a pallet or overpack for subsequent manual or mechanical handling;
- packaging must be constructed and closed to prevent any loss of contents when prepared for transport (i.e. caused by vibration, or by changes in temperature, humidity, or pressure);
- no dangerous residue must be allowed to adhere to the outside of packaging during carriage.

Dangerous goods packages must be marked with their designation, the UN number (preceded by the letters 'UN'), the danger sign and any relevant subsidiary hazard signs. Markings must be clear (i.e. being indelibly marked on the package or printed on a label securely fixed to the package), in English, or the official language of another EU Member State if being supplied to that state, so they can be read easily, and so they stand out from the background to enable them to be readily noticeable.

17.1.6 Vehicles, tanks, and pressure receptacles

In general terms, all containers, tanks, and dangerous goods vehicles must be suitable for the purpose, in particular for the hazardous properties of the goods and journey to be undertaken, and must be adequately maintained to comply with the constructional and approval requirements of ADR. Designers, manufacturers, importers, or suppliers of transportable pressure receptacles (i.e. gas receptacles, tank containers, and other forms of gas storage vessel) must ensure they are safe and suitable for their purpose and comply fully with ADR. Road tanker vehicles and demountable tanks (i.e. tanks contained within a strong metal frame and capable of being lifted on and off road vehicles and road wagons) used for the carriage of dangerous goods by road under ADR must be examined annually and certified by the relevant authority in their country of registration. In the UK, road tankers are examined by the Vehicle Operator and Services Agency (VOSA) at Goods Vehicle Testing Stations, while examination of demountable tanks is carried out by a number of commercial testing organizations. On satisfactory completion of the examination an ADR certificate is issued and this must be carried on the vehicle when undertaking a dangerous goods journey under ADR.

17.1.7 Vehicle marking and placarding

Containers, tanks, and vehicles used for carrying dangerous goods must display relevant information as described below:

- All panels and danger signs must be kept clean and free from obstruction.
- Signs and panels relating to dangerous goods no longer being carried must be covered or removed.
- Where an orange-coloured panel is covered, the covering material must remain effective after 15 minutes engulfment in fire.
- It is an offence to display information when the container, tank, or vehicle is not carrying dangerous goods, and to cause or permit the display of any information likely to confuse the emergency services.

- It is an offence to remove panels or signs from a container, tank, or vehicle carrying dangerous goods (except for updating the information) and to falsify information on any panel or sign.

17.1.7.1 Hazard warning panels

A reflectorized orange-coloured, black-bordered panel (plain with no letters or figures) must be displayed at the front and rear of vehicles carrying dangerous goods. Where a *single load* of dangerous goods is carried in a container, tank, or vehicle separate orange-coloured panels showing the appropriate UN number and hazard identification number (HIN) must be displayed at the front and rear of the vehicle. Where a vehicle is carrying a *multi-load* in tanks, or in bulk in separate compartments of the vehicle, or in separate containers, an orange-coloured panel showing the appropriate emergency action code (EAC) only must be displayed at the rear of the vehicle. Additionally, orange-coloured panels, as follows, are required on both sides of each tank or (if it has multiple compartments) on each compartment of the vehicle, or on each container on the vehicle:

- at least one on each side showing the appropriate UN number and HIN;
- the remainder showing only the appropriate UN number.

Alternatively, for dangerous goods carried in a tank, the panels may be displayed on both sides of the frame of each tank, or on the vehicle positioned immediately below the tank or tank compartment concerned. Where a mixed load of diesel fuel and/or gas oil or heating oil (UN 1202), petrol, motor spirit or gasoline (UN 1203), and/or kerosene (UN 1223) is carried in a multi-compartment road tanker it may be labelled as a *single load* only, showing the UN number and HIN for the most hazardous of the products carried.

The orange-coloured panels must be either of the following:

- A rigid plate fitted as near vertical as possible.
- In the case of a vehicle carrying dangerous goods in a tank container or in bulk in a container:
  - orange-coloured self-adhesive sheets;
  - orange-coloured paint (or equivalent), provided the material is weather-resistant and ensures durable marking.

UN numbers and HIN must be shown in black, but where the EAC is white on a black background, it must appear as orange on a black rectangle. Except where panels comprise self-adhesive sheets or are applied by paint, UN numbers and EACs must be indelible and remain legible after 15 minutes' engulfment in fire.

17.1.7.2 EACs

Where, under ADR requirements, orange-coloured plates bearing HIN must be displayed on dangerous goods carrying vehicles, if the carriage is within GB then the HIN must be replaced with the appropriate EAC specified in the Dangerous Goods EAC List (see Figure 17.4).

The EACs are as follows:

- By numbers 1 to 4 indicating the equipment suitable for fire fighting and for dispersing spillages (i.e. 1 = coarse spray, 2 = fine spray, 3 = foam, 4 = dry agent).
- By letters indicating the appropriate precautions to take.

Letter	Danger of violent reaction or explosion	Protective clothing and breathing apparatus	Measures to be taken
P	Yes	Chemical protective clothing	Dilute
R	No	Chemical protective clothing	Dilute
S	Yes	Breathing apparatus	Dilute
T	No	Breathing apparatus	Dilute
W	Yes	Chemical protective clothing	Contain
X	No	Chemical protective clothing	Contain
Y	Yes	Breathing apparatus	Contain
Z	No	Breathing apparatus	Contain

Notes: Dilute means that dangerous goods spillages may safely be washed into drains provided large quantities of water are provided for dilution. Contain is used, this indicates a need to prevent spillages from entering drains or watercourses. Chemical protective clothing includes breathing apparatus. The letter 'E' shown at the end of an emergency action code means that consideration should be given to evacuating people from the neighbourhood of an incident.

Fig. 17.4 EACs for action in dangerous goods incidents/accidents.

17.1.7.3 Display of telephone numbers

A contact telephone number, comprising black digits on an orange background, must be shown on vehicles carrying single or multi-loads of dangerous goods in tanks, positioned:

- at the rear of the vehicle;
- on both sides of the tank (or each tank if more than one), the frame of each tank, or the vehicle;
- located in the immediate vicinity of the orange-coloured panels.

Instead of a telephone number, the words '*consult local depot*' or '*contact local depot*' may be substituted, but only if:

- the name of the operator is clearly marked on the tank or the vehicle;
- the fire chief for every area in which the vehicle will operate has been notified in writing of the address and telephone number of that local depot, and has confirmed in writing that he is satisfied with the arrangements.

17.1.7.4 Display of hazard labels and subsidiary hazard signs

Where a vehicle is carrying:

- packaged dangerous goods in a container, any hazard label or subsidiary hazard sign required on the packages must also be displayed on at least one side of the container;
- dangerous goods in a tank container or in bulk in a container, any hazard label or subsidiary hazard sign required on the packages containing such goods must be displayed on each side of

the tank container or container, and where such signs are not visible from outside the carrying vehicle, the same signs must also be shown on each side of and at the rear of the vehicle;

- dangerous goods in a tank, other than a tank container, or in bulk in a vehicle, but not in bulk in a container on a vehicle, any hazard label or subsidiary hazard sign required on the packages containing such goods must be displayed on each side of and at the rear of the vehicle.

17.1.8 Transport information and documentation

Consignors of dangerous goods must provide the transport operator with a transport document showing:

- the designation, classification code, and UN number for the goods;
- any additional information needed to determine their transport category and their control and emergency temperatures;
- for packaged goods, the number and weight or volume of individual packages, or the total mass or volume in each transport category;
- for bulk loads, the weight or volume in each tank or container and the number of tanks/containers;
- the name and address of both consignor and consignee;
- any other information which the operator must give to the driver;
- a 'consignor's declaration' that the goods may be carried as presented, that they, their packaging and any container or tank in which they are contained is fit for carriage and is properly labelled.

It is an offence to provide false or misleading information and where one transport operator subcontracts a dangerous goods consignment to another operator, he must pass on the information provided by the consignor.

17.1.8.1 Documentation to be carried during carriage

Drivers of vehicles carrying dangerous goods must be provided with the following 'transport documentation', in writing:

- The information provided by the consignor.
- The weight or volume of the load.
- The HIN or the EAC (where appropriate).
- Emergency information comprising:
  - the dangers inherent in the goods and safety measures;
  - what to do and the treatment to be given should any person come into contact with the goods;
  - what to do in the event of fire and what fire-fighting appliances or equipment must not be used;
  - what to do in case of breakage or deterioration of packagings or of the goods, particularly where this results in a spillage onto the road;
  - what to do to avoid or minimize damage in the event of spillage of goods likely to pollute water supplies.
- Any relevant additional information about the particular type of dangerous goods being carried.

It is an offence to provide false or misleading information to drivers about the particular type of dangerous goods being carried.

Drivers must keep the transport documentation readily available during dangerous goods journeys and produce it on request by the police or a goods vehicle examiner. Where a dangerous goods carrying trailer is detached from the towing vehicle the transport documentation (or an authenticated copy) must be given to the owner/manager of the premises where it is parked, or attached to the trailer in a readily visible position. However, it is important to note that documentation relating to dangerous goods no longer on a vehicle must be either removed completely, or placed in a securely closed container clearly marked to show that it does not relate to dangerous goods still on the vehicle. Operators must keep a record of journey transport documentation for at least 3 months.

17.1.8.2 Transport emergency cards (Tremcards)

Tremcards are a widely used system (devised by the *European Council of Chemical Manufacturers' Federation – CEFIC*) by which the emergency and other relevant information that a driver must carry, as detailed above, is provided in a convenient, ready-made format (covering single and multi-substance loads – for use where combinations of substances are loaded on the same vehicle). Tremcards are available in paper format for single journey use or laminated for repeated use with the relevant text in a single language, or four or eight languages from CEFIC's range of language versions. Specifically, they:

- detail information on the substance(s) to which they relate;
- specify any protective devices needed when such substances are handled;
- provide information on the action to be taken in the event of an emergency;
- any other additional information provided by the manufacturer.

17.1.9 Emergency procedures and road traffic accidents

Vehicles carrying dangerous goods must be equipped so the driver can take emergency measures and, where toxic gases are carried, they must carry respiratory equipment to enable the crew to escape safely.

Vehicles must carry the following:

- At least one portable fire extinguisher with a minimum capacity of 2 kilograms of dry powder, suitable for fighting a fire in the engine (unless the vehicle has an automatic extinguisher system) or cab, and not likely to aggravate a fire in the load.
- At least one portable fire extinguisher with a minimum capacity of 6 kilograms of dry powder, suitable for fighting a tyre or brake fire, or a fire in the load, and not likely to aggravate a fire in the engine or cab.
- Fire extinguishers with a minimum total capacity of 12 kilograms are required.

Vehicles with a maximum permissible mass (i.e. weight) of more than 3.5 up to 7.5 tonnes must carry, in addition to the 2-kilogram extinguisher referred to above, one or more portable extinguishers with a minimum total capacity of 8 kilograms with one at least of 6 kilograms capacity. Portable fire extinguishers must not be liable to release toxic gases into the driver's cab, or under the heat of a fire. They must be marked in compliance with a recognized standard; fitted with a seal verifying they have not been used; and, where manufactured after 31 December 1996, be inscribed with the date for their next inspection – it is an offence to carry an extinguisher with an overdue inspection date.

17.1.9.1 Emergency procedures

In accident or emergency situations, drivers must comply with the emergency information given to them and apply the emergency procedures for which they are trained and which are stated on the emergency action information given to them (e.g. on the Tremcard), including:

- notifying the relevant emergency services by the quickest practical means;
- taking steps to prevent chemical contamination of the environment (e.g. water courses);

- taking steps to prevent the risk of fire;
- keeping the public away from the scene;
- applying first aid where necessary.

17.1.10 DGSAs

Any firm, organization, or individual who loads, unloads, or transports dangerous goods is required to appoint a trained and certificated DGSA under ADR. Specifically, it is illegal for:

- any firm (or organization) to load, unload, or transport dangerous goods unless they have appointed a qualified person as a DGSA;
- any individual or self-employed person involved in the transport of dangerous goods to have failed to qualify as a DGSA.

A DGSA is legally defined as a person who is self-appointed or is appointed by the head of an undertaking (i.e. the employer), and whose job is to carry out legally specified tasks and functions relating to the loading, unloading, and transport of dangerous goods. Any person whose employer requires them to take up the role of DGSA may do so provided they:

- pass the necessary examination to gain a DGSA VTC which will allow them to take up such a position;
- are prepared to carry out the duties of a DGSA in an effective manner and bear the legal responsibilities as set out in the legislation.

This applies to:

- any existing or new, full- or part-time, employee;
- any person employed as a full- or part-time consultant;
- any employer or self-employed person;
- any individual working for themselves who load, unload, and transport dangerous goods (e.g. such as freelance sales or technical representatives and consultants, or self-employed owner-driver road hauliers) who must qualify and appoint themselves.

When dangerous goods are loaded, unloaded, or transported responsibility for appointing a DGSA rests with:

- the employer in the case of legally constituted firms (i.e. limited companies) and other bodies who must appoint a qualified person, or achieve the qualification and appoint themselves;
- the self-employed person in the case of sole proprietor businesses who must be qualified and appoint themselves;
- at least one partner in partnership businesses who must be qualified and appoint themselves.

17.1.10.1 DGSA Certification

DGSA Certification is the issue of a VTC to an individual who has achieved a satisfactory pass mark in the three official qualifying examinations (namely the core, a mode subject, and a class subject) within a 12-month period.

The VTC should cover the DGSA for:

- the particular mode of transport (i.e. road, rail, or inland waterway) used for the carriage of dangerous goods in their own or their employer's business;
- the class or classes of dangerous goods loaded, unloaded, or transported within their own or their employer's business.

A VTC is valid for 5 years from the date of issue. For an appointed safety adviser to remain qualified they must, within the 12 months preceding the expiry date of their current certificate, re-take and pass the full official DGSA examination to enable their certification to be extended for a further 5 years.

17.1.10.2 The tasks and functions of a DGSA

The appointment of a DGSA is intended to protect their employer, the firm's employees, and the general public – as well as themselves if acting in an individual capacity – from any risk of danger, and property and the environment from risk of damage, resulting from the transport and related loading and unloading of dangerous goods. Statutory duties and responsibilities of a DGSA include the following:

- Monitoring compliance with the law on the transport of dangerous goods.
- Advising their employer on the transport of dangerous goods.
- Preparing an annual report to their employer on the firm or organization's activities in the transport of dangerous goods. These annual reports must be retained for 5 years and made available to the authorities for inspection on request.

The DGSA must also monitor the following practices and procedures within their firm or organization insofar as they relate to the transport of dangerous goods:

- Procedures for complying with the rules governing the identification of dangerous goods being transported.
- Practices, when purchasing means of transport, for taking account of any special requirements in connection with the dangerous goods being transported.
- Procedures established for checking the equipment used in connection with the transport, loading, or unloading of dangerous goods.
- Proper training of employees and maintaining records of such training.
- Implementation of proper emergency procedures in the event of any accident or incident that may affect safety during the transport, loading, or unloading of dangerous goods.
- Investigation of and, where appropriate, preparation of reports on serious accidents, incidents, or serious infringements recorded during the transport, loading, or unloading of dangerous goods.
- Implementation of appropriate measures to avoid the recurrence of accidents, incidents, or serious infringements.
- Taking account of the legal prescriptions and special requirements associated with the transport of dangerous goods in the choice and use of sub-contractors or third parties.
- Verification that employees involved in the transport, loading, or unloading of dangerous goods have detailed operational procedures to follow and have received appropriate instructions.
- Introduction of measures to:
  - increase awareness of the risks inherent in the transport, loading, and unloading of dangerous goods;
  - implementation of verification procedures to ensure that vehicles carry the required documents and safety equipment and that they comply with the law;
  - implementation of verification procedures to ensure that the rules governing loading and unloading are complied with.

17.1.11 Certification of dangerous goods vehicle drivers

Drivers of vehicles carrying dangerous goods must be instructed and trained in accordance with ADR, and as appropriate to the goods and type of vehicle in question, so they understand the dangers of the particular goods being carried and the emergency action to be taken, as well as their duties under current ADR requirements.

Minimum training for drivers must cover at least:

- general requirements on dangerous goods carriage;
- main types of hazard;
- environmental protection in the control of the transfer of wastes;
- preventive and safety measures appropriate to various types of hazard;
- what to do after an accident (first aid, road safety, the use of protective equipment, etc.);
- labelling and marking to indicate danger;
- what to do and not do when carrying dangerous goods;
- the purpose and operation of technical equipment on vehicles used for carrying dangerous goods;
- prohibitions on mixed loading in the same vehicle or container;
- precautions during loading and unloading of dangerous goods;
- civil liability;
- multimodal transport operations.

For drivers of vehicles carrying packaged dangerous goods, training must also cover the handling and stowage of packages, and for road tanker or tank container drivers training must cover the behaviour of such vehicles on the road, including load movement during transit.

Transport operators must ensure their drivers hold valid VTCs appropriate to the dangerous goods work they are employed in (gained by being trained as described above) and drivers must carry their VTCs on all dangerous goods journeys and produce them on request by the police or by a goods vehicle examiner.

17.2 The international carriage of dangerous goods by rail: RID

The transport of dangerous goods by rail in Europe and in certain other countries (see list above for ADR) is subject to the provisions of what are known as RID (the International Carriage of Dangerous Goods by Rail). These are contained in Annex 1 to the CIM Uniform Rules which in turn, form Appendix B to the COTIF. These RID Rules lay down minimum standards for the safe packing and transport of various types of dangerous goods travelling to or through another country, and parties to the Convention (which are the countries listed above) guarantee to permit the carriage of goods meeting these standards through their territory, even when their own domestic requirements may be different.

RID comprises three parts, broadly similar in structure and content to ADR (although historically, in fact, it is ADR which followed on from RID) as described above:

Part I General requirements, the structure of the regulations, definitions, and a list of units of measurement.

Part II Special requirements for the various classes of substances and articles which are categorized according to type of hazard into nine main classes (listed below, some of which are further subdivided, making a total of 13 in all. The numbering system for the classes follows UN Recommendations (in the Orange Book) mentioned previously.

Part III Appendices providing details of technical requirements for testing and approval of tanks, receptacles and equipment, and other matters.

17.2.1 Substance classes

The nine classes (and sub-classes) are as follows:

Class 1: Explosive substances and articles
Class 2: Gases: compressed, liquefied or dissolved under pressure

Class 3: Flammable liquids
Class 4 4.1: Flammable solids
4.2: Substances liable to spontaneous combustion
4.3: Substances which, in contact with water, emit flammable gases
Class 5 5.1: Oxidizing substances
5.2: Organic peroxides
Class 6 6.1: Toxic substances
6.2: Infectious substances
Class 7: Radioactive material
Class 8: Corrosive substances
Class 9: Miscellaneous dangerous substances and articles

With the exception of Class 7 (radioactive substances), for which a different layout is used, the structure of each class as set out in RID is similar, comprising the following:

1. List of substances and articles
2. Conditions of carriage
   - A Packages
     - (i) General conditions of packing
     - (ii) Packing of individual substances and articles
     - (iii) Mixed packing
     - (iv) Marking and danger labels on packages
   - B Method of despatch and restrictions on forwarding
   - C Particulars in the consignment note
   - D Transport equipment
     - (i) Conditions relating to wagons and loading
     - (ii) Marking and danger levels on wagons, tank wagons, tank-containers, and small containers
   - E Prohibitions on mixed loading
   - F Empty packagings
   - G Other requirements

Within each class the various requirements are divided into three-digit 'marginals' or paragraphs (the paragraph numbers being shown in the margins of each page), hence use of the term marginals when referring to the contents of RID (and ADR).

The list of substances and articles within each class comprises a certain number of items, each consisting of a group of substances which have, subject to prescribed limitations, the same physical and chemical properties. In assigning a particular substance or article to one of the classes and items the following basic properties are taken account of: vapour pressure, critical temperature, flash-point, flammability, toxicity, corrosivity, explosivity, radioactivity, stability, oxidizing properties, chemical structure, viscosity, reactivity with water, and whether organic or inorganic.

17.2.2 Packaging and labelling

RID sets out the forms of packaging acceptable for any substance or article (using the concept of packaging groups). Within the classes concerned, the same type of packaging can be used for all substances in the same packing group. Similarly, RID lays down specific provisions for the labelling and marking of packages, wagons, and containers containing dangerous goods. The labels are based on those in the Orange Book (see p. 223).

17.2.3 Consignment notes

Under the Uniform Rules of COTIF, an appropriate consignment note must accompany all goods. For dangerous goods subject to RID, the description of the goods in the consignment note must conform to one of the names in the list of substances at the beginning of the relevant class. In a non-restrictive class, where the substance is not mentioned by name, the chemical name must be entered – trade names alone are not allowed. The description of the goods must then be followed by the class number, the item number (adding the letter where given) and the initials RID. In addition, a cross must be inserted in the appropriate box in the consignment note. For certain substances, such as those which are chemically unstable, additional declarations are required. The consignor is expected to certify that appropriate measures have been taken to prevent dangerous decomposition or polymerization during carriage, and substances carried in the molten state must be declared as such.

17.2.4 Empty packagings

Since empty uncleaned packages can be as dangerous as, or even more dangerous than, full ones, they must be treated in exactly the same way as when full, with appropriate labelling, documentation, and declaration.

17.2.5 General requirements

In the list of general requirements to RID, specific mention is made of piggyback and combined road–rail transport. In the first case, RID states that vehicles handed over for carriage by piggyback (and their contents) must comply with ADR, although certain substances such as organic peroxides are not accepted for piggyback transit. Rail wagons carrying piggyback lorries with dangerous goods must display danger labels but this is not necessary in the case of rolling motorway systems such as Eurotunnel's freight shuttle.

17.3 The carriage of dangerous goods through the Channel Tunnel (IGC)

The over-riding aim of Eurotunnel's hazardous goods policy is to ensure a safe transport system and the measures adopted make it one of the world's safest transport systems. This policy has been formulated with two important groups in mind:

1. The public with its concern for safety.
2. Industry with its concern for safeguarding the commercial interest.

Safety procedures for the Tunnel were devised to conform to the IGC's rigorous requirements, and both the safety measures and the list of hazardous goods authorized for transit in the Tunnel have been approved by the IGC, which is made up of representatives from the British and French governments. These safety procedures are designed into the system to minimize the risks of incidents in the Tunnel, so enabling certain types of hazardous goods to be carried. The ADR regulations on the carriage of dangerous goods by road and the RID regulations applying to rail transport were used as the basis of Eurotunnel's policy.

A common policy is applied to the carriage of heavy goods vehicles on the freight shuttle service and to rail transit, and while access procedures for heavy goods vehicles and their drivers to the terminals and Tunnels are different from those of rail-freight trains, no difference is allowed between the types of goods which may be carried. In consequence of this policy, the carriage of hazardous goods through the Channel Tunnel must comply with the ADR (see above). Hazardous goods definitions are subject to the ADR regulations and declarations must be made using these. Eurotunnel will accept, restrict, or refuse goods only on the basis of ADR definitions, a policy that has been decreed by a bye-law in the UK and an *Arrêté Ministériel* in France, and which forms an integral part of Eurotunnel's conditions of carriage. In the case of any conflict between the ADR regulations and Eurotunnel's own regulations, the more stringent regulation shall apply.

Road hauliers are responsible for ensuring that the goods they are carrying comply with these regulations. It is also the carrier's responsibility to provide all the obligatory information on the substances being carried.

17.3.1 Quantities of dangerous goods accepted

The quantity of goods that will be accepted for transit through the Tunnel is divided into three categories:

1. Carried with no restrictions on quantity.
2. Carried subject to restrictions on quantity:
   - Product may be carried in unit packages of 250 litres or 400 kilograms: no restriction on quantity carried per unit of transport (i.e. lorry or trailer), but only in containers with a maximum capacity of 250 litres (for liquids) or 400 kilograms (for solids).
   - Maximum quantity per transport unit (i.e. lorry or trailer):
     carriage is restricted to a maximum of 250 litres or 400 kilograms per transport unit.
     Note: The number of packages and their volume is unrestricted provided that the overall 250-litre or 400-kilogram limit per transport unit is not exceeded. If several products of the same class in any one transport unit fall into this category, the sum of all their quantities is limited to 250 litres or 400 kilograms per transport unit.
   - Other maximum quantities allowed: the authorized limit is in net kilogram per transport unit unless otherwise indicated.
3. Not permitted.

The following list of substances describes those which are accepted for carriage through the Tunnel, with or without restrictions. Other goods, classified under ADR as hazardous, are *not* accepted for transit.

Class 1: Explosive substances and articles
Apart from fuse cord used in explosives, most goods from this class are excluded from the Tunnel.

Class 2: Gas
Although natural gas is not accepted, several products in this class can be accepted. These include gases such as nitrogen, oxygen, argon, helium, and neon. Toxic gases such as liquefied ammonia, or inflammable gas, such as heavy hydrogen, are not accepted.

Class 3: Inflammable liquids
Several commonly used household products are accepted including alcohol, glue, varnish, and paint. The determining factor is packaging.

Class 4.1: Inflammable solids
Household goods such as matches, camphor, and sulphur are accepted without restriction since they represent a minor risk. Other products classified hazardous by ADR but which will be carried include film. All solids with a greater risk are, in principle, excluded.

Class 4.2: Substances liable to spontaneous combustion
Goods in this class are high risk and are excluded entirely.

Class 4.3: Substances which give off inflammable gases on contact with water
Goods in this class are high risk and are excluded entirely.

Class 5.1: Oxidizing substances
Ammonium nitrate based fertilizers are accepted but packaging is a determining factor, bulk, or tanker loads being excluded.

Class 5.2: Organic peroxides
No products in this class are accepted.

Class 6.1: Toxic substances
Packaging is the key for determining whether goods in this class may be carried. Chloroform and arsenic are no risk when correctly packaged and pesticides made from dichlorodiphenyltrichloroethane (DDT) are also relatively safe when properly packaged.

Class 6.2: Repugnant substances and substances liable to cause infection
These goods represent no danger and are accepted.

Class 7: Radioactive substances
No radioactive products such as radioactive waste are accepted, although certain medical products (e.g. cardio-stimulators) with radioactive components that represent no danger are accepted.
Note: Goods in this class are not authorized for transit without prior agreement between the manufacturer and Eurotunnel.

Class 8: Corrosive substances
Highly corrosive products are excluded, but household goods such as caustic soda, ammonia, and sodium hydrochloride (bleach), being only marginally toxic, are accepted for carriage without restriction.

Class 9: Miscellaneous dangerous substances and articles
Goods such as lithium batteries are in this class and are accepted since they represent only minimal risk.

17.3.2 Action prior to and on arrival at the Channel Tunnel

All goods classified as hazardous under ADR must be declared to Eurotunnel prior to transit under the following procedure.

1. *Before arrival*: At the point of loading, check that the driver is issued with all the necessary documentation for registering his goods on arrival at the freight terminal (see item below).
2. *On arrival at the freight terminal*: The driver should follow the signs to the Freight Terminal and then follow the route identified to the Hazardous Goods Control Point.
3. *On arrival at the Hazardous Goods Control Point*: The driver should park his vehicle adjacent to the Control Point building and report to the Hazardous Goods Agent in the building.
4. *Documentation required*: The driver must give the Hazardous Goods Agent, the documents (prepared in English or French) carrying the information required by ADR and specifically for registration of goods by Eurotunnel as follows:
   - UN number.
   - ADR references (class, number, and letter) of all the hazardous goods being carried.
   - The official ADR name for the substances carried, including their volume and weight.
   - Details of packaging.
5. *Checks*: The Hazardous Goods Agent will check that the goods being carried comply with Eurotunnel's regulations as set out in its *Hazardous Goods Handbook* and will make an external examination of the vehicle. (This examination has no bearing on possible customs or police examinations that may take place before departure.)
6. *Identification*: When the documentation and vehicle have been checked, the Hazardous Goods Agent will issue the driver with a label showing a reference number for the vehicle transit and will create a file with this number as a reference:
   - These reference numbers enable Eurotunnel to follow the progress of the vehicle in the system and, should the need arise, to react swiftly according to the nature of the goods being carried.
   - The label issued to the driver, which identifies the vehicle as carrying hazardous goods throughout the journey, is placed on the windscreen in a visible position, and must not be removed until the vehicle has left the destination terminal at the end of its journey.
7. *Journey through the system*: Once the Hazardous Goods Agent has authorized the transit, the driver can return to his vehicle and follow the normal route through the terminal.

18

Safety in Transport

By their very nature, transport operations, whether by road, rail, inland waterway, sea, or where they involve a combination of modes, are inherently dangerous, involving as they do the movement of heavy vehicles, rail rolling stock, and waterway vessels, and the loading, unloading, lifting, and transfer of heavy-load units such as shipping containers and swap bodies, often in confined and congested premises. In the absence of great care and diligent observation of safety procedures, the risk of accident is ever present. Hence the reason that, as this chapter shows, extensive regimes of legislative control have been established to safeguard, so far as is reasonably possible, against every such eventuality. Furthermore, in the interest of cross-border liberalization and fair competition throughout the European Union (EU), there is concern to identify and harmonize safety and technical standards across all forms of equipment (including transport equipment) used in international trade, all of which spells yet more complex regulation and stringent standards which the transport operator must observe, or be heavily penalized should he unwittingly fail to do so. The important point to remember is that irrespective of the circumstances and irrespective of policy aims and objectives, concern for safety must remain paramount.

Within the operations of combined road–rail and the wider aspects of intermodal transport there must be no exception to the attention given to safety issues. In fact, the extensive portfolio of safety and technical standards that govern the construction, certification, and use of transport units such as road vehicles and trailers, swap bodies and containers is testimony to the concern devoted to safety issues. Most of these standards are established by various national and EU legislative measures, by the International Standards Organization (ISO), by other organizations such as Comité Européen de Normalisation (CEN) (a combination of national standards organizations) which sets the Euro-wide standard for swap bodies; and by the Union International des Chemins de Fer (UIC) which sets standards for swap bodies which are to be carried by rail.

Transport operators, along with all other sectors of business and industry, are under constant pressure to become ever more conscious of the need for safety in their operations, in the provision of facilities for safeguarding the health and safety of employees, and in the way employees work and conduct themselves in a safe manner in the work premises. The problems of safe transport operation in general and of safe loading and the avoidance of vehicle overloading in particular are not new and there is concern also on the wider front of safety in load handling and in the use of loading aids (e.g. using forklift trucks), and with regard to vehicle manoeuvring in depots and works premises where too many unfortunate deaths are caused by lack of attention to proper (i.e. safe) procedures. For example, in the UK alone it is estimated that nearly 100 people are killed every year in workplace accidents involving vehicles (including cars, heavy-goods vehicles, forklift trucks, and tractors) and according to the Health and Safety Commission in 2000/2001, 2490 major injuries were sustained at work and almost 6000 people lost work because of injury. Imagine what those numbers would amount to taken on a Euro-wide and then on a worldwide basis, and one can begin to appreciate how serious and significant a problem it is, and one that demands the utmost care and vigilance at all times. The Health and Safety Executive (HSE) reports that the four main types of workplace transport accidents are: moving vehicles hitting or running over people; people falling off vehicles; vehicles overturning; and objects falling off vehicles.

This chapter examines some of the key safety issues such as the regulations applying to road vehicles and trailers, the safe loading of vehicles, the use and marking of freight containers and international standards for swap bodies, and certain rail and waterway transport safety issues.

18.1 Safety in road freighting

As we are all aware, the toll of road accidents throughout the whole of the EU is nothing short of horrendous. More than 41 000 deaths and 1.7 million injuries are reported annually by Community Road Accident Database (CARE), which is the European centralized database on road accidents hosted by the European Commission at the Luxembourg data centre. Many actions are taken at national level among Member States to ameliorate this dreadful toll on human life and misery – accepting perhaps that it is something that we shall, unfortunately, never be rid of. In the UK for instance we have a powerful body of road traffic legislation, which determines vehicle speeds, sets many other traffic controls, and provides the basis for strict enforcement and for penalizing convicted offenders. We also have our long established Highway Code (first published as long ago as 1931) that all vehicle drivers are taught to follow from their first driving experience. Other European Member States have their own traffic rules and regulations as witness the speed limits and parking restrictions that we encounter when driving and holidaying abroad. Additionally, throughout the EU, legislative standards in one form or another are imposed to ensure the safe operation, construction, use, and testing of goods vehicles and relevant enforcement regimes are applied to ensure compliance.

Unfortunately, despite standardized rules there is a common complaint that enforcement of such standards varies significantly between Member States (those in the north of the EU, i.e. the UK, Germany, France, the Netherlands) generally being believed to apply more stringent regimes than their southern European counterparts (i.e. Spain, Greece, Italy). This leads to suggestions of unfair competition in a European road transport market that is supposed to provide a harmonized ('level playing field') scenario. The UK, for its part, has always been held to be in the forefront when it comes to goods vehicle controls, having an extensive and complex body of domestic legislation relating to safety and mechanical standards for vehicles, and applying rigorous enforcement of those controls.

18.1.1 European legislation: Goods Vehicle Type Approval

As the basis for common technical standards in goods vehicle construction, Goods Vehicle Type Approval is a compulsory scheme, originating in the EU, which requires vehicle manufacturers and importers to submit new vehicles (i.e. new designs, new models, and changes of specification for existing approved models) for official approval before they are sold for use on European roads. This is to ensure they meet prescribed road safety and environmental standards, and to determine the maximum weights (i.e. gross, train, and individual axle weights) at which they may legally operate on the road. The manufacturer or importer must demonstrate, by submitting a sample vehicle for inspection, compliance with each safety or environmental standard (i.e. individual system approval) before the vehicle model can receive overall vehicle type approval. Following approval (including approval of the manufacturing process to ensure conformity of production) any number of models to that exact specification, each carrying an individual certificate of conformity, may be manufactured/imported, sold, and used on the road. In effect this system makes it illegal for a manufacturer or importer to sell a vehicle which does not meet Type Approval Standards, and in the UK, for example, it is actually impossible to register a new vehicle without producing a valid Type Approval Certificate.

EU directives, and regulations of the United Nations Economic Commission for Europe (ECE), of which there are many going back as far as 1970, set out Type Approval standards for vehicles and components. Principal among these are Council Directives 70/156/EEC as amended by Directives 87/403/EEC and 92/53/EEC which deal with 'the approximation of the laws of Member States relating to the Type Approval of motor vehicles and their trailers'.

In the UK, the requirements of Council Directives and ECE Regulations have been incorporated into domestic legislation as follows:

1. The Motor Vehicles (Type Approval) Regulations 1980 (as amended), which provide, in accordance with EU Council Directive 70/156/EEC for the type approval of motor vehicles, other than land tractors, made after 1 July 1973 which have four or more wheels and are capable of exceeding a speed of 25 kilometres/hour on the level under their own power, and trailers made after this date which are drawn by such vehicles.
2. The Motor Vehicles (Type Approval for Goods Vehicles) (Great Britain (GB)) Regulations 1982 (as amended), which, in accordance with the Road Traffic Act 1988 require most goods vehicles and bi-purpose vehicles made since 1 October 1982 and used since 1 April 1983 to be type approved. There are exemptions for certain specialized vehicles.

18.1.2 UK legislation: the C&U regulations

In the UK, domestic legislation in the form of The Road Vehicles (Construction and Use) Regulations 1986 (as amended) – commonly referred to as the C&U regulations – specifies statutory requirements regarding the safety of vehicles and loads. Specifically, these regulations require that all vehicles and trailers, and all their parts and accessories, and the weight, distribution, packing, and adjustment of their loads, shall be such that no danger is caused or likely to be caused to any person in or on the vehicle or trailer or on the road. Additionally, the regulations state that no motor vehicle or trailer must be used for any purpose for which it is so unsuited as to cause or be likely to cause danger or nuisance to any person in or on the vehicle or trailer or on the road. Loads must be secured, if necessary by physical restraint, to stop them falling or being blown from a vehicle. This is particularly relevant to the carriage of shipping containers, which erroneously, are often thought to be secure by virtue of their weight alone; numerous serious and fatal accidents in the UK alone have shown this not to be the case, and any vehicle operator failing to secure such loads effectively may expect to be prosecuted and heavily penalized on conviction. Specific aspects of the legislation regulate the weight and dimensions of vehicles and their loads.

18.1.3 Maximum weights and dimensions for road vehicles in the UK

The sheer weight and size of road vehicles, and particularly the largest, which are those most likely to be used in combined and intermodal transport operations, create risks of danger. In the context of the types of vehicle likely to be found in combined transport operations, the normal maximum weight limit for heavy-goods vehicles in the UK is 44 tonnes, provided they are 'plated' (i.e. approved) for operation at this weight. Maximum weights for vehicles and trailers are indicated by a 'plate' issued by the Department for Transport showing the maximum permissible gross weight and maximum weights for the individual axles at which they may be operated on roads in GB. The requirement for annual plating and testing of goods vehicles is contained in the Road Traffic Act 1988 and is detailed in the Goods Vehicles (Plating and Testing) Regulations 1988 as amended.

The 44-tonne maximum weight limit approved for this particular purpose aligns with the EU standard for articulated vehicles and drawbar trailer combinations used in intermodal transport. The EU maximum weight limit for goods vehicles used in other transport operations is 40 tonnes. The EU requirements for harmonized vehicle, and trailer weights and dimensions are to be found in Council Directive 85/3/EEC as amended by Directives 89/338/EEC, 89/461/EEC, and 91/60/EEC.

For safety reasons, the C&U regulations require the 'travelling height' of a vehicle or trailer carrying a container to be indicated to the driver when this exceeds 3.66 metres. This is to help avoid the many instances of 'bridge bashing' that occur causing great danger to the lorry occupants, other road users, and especially to train passengers on the overhead rail tracks which become dislodged. The 'travelling height' is measured from the ground to within ±25 millimetres of the highest point of the container, and must be indicated by means of a notice, positioned where the driver can see it from the driving position, stating the height in feet and inches in figures at least 40-millimetre tall. For these purposes, a container is defined in the legislation as an article of equipment having a volume of at least 8 cubic metre, constructed wholly or mostly of metal and intended for repeated use for the carriage of goods or burden.

Heavy-goods vehicles are restricted in width to 2.55 metres both in the UK and by European standards (except for refrigerated vehicles which may be 2.6-metre wide subject to meeting requirements as to the thickness of the sidewalls, etc.). This dimension includes both the width of the vehicle itself (i.e. between its extreme projecting points) and any demountable load-carrying body or container (referred to in the legislation as a receptacle and defined as being of permanent character and strong enough for repeated use) carried on it. However, items such as driving mirrors and equipment to assist with transferring the load-carrying receptacle to or from a railway vehicle or securing the receptacle to a rail wagon are specifically excluded from this width limit.

The maximum length permitted for heavy-articulated and drawbar vehicles is specified in a number of ways, covering both the complete vehicle and its constituent parts. The maximum overall length for an articulated vehicle is 16.5 metres (apart from certain low loader articulated vehicles which may be up to 18-metre overall length) provided the combination:

1. Can turn within minimum and maximum swept inner and outer concentric circles of 5.3-metre and 12.5-metre radius, respectively. If this requirement cannot be met the overall maximum length for the vehicle remains at the old limit of 15 metres.
2. Has a semi-trailer with a distance from the centre line of the kingpin to the rear of the trailer which does not exceed 12 metres and where the distance from the kingpin (or rearmost kingpin if it has more than one) to the furthest point on the front corner of the trailer does not exceed 2.04 metres.

In practical terms this provides a load space length of approximately 13.6 metres (including front and rear wall thicknesses, and headboards) with flat platform or dry-freight trailers (at 2.55-metre wide). Dry-freight semi-trailers built to this length can accommodate 26 metric pallets (i.e. 1000 millimetres times 1200 millimetres) or 33 Euro-pallets (i.e. 800 millimetres times 1200 millimetres).

It is important to mention here that the carriage of 45-foot long ISO containers on standard 16.5-metre length articulated vehicles (a practice which is on the increase, especially in Europe) results in a small (i.e. 4.75 inches), but nevertheless illegal, overhang beyond the maximum length permitted (i.e. the container is 13.72-metre long against the semi-trailer maximum length of 13.6 metres). This is overcome by the use of ISO containers fitted with the Geest-patented tapered twistlock fitting which permits the front offside of the container to turn within the specified maximum swing clearance.

When a rigid motor vehicle is drawing a trailer the maximum overall length for the combination is 18.75 metres (as per Directive 91/60/EEC) subject to certain other minimum dimensional requirements being met as follows:

- Maximum load space is limited to 15.65 metres; to be shared between the two bodies (previous close-coupled drawbars operating within the old 18-metre length limit could achieve up to 16.2-metre load space).

- A minimum coupling dimension of 0.35 metres is specified (this allows a 16-metre 'envelope' of load and coupling space).
- A minimum cab length of 2.35 metres is required.

In measuring vehicle or trailer length account must be taken of any load-carrying receptacle (e.g. swap body or container) used with the vehicle. Excluded from the overall length measurement are such things as rubber or resilient buffers and receptacles for customs seals. In the case of drawbar combinations the length of the drawbar coupling is excluded from overall length calculations. With dropside-bodied vehicles the length of the tailboard in the lowered (i.e. horizontal) position is excluded unless it is supporting part of the load, in which case it must be included in the overall length measurement and for the purposes of establishing overhang limits (see below).

Overhang is the distance by which the body and other parts of a vehicle extend beyond the rear axle. The maximum overhang permitted for rigid goods vehicles (i.e. motorcars and heavy motorcars) is 60 per cent of the distance between the centre of the front axle and the point from which the overhang is to be measured. The point from which overhang is measured is, in the case of two-axled vehicles, the centre line through the rear axle, and in the case of vehicles with three or more axles, two of which are rear axles, 110 millimetres to the rear of the centre line between the two rear axles.

18.1.4 Annual testing of goods vehicles

Most goods vehicles are required to be tested annually to ensure they are safe to operate on the road and meet the legal requirements relating to mechanical condition. This applies both in the UK where stringent standards are applied, and elsewhere in Europe, where standards vary. In the UK, the annual testing of goods vehicles and trailers is intended to determine whether they meet the safety and technical standards specified in the C&U regulations. Testing of goods vehicles is carried out on behalf of the Department for Transport (DfT) at testing stations run by the Vehicle Operator and Services Agency (VOSA). It is an offence for a vehicle or trailer to be operated on the road if a valid test certificate is not in force.

18.1.5 'Safety of Loads on Vehicles': DfT Code of Practice

A Code of Practice, *The Safety of Loads on Vehicles*, published in the UK by the DfT (3rd edition 2002), sets out general requirements in regard to the legal aspects of safe loading, information on the forces involved in restraining loads, the strength requirements of restraining systems and load securing equipment, and then details specialized requirements for containers, pallets, engineering plant, general freight, timber, metal, and loose bulk loads. Of particular note is the basic principle on which the Code is based which is that:

> the combined strength of the load restraint system must be sufficient to withstand a force not less than the total weight of the load forward and half the weight of the load backwards and sideways.

The Code makes special mention of the hazards when a road vehicle is carried on a roll-on/roll-off ferry ship where both the vehicle and its load will be subject to different forces than normally experienced in road freighting due to the rolling and pitching motions of the vessel, and hence it advises that a restraint system that is suitable for road use will often not be adequate at sea.

One section of the Code deals specifically with containers, giving advice on how they should be secured and how goods should be stowed in them. It refers to both closed 'box' type containers and those of open frame structure enclosing the load of a bulk liquid or dry powder tank, which may in either case be lifted on and off the vehicle as a single unit comprising container and load. ISO freight containers, it says, may be secured to the vehicle by means of container locks, commonly known as twist locks. These twist locks should be inspected regularly for wear, damage, and correct operation. Locking devices, which are intended to prevent the operating levers from moving during transit, should be given special attention. A minimum of four twist locks should be provided (and used) for each container carried. As the Code says, provided the twist locks are fully engaged and locked in position, the container will be adequately secured and no further restraint will be necessary. If not carried on a vehicle equipped with twist locks, the container will need to be secured with chains and tensioners, although this method of carriage is not recommended. The use of ropes is totally unsuitable, the Code warns, and may lead the vehicle operator and driver into conflict with the law both for having an insecure load and for causing danger. It says that containers that do not have ISO corner castings may be fitted with special attachment brackets or lashing rings which should be in good condition and sufficiently sound for their purpose. Lashings should only be secured to those special points and to suitable points on the vehicle (not rope hooks on the side raves which are intended only for securing load covering sheets).

The Code warns that incorrect loading of containers can result in dangerous situations during transit through instability of the vehicle and cause damage to the goods themselves. The following special points are advised:

1. The load should not exceed the permitted payload of the container.
2. The load should be evenly distributed across the floor area of the container with not more than 60 per cent of the load in less than half the length of the container.
3. Heavy goods should not be stowed on top of lighter goods and the centre of gravity of the load should be below the midpoint of the container height.
4. The load should be secured inside the container sufficient to withstand reasonable forces which may be imposed on it during transit. A tightly packed load will be less likely to move than one which is loosely packed.
5. When packing is complete the rear of the container should be secured so that no part of the load or dunnage falls out on opening the doors at the end of the transit. Webbing lashings or nets are recommended for use, or a timber gate-like structure.

18.2 Rail safety

Rail safety regulatory controls and standards are, of necessity very extensive, extremely complex and highly specialized, too much so to attempt to detail here. But then the likely reader of this book, who may well be able to influence and control safety issues as they relate to one or more heavy-goods vehicles, is not likely to be in a position to impact in any meaningful way, as an individual or as a sole employee, on railway safety issues. These are matters for powerful committees, panels of technical experts, and boards of directors. In fact, organizations like the Rail Safety and Standards Board (RSSB), which was established on 1 April 2003, implementing one of the core sets of recommendations from the Cullen's enquiry into the Ladbroke Grove (London) rail accident of 1999. The RSSB's prime objective is to:

> lead and facilitate the railway industry's work to achieve continuous improvement in the health and safety performance of the railways in Great Britain, and thus to facilitate the reduction of risk to passengers, employees and the affected public.

In a document published in February 2005, 'How Safe is Safe Enough' – an overview of how Britain's railways take decisions that affect safety, the Railway Society of Southern Africa (RSSA) said, by way of an introduction;

> Britain's railways are very safe, comparable with other modes of public transport and with other countries' railways. But they can never be completely safe; neither engineering

> nor human systems are perfect and safety has to be balanced with the demands of performance and cost. Every decision that affects the railway is a safety decision because it should reflect that balance – whether it is taken by a company, government, or the regulators, and whether it concerns policy, investment, or operations. Safety decisions are business decisions, and a responsible company follows a similar process for both.

As well as the new RSSA mentioned above, also in 2003 the Rail Accident Investigation Branch (RAIB) was established under the provisions of the Railways and Transport Safety Act 2003. The Branch is due to be operational in spring 2005, and it will be the independent railway accident investigation body for the UK, as required by the European Railway Safety Directive. This Directive (2004/49/EC of 29 April 2004) is intended, among other purposes, to ensure the development and improvement of safety on the Community's railways by harmonizing the regulatory structure in the Member States, define responsibilities among the actors, develop safety targets and common safety methods with a view to greater harmonization of national rules, require the establishment in every Member State, of a safety authority and an accident and incident investigating body (see above regarding the RSSA and the RAIB in the UK), define common principles for the management, regulation, and supervision of railway safety.

18.3 Freight container safety regulations

Shipping containers present various problems of safety, they are large and very heavy when being lifted and loaded, thus presenting a variety of risks. Other risks present themselves when containers are stacked – as often they are, to as many as 8 or 10 units high. For these reasons, owners and lessees, and others in control of freight containers used at work or supplied for use at work must ensure that they comply with the International Convention for Safe Containers, Geneva 1972 (ratified by the UK 8 March 1978). The Freight Containers (Safety Convention) Regulations 1984, which came into force on 1 January 1985 and implemented this international agreement in the UK, apply to containers used within GB as well as to those used in international trade.

For the purposes of these regulations, 'containers' are those which are of the following:

1. Containers are of a permanent character and accordingly strong enough for repeated use.
2. Designed to facilitate the transport of goods by one or more modes of transport without intermediate reloading.
3. Designed to be secured or readily handled or both, having corner fittings for these purposes.
4. Are of a size such that the area enclosed by the outer bottom corners is either:
   (a) if the container is fitted with top corner fittings, at least 7 square metres;
   (b) if it does not have top corner fittings, a bottom area of at least 14 square metres.

Containers must have a valid approval issued by the HSE or a body appointed by the HSE (or under the authority of a foreign government which has acceded to the Convention) for the purpose confirming that they meet specified standards of design and construction, and should be fitted with a safety approval plate to this effect. If they are marked with their gross weight such marking must be consistent with the maximum operating gross weight shown on the safety approval plate. Containers must be maintained in an efficient state, in efficient working order, and in good repair. Details of the arrangements for the approval of containers in GB are set out in a document *Arrangements in GB for the Approval of Containers*, available from the HSE.

The safety approval plate (issued by the HSE) as described in the regulations must consist of a permanent, non-corroding, fireproof rectangular plate measuring not less than 200 millimetres by

100 millimetres and be permanently fitted to the container where it is clearly visible, adjacent to any other officially approved plate, and not capable of being easily damaged.

The nine lines on the plate must contain the following information:

Line 1: The country of approval and the approval reference.
Line 2: The month and year of manufacture.
Line 3: Manufacturer's identification number, or where the number is unknown, the owner's identification or a number allotted by the government or organization which granted the approval.
Line 4: The maximum gross weight in kilograms and pounds.
Line 5: The allowable stacking weight for 1.8 grams in kilograms and pounds (i.e. the designed maximum superimposed static stacking weight).
Line 6: The transverse racking test load value in kilograms and pounds.
Line 7: The end wall strength value as a proportion of the maximum permissible payload, which shall not be entered unless the end walls are designed to withstand a load of less or more 0.4 times the maximum permissible payload.
Line 8: The sidewall strength value as a proportion of the maximum permissible payload, which shall not be entered unless the sidewalls are designed to withstand a load of less or more 0.6 times the maximum permissible payload.
Line 9: If the approved examination scheme or programme so requires:
(a) a legend indicating that the container is subject to a continuous examination programme, or
(b) the date (month and year only) before which the container shall next be thoroughly examined.

Lines 7 and 8 may be used for the above purposes (a) and (b) if they are not required to contain other information.

18.4 International standards for swap bodies

European standards relating to the dimensions, structural strength, and safety of intermodal swap bodies are set by the CEN. This organization is made up of the respective national standards organizations of the EU Member States plus Iceland, Norway, and Switzerland. The whole concept of the CEN standard is to define a single European standard that will allow the unimpeded circulation of swap bodies throughout the road–rail systems of the EU (and beyond), and that is accepted, without alteration, by all the CEN member nations. In the UK, the British Standards Institute (BSI) publishes English language versions of the CEN standards under the title *Swap Bodies – Testing* (Code BS EN283: 1991).

Much of the testing and study work on defining the standards has been carried out by the German motor vehicle standards organization Normenausschuss Kraftfahrzeuge (FAKRA), which is part of the German standards institute DIN (Deutsches Institut fur Normung).

Under the CEN standard, swap bodies suitable for use in combined road–rail operations are defined (with sub-definitions for box bodies, open-siders, curtain-siders, and dropside types) and placed in one of three categories A, B, or C, with dimensions specified for each category and the fittings they must have to allow them to move freely on intermodal journeys within Europe. Individual standards are specified for the testing of each of these three categories of swap body to ensure they meet minimum standards of compliance. Static and dynamic tests are carried out, together with strength tests of the structure including side and end walls, the floor, fittings provided for lifting and the stability of support legs. Tests are also carried out to ensure adequate weather protection is provided where necessary.

A UIC (the International Railway Union) certification plate may be affixed to an approved swap body showing technical details of the approval and a registration number for the body for identification purposes.

The UIC sets its own standards for combined transport load units that are to be carried by rail. The following are the relevant standards:

1. 1985 UIC 592-4: Technical conditions for swap bodies that can be handled by grabs.
2. 1986 UIC 596-6: Traffic of road vehicles on wagons technical organization – conditions for coding load units in combined transport and combined transport lines.

18.5 Maritime safety

The principal agencies in the UK charged with regulatory control relating to waterborne transport are the HSE and the Maritime and Coastguard Agency (MCA) between which a Memorandum of Understanding exists. The work of the marine section of the HSE includes overseeing legislation dealing with the health and safety of people working in docks, shipyards, and inland waterways. Under the Memorandum of Understanding with MCA, HSE's enforcement role generally ends at the gangway. However, HSE enforces The Health and Safety at Work Act (1974) and the *Docks* Regulations (1988) where they apply to shore-based dock workers who may work on ships. HSE also enforces the Dangerous Substances in Harbour Areas Regulations 1987.

Following the Thames Safety Inquiry (i.e. of December 1999 on the 'Marchioness' disaster on the River Thames in August 1989) there was a review of HSE responsibilities in relation to 'inland waterway safety'. HSE policy has remained that it would have no involvement in matters concerning the design and safety standards incorporated in the construction, testing and operational use of ships, and any navigation or similar issues. Matters of manning, navigation, design and structural integrity of vessels, protection of workers against navigational or other risks to health or safety arising from the marine environment would be the responsibility of the Secretary of State for Transport through MCA. Local Authorities are also responsible for enforcing health and safety on certain activities on, or adjacent to, inland navigations, which are prescribed in the Health and Safety (Enforcing Authority) Regulations 1998. These principally relate to leisure activities, catering services, and the hiring out *of* pleasure craft.

The MCA enforces health and safety on ships. The MCA's main functions are to develop, promote, and enforce high standards of marine safety to minimize the loss of life among seafarers and coastal users, and to minimize pollution from ships of the sea and coastline. The MCA's statutory powers and responsibilities derive primarily from the Coastguard Act 1925, the Merchant Shipping Act 1995 and the Merchant Shipping and Maritime Security Act 1997 and associated secondary legislation. The Director of Logistics and Maritime Transport at the DfT is responsible for policy oversight and co-ordination of the MCA and the Marine Accident Investigation Branch (MAIB).

MCA is responsible for enforcing all merchant shipping regulations in respect of occupational health and safety: safety of vessels, safe navigation, and operation (including manning levels and crew competency). For inland and coastal waters purposes, a vessel is considered to be used in navigation, and therefore under the jurisdiction of MCA, if it operates on waters listed in the Annexe to Merchant Shipping Notice M5N 1776 (M) Categorization of Waters, and its subsequent revisions. Currently, the MCA is consulting on the 'Technical Standards for Inland Waterway Vessels', the harmonization of Boatmaster's licences for inland waterway operations, and implementation of the Working Time Directive (WTD) for mobile workers on inland waterways. The 'Technical Standards' comprise regulations regarding shipbuilding requirements, steering systems, engine design, electrical equipment, rigging and outfitting, accommodation, and other ancillary equipment. These standards will apply solely to new-build vessels in the first instance. In the area of crew competence, it is proposed that Boatmaster's licences would cover such matters as navigation rules, vessel manoeuvring and handling, knowledge of vessel construction and stability, engines, accident prevention, actions in the event of damage or danger, use of rescue and first-aid equipment, and prevention of fire and pollution.

To ensure safe operations on its waterways, British Waterways (BW) is presently implementing its own registration scheme for vessels and certificates of crew competence. *British Waterways Freight Vessel Conditions 2003* and *British Waterways Carriage* of *Freight Conditions* came into force in April 2003. The Freight Vessel Conditions require owners or operators of freight vessels operating on BW waterways to register the vessel with BW. However, to be able to register, the vessel will be subject to an annual fitness-for-purpose safety inspection, and it must have adequate public liability insurance. The Conditions of Carriage of Freight require the freight contractor to agree an operational schedule with BW, including routes to be used, competent crew required, lock/bridge/tide times and manning, carriage of cargo risk assessment, and risk assessment for load and discharge of goods where this is across BW owned or controlled property.

Other Navigation Authorities have a safety role on their own waters, in part to fulfil their duty of care under Section 3 of the Health and Safety at Work Act. Port and Harbour Authorities may also have byelaws which cover various aspects of operational or vessel safety. The Port Marine Safety Code is an important document in this respect because it sets out a summary of the legal duties and powers of harbour authorities relating to marine safety and it aims to promote good practice.

On a broader scale, according to the European Maritime Safety Agency (EMSA), which was established in 2002 under provisions contained in European Council Regulation 1406/2002/EC (of 27 June 2002), maritime transport is of fundamental importance to Europe and the rest of the world. In its published rationale it states that:

> To put this in perspective, over 90 per cent of European Union external trade goes by sea and more than one billion tonnes of freight a year are loaded and unloaded in EU ports. This means that shipping is the most important mode of transport in terms of volume. Furthermore, as a result of its geography, its history and the effects of globalisation, maritime transport will continue to be the most important transport mode in developing EU trade for the foreseeable future.
>
> In this context, European citizens have the right to expect their maritime passenger and goods transport to be safe, secure and clean. So, in support of these goals, and particularly in the wake of the recent *Erika* and *Prestige* oil tanker accidents, the set up of EMSA (under Regulation 1406/2002/EC) is one of the key EU level initiatives aimed at improving the situation.
>
> Although it has recently been given an operational task in oil pollution preparedness and response, the Agency's main objective is to provide technical and scientific assistance to the European Commission and Member States in the proper development and implementation of EU legislation on maritime safety, pollution by ships and security on board ships.

18.6 Duty of care

It is useful to note here the requirements under the Health and Safety at Work Act in the UK for employers to have a general duty of care towards their employees to ensure their health, safety, and welfare irrespective of the mode of transport in which they are engaged or ancillary activity. They also have a duty to ensure their premises are safe and without risk to the health of those who are not employees but who use the premises as a place of work. Employees must take reasonable care for their own health while at work. Employers and self-employed persons must ensure that their work activities do not expose other persons (i.e. not in their employ) to risks to their health and safety through their work activities.

Glossary of terms

The following terms are in common use in intermodal transport operations; many are used in this book.

Accompanied vehicles Road vehicles travelling on RO-RO ferry ships or on Eurotunnel's freight shuttle service through the Channel Tunnel where the driver travels with his vehicle so as to be available to drive it through to it final destination.

ADR Accord Dangereux Routier International agreement concerning the international transport of dangerous goods by road.

Alternative transport modes Road, rail, inland waterway, sea and air transport.

Articulated lorry Heavy lorry combination comprising a powered tractive unit and a semi-trailer attached so that at least 20 per cent of the weight of the load is borne by the tractive unit.

Berne Gauge European standard rail loading gauge. Exceeds normal UK standards.

Bimodal transport The carriage of goods by two separate modes of transport, usually road and rail.

Bimodal trailer Specially built road-going semi-trailer with retractable running gear that can be mounted on rail bogies for transport by rail.

Block trains Trains that run between two terminals without intermediate marshalling or transhipment of loading units.

Bogie The wheeled undercarriage of a road or rail vehicle, usually comprising a number of axles (e.g. two or three on a road vehicle) joined together by a compensating suspension arrangement.

Bulk freight Generally, grain and granular products, sand and gravel, coal and other extracted minerals, chemicals and other liquid products, oils and petroleum, etc. carried in tanker, or in some cases tipper road and/or rail vehicles or inland waterway vessels.

C and D Collection and delivery services, usually by road, into and out of sorting depots, or delivery hubs.

Cabotage The collection and delivery of goods within a country other than that in which the vehicle is registered.

Catenery High-voltage electric cable suspended above the rail track from which trains draw their power.

CEN Comité Européen de Normalisation (European Committee for Standardization).

CIM International convention on the movement of goods by rail.

CMR Convention Marchandises Routiers. Contract for the International Carriage of Goods by Road.

Combined transport A system in which a number of different modes are combined to provide the most efficient/economic transport of goods. Most commonly combining the short-haul collection and delivery of goods by road and the long trunk haul by rail.

Congestion Where the road system cannot accommodate the numbers of vehicles trying to use it at key (i.e. peak) times.

Container Rigid box (or frame containing tank) used in unit load transport which conforms to standard dimensions (i.e. to ISO standards) usually of 20, 30, 40 and 45 feet lengths by 8 feet or 2.5 metres wide and 8 feet 6 inches or 9 feet 6 inches high. Built to withstand the stresses of lifting under load and stacking to as many as 8/10 units high. Fitted with corner castings top and bottom for securing to road–rail vehicles and for lifting with the aid of twistlock attachments. Also known as ISO/maritime/shipping/freight container.

Consignee The person/firm to whom goods are sent (i.e. the recipient of the goods).

Consignor The person/firm who send goods.

Corner casting Heavy cast metal attachments fitted to the top and bottom corners of shipping containers, and to the bottom corners of swap bodies – also some located in intermediate positions. They match up with twistlocks for securing the container/swap body to a road vehicle skeletal chassis, or to a rail wagon. For lifting, the top castings are located by the gantry crane spreader.

C&U Construction and Use. *The Road Vehicles (Construction and Use) Regulations 1986* (as amended). Establishes constructional and use limitations, and legal requirements for road vehicles in UK.

Cube-out An expression that describes when a vehicle body or container volumetric capacity is fully utilized before its maximum weight-carrying capacity is achieved.

Curtain sides Plastic/flexible side curtains fitted to semi-trailers and some swap bodies to provide weather-proof and (relatively) secure protection for the load but which can be drawn back to allow side access for loading/unloading.

Demountable body Detachable road vehicle body fitted with folding/retractable landing legs for demounting. Interchangeable between vehicles/trailers and can be loaded/unloaded when freestanding. Not to be confused with the more rigidly built swap body used in intermodal transport.

Demurrage A charge payable in respect of delay to goods or a vehicle.

DfT Department for Transport. UK governmental transport authority.

DIRFT Daventry International Rail Freight Terminal. A freight terminal providing road to rail interchange facilities.

Drawbar trailer combination Road vehicle combination comprising a rigid towing (i.e. drawing) vehicle and a trailer that has either a front (steerable) and rear axle(s), or centre axles. Not to be confused with an articulated semi-trailer.

Drivers' hours rules Limitations placed on the working times of hgv drivers in the interests of public safety by *EC Regulation 3820/85/EEC* which set maximum daily and weekly driving times, and minimum break and rest periods.

DRS Direct Rail Services. Rail-freight operator.

EC European Commission. Official body within EU structure which makes and enforces European legislation.

ECMT European Conference of Ministers of Transport. Inter-governmental organization established in 1953.

EDI Electronic data interchange. System for exchanging data by electronic means (i.e. by computer).

EIA European Intermodal Association.

Europallet Pallet conforming to European standard dimensions; namely, 800 millimetres × 1200 millimetres (usually loaded up to 1000 kilograms gross weight).

EWS English Welsh and Scottish Limited. Rail-freight operator.

EU European Union. The Union of EU Member States comprising 25 members since May 2004.

Euroterminal Terminology for UK freight terminals that handle intermodal road–rail traffic mainly destined for Europe.

Fifth wheel (5th wheel) Part of articulated vehicle combination; that is turntable plate mounted on tractive unit chassis on which the semi-trailer rests and turns, and is secured by the kingpin.

FOC Freight operating company (i.e. rail-freight company).

Freight Facilities Grant (FFG) UK government grant towards the provision of rail facilities in the interest of converting freight from road to rail.

Freight grants Financial aid provided by government, via the Strategic Rail Authority, to support projects which aim to switch freight traffic from road to rail or to inland waterway or short-sea shipping.

Freightliner UK domestic container operator.

Freight lifted Statistical figure which represents the tonnage of freight transported.

Freight moved Statistical figure which represents the tonnage of freight multiplied by the distance it is transported (e.g. in kilometres).

Freight village Intermodal terminal with provision for road–rail transfers, storage and distribution operations.

Gantry crane Large, usually rail-mounted, crane (with spreader lifting attachments) seen in ports and freight terminals which straddles roads and rail tracks to allow rapid transfer of containers, swap bodies or complete road vehicles from one mode to another, or to/from ground storage, or between shore-based facilities and the hold of a berthed ship (e.g. a containership).

GBRf GB Railfreight. Freight train operator.

Grappler arms Attachments fitted to crane-mounted spreader to enable bottom lift swap bodies and/or piggyback semi-trailers to picked up. The arms locate in lifting pockets in the strengthened underframe of the body/semi-trailer chassis.

Gross vehicle weight The total weight of a vehicle (gvw), its load and the driver – the maximum allowed by law.

HGV Heavy goods vehicle. Goods vehicle with a gross weight exceeding 3.5 tonnes. Also referred to as a large goods vehicle (LGV).

High cube (or Hi-Cube) Containers and lorry trailers with extra height to provide capacity for high-volume, low-weight products (e.g. mattresses or polystyrene products).

Highways Agency UK Government agency responsible for managing the road network.

Hub Central point to which collected loads are brought for transhipment, sorting, and re-distribution.

Hub and spoke Distribution system where collected goods are centralized (i.e. at the hub) and then re-allocated for regional delivery on routes radiating out from the central hub (i.e. as in the spokes of a wheel).

Incoterms Standard terms established by the International Chamber of Commerce to facilitate international trade.

Intermodal transport The transport of unitized goods on a single journey by a number of different modes (e.g. road, rail, inland waterway, sea and air).

ICC International Chamber of Commerce.

ISO International Standards Organization, which in the context of this book sets structural and dimensional standards for shipping/freight containers, hence use of the term 'ISO container'.

ITU Intermodal transport unit – unit load device used in intermodal transport, for example swap body, container, piggyback semi-trailer.

JiT Just in time. Logistics system whereby goods are scheduled for delivery only as required (e.g. to match production schedules) to obviate the need for stockholding.

Kangaroo Continental name for piggyback road–rail system where semi-trailers are carried by rail (principal operator being French company Novatrans).

Kingpin Steel pin mounted on front underside of articulated semi-trailer which locates into the fifth-wheel plate on a tractive unit.

Landing legs The support legs on an articulated semi-trailer which are retracted when travelling, but are wound down when the trailer is detached from the towing unit.

Lift truck Loading vehicle with forks (i.e. forklift truck) or container loading/stacking attachment (i.e. stacker/reach stacker).

Load factor The percentage of available load space occupied by a load. A ratio of the used space against the total available space – or weight-carrying capability.

Loading Gauge Coded representation for the profile through which a rail wagon and load must be able to pass – limited by tunnel and bridge dimensions, platform clearances and the height of catenary (i.e. suspended power cables) above the track. For example, W6 gauge for conventional freight wagons and W9 (SB1c), and W10 gauge clearance for tall containers, piggyback freight and swap body traffic.

Loading unit General terminology for a container/swap body/semi-trailer in which loads are placed for unitized handling.

Lo-Lo Lift-on/Lift-off – loading method involving lifting (i.e. on to a ship) as opposed to rolling on wheels (see Ro-Ro).

Marco Polo EU programme which aims (through the provision of grant aid) to reduce road congestion and improve the environmental performance of freight transport within the Community and to enhance intermodality, thereby contributing to an efficient and sustainable transport system.

Maritime container Another term for a container which conforms to ISO standard dimensions for use in cellular container ships and in road–rail transport – also known as ISO/shipping/freight container.

Modal split The percentage split between different modes of transport comprising an intermodal operation.

Multifret wagon Low platform rail wagon used for carrying unit load traffic (i.e. swap bodies and containers) on intermodal rail services (e.g. through the Channel Tunnel).

Multimodal transport The use of a number of (i.e. multiple) different modes in the transport of goods; generally, it implies more than just two modes.

Network rail Owner and operator of the UK rail infrastructure (i.e. track, signalling and associated equipment); train operators pay track access charges for its use.

Open access EU direction (i.e. under *EC Directive 91/440/EEC*) that railway systems must be open to access by private (i.e. competing) operators.

PACT Pilot Actions for Combined Transport. EU Grant Aid Scheme to encourage intermodal transport projects; predecessor to Marco Polo programme.

Pallet Rigid base (of wood or composite material) on which goods are loaded to form an easily lifted and stacked unit load. Usually conforms to standard dimensions (ISO) 1200 millimetres × 1000 millimetres, or (Euro) 800 millimetres × 1200 millimetres.

Pallet-wide container Shipping (freight) container with 2.44-metre internal width to allow 1200 millimetre pallets to be loaded two across.

Payload The (maximum) weight of goods that can be legally carried on a vehicle.

Piggyback transport The carriage by rail of specially strengthened and equipped road-going semi-trailers on purpose-built rail wagons with pockets to accommodate the trailer bogie, thus providing minimum height for gauge clearance. Also using spine wagons which give low height and narrower width clearance.

Piggyback semi-trailer Road-going semi-trailer specially strengthened to withstand lifting under load and with folding sideguards and rear under-run bumper for transit on piggyback rail wagon.

Pocket wagon Purpose-built rail wagon with pockets to accommodate the two/three-axle bogie of road semi-trailers to provide low height for loading gauge clearance.

RDC Regional Distribution Centre.

Reach stacker Forklift type truck used for lifting/loading containers or swap bodies, etc. Fitted with spreader attachment for top-lifting containers and grappler arms for bottom lifting of swap bodies and piggyback semi-trailers.

RID Uniform rules on the carriage of dangerous goods by rail, part of CIM (see above).

RoadRailer Brand named bimodal transport system (from North America) currently on trial in the UK and Europe.

Rolling motorway Intermodal transport system where complete road vehicles are carried on special drive-on/drive-off low-height rail wagons (i.e. as with Eurotunnel's freight shuttle service through the Channel Tunnel and rail services through the Swiss Alps).

Rolling stock Collective term usually applied in rail operations for locomotives and freight wagons.

Ro-Ro Roll-on/Roll-off. The loading method whereby road vehicles are driven on/off a ferry ship.

RTD Road Transport Directive. *EC Directive 2002/15/EC* which sets working time limits for lorry drivers who were exempted from the original working time directive.

SAD Single Administrative Document. Multipart customs document used in inter-European trade.

Semi-trailer A trailer coupled to a road vehicle (i.e. the tractive unit) to form an articulated vehicle in such a way that a proportion (i.e. a minimum 20 per cent) of the weight of the load is borne by the motor vehicle.

Shipper Person/firm who consigns goods for despatch.

Skeletal Used in specialized container/swap body transport. Describes a sub-frame or chassis comprising only a skeleton framework (to reduce tare weight) mounted on a vehicle/trailer/semi-trailer in place of a normal freight body and equipped with twist locks on which the unit load is placed and secured.

SNCF Societe Nationale des Chemins de Fer. French national railway company. Operates freight division as SNCF Fret.

Spine wagon Type of rail wagon designed primarily for piggyback operations (i.e. carrying semi-trailers). Constructed with substantial spine to which the rail bogies are attached and with outrigger platforms for carrying road semi-trailer bogies. Has the advantage of being narrower (and lighter in weight) than conventional piggyback pocket wagons.

SRA Strategic Rail Authority. UK Governmental rail authority, due to be replaced in 2005/2006.

Straddle carrier Rubber-tyred mobile container lifting and transporting equipment wide and tall enough to straddle a number of lanes/container stacks/vehicles in a container port. Lifts containers from the top by means of a spreader.

Supply chain Describes the flow of goods from their original source through all processes to their final delivery to the end consumer, including where appropriate returns and disposal of waste.

Sustainable distribution Distribution system that causes minimal impact on the environment.

Swap body Load-carrying unit designed, to standard dimensions, for lifting under load and transfer between road and rail vehicles. Cannot be top lifted or stacked like a container. Known in French as a *caisse mobile*.

Tachograph Instrument designed to record the working time, breaks, and rest periods of road vehicle drivers to ensure they are conforming to statutory limits. Mandatory use as required by *EC Regulation 3821/85/EEC*. Digital version to be introduced from 2005/2007.

Tare weight The unladen weight of a vehicle/trailer or loading unit, etc. When subtracted from the gross weight it will indicate maximum payload potential.

TEN-T Trans-European Transport Network. A system of strategic routes across Europe defined by the EC.

Terminal Purpose-built and equipped premises where unit loads are transferred between modes (e.g. road to rail), where block trains are formed and where any necessary customs or security checks are carried out.

TEU Twenty-foot equivalent unit. Term of measurement for containers (i.e. 40 foot container is 2 × teus).

Tilt Tarpaulin (or reinforced plastic) sheet used to cover loads. Usually placed over a supporting frame on a semi-trailer/swap body.

Through train Rail train which runs directly through from the originating terminal to the destination terminal without intermediate marshalling, also known as a block train. Provides fast and reliable transit times.

TIR Transport International Routier, international road transport convention which permits the cross-border movement of sealed loads under simplified customs procedures.

Tonne Kilometres The unit by which the movement of freight is measured (i.e. tonnes lifted multiplied by the distance it is carried).

Top lift System whereby ISO containers are lifted from the top, via twistlocks mounted on the top corners, by gantry or container cranes. Obviously requires adequate built-in structural strength to withstand maximum container capacity loads.

Track Access Grant (TAG) UK government grant to defray Network Rail's charges for use of the rail track to encourage a switch of traffic from road to rail.

Track gauge The distance between rail lines. For the UK and Europe the dimension is 1435 millimetres (i.e. 4 feet 8½ inches); Eire, Spain, Portugal, Russia and Finland operate on different gauges.

Traction The provision of motive power to move loads – a term usually found in rail freight, but also occasionally in road haulage circles.

Tractive unit The motive power unit used in an articulated vehicle combination.

Trailer A wheeled load-carrying vehicle drawn by a motor vehicle to form a drawbar combination. Not to be confused with an articulated semi-trailer.

Train path An available time slot into which freight trains can be scheduled to run.

Transhipment The transfer of freight between vehicles, between transport modes, or between distribution centre and vehicle.

Transit The act of crossing, by vehicle, from one country to another.

Trunking The haulage of freight over long distances (e.g. the trunk haul between load collection and delivery).

Twistlock Standard fixing device, which locates in a corner casting, for securing container/swap body to a road vehicle/rail wagon or attached to spreader for top lifting of containers.

UIC Union Internationale des Chemins de Fer, International Union of Railways. Body representing and setting technical standards for most of the world's railways. Also established UIC system of loading gauges.

UIRR International Union of Combined Road–Rail Transport with 19-member companies throughout Europe.

Unaccompanied transport The intermodal transport of semi-trailers without the driver (i.e. on ferry ship or by rail); collected at the arrival port/terminal by another tractive unit and driver.

UNCTAD United Nations Conference on Trade and Development.

Unitization System whereby goods are assembled (consolidated) and loaded into standard loading units (e.g. pallets/containers/swap bodies).

UTI Unité de Transport Intermodal, French term for intermodal transport unit.

WCML West Coast Main Line. The UK's main rail route from London to the North West and Glasgow in Scotland.

WTD Working Time Directive. *EC Directive 93/104/EC*, which sets working time limits for most employees in Europe. Intended to be in the interests of improving working environments and ensuring better protection for workers.

Bibliography

The following lists of books, booklets, conference papers, and journal and newspaper articles, as well as government regulations, and international agreements and standards provide sources of further information for interested readers. Also included in this section is a list of the principal UK transport journals that carry news and feature items on intermodal and combined road–rail transport operations.

Legislation, International Standards and Official Reports

Commission of the European Communities (2003) *Study on Freight Integrators: Final Report*. Berlin. Consortium of five partners: ZLU, ISF, EIA, Kravag-Logistic, ELA.

European Commission, Directorate General for Energy and Transport (2003) *Inland Waterways: A Transport Solution That Works*. Brussels. ISBN 92-894-4344-8.

European Commission (2002) *Trans-European Transport Network: TEN-T Priority Projects*. Luxembourg.

European Conference of Ministers of Transport (1995) *Transforming the Structure of the Freight Transport Sector, 99th Round Table Conference on Transport Economics*, Economic Research Centre, Paris.

International Road Transport Union (IRU) and International Union of Combined Rail–Road Companies (UIRR) (May 1995) *Joint Declaration on the Promotion of Combined Transport*.

UK Secretary of State for Transport (1995) *Transport and the Economy*. Speech to the CBI.

Royal Commission on Environmental Pollution (1994) *Transport and the Environment*, Report.

Planning Policy Guidance Note 13 – PPG 13 (1995) HMSO, London.

Sustainable Distribution: A Strategy (1999) DETR – now the DfT.

Transport Ten Year Plan, DfT (2000) Dubbed Transport 2010.

The Piggyback Consortium (1994) *Final Report*.

HMSO, London (1994) *The Road Vehicles (Construction and Use) (Amendment) (No 2) Regulations 1994*.

Council Regulation EEC 3315/94 (1994) Amending Council Regulation 3118/93/EEC dealing with cabotage and cabotage permit allocations.

Freight on Water – A New Perspective (2002) DEFRA.

Strategic Rail Authority (SRA) *Freight Strategy* (2001) *The Way Forward for Britain's Railway* (2002).

Everyone's Railway – The Wider Case for Rail (2003) and *Platform for Progress* (2003).

European Commission (1992) *The Future Development of the Common Transport Policy* – COM 92/494 (final).

European Council Directive 92/106/EEC (1992) Establishing Common Rules for the Combined Transport of Goods.

Intermodality and Intermodal Freight Transport in the European Union (COM 97/243) (1997) EC.

White Paper, *European Transport Policy for 2010: Time to Decide* (2001) EC.

European Commission proposal for a Directive on *Intermodal Loading Units* in April 2003 (COM 2003 155 final).

Directive 91/440/EEC (1991) The 'Railway Directive'.
UK Marine Motorways Study (2002/2003) DfT with the Engineering and Physical Sciences Research Council (EPSRC) funded the under the LINK Future Integrated Transport Programme.
Department of Transport (1994) *Freight Facilities Grants.* Guide for applicants, London.
Department of Transport (1994) *Track Access Grants.* Guide for applicants, London.
Commission of the European Communities, Brussels (1994) *Pilot Action for Combined Transport.* Paper.
European Commission (1993) *Trans-European Networks: Towards a Master Plan for the Road Network and Road Traffic.* Motorway Working Group Report. Brussels and Luxembourg.
Council Regulation EEC 3118/93. Specifying the permanent cabotage regime.
Eurotunnel, London (1993) *The Loading Gauge Issue.* Report.
Department of Transport (1992) *Moving Freight by Rail: Loading Gauge.* London.
Commission of the European Communities, Brussels (1992) *The Future Development of the Common Transport Policy*, White Paper.
European Commission, Brussels (1992) *Adoption of a Communication on the Establishment of a European Combined Transport Network,* Information Memo.
HMSO, London (1992) *The Goods Vehicles (Community Authorisations) Regulations 1992.* Brings into force in the UK *Council Regulation EEC/881/92.*
Council Regulation EEC 881/92. Deals with Community Authorisations.
International Road Transport Union (IRU), Geneva (1992) *The Transport of Goods by Road and Its Environment in the Europe of Tomorrow.*
European Conference of Ministers of Transport, Paris (1991) *Freight Transport and the Environment.*
European Commission, Brussels (1991) *High Level Working Group: Combined Transport Report.*
International Union of Railways (UIC) and the International Road Transport Union (IRU) (1991) *The Conditions for the Development of Combined Transport in Europe.* Joint presentation.
British Standards Institution (1991) *Swap Bodies – Testing* (BS EN283:1991 BSI).
British Standards Institution. *Stowage of Goods in Containers* (BS 5073 BSI).
British Standards Institution. *Freight Containers Parts 1 and 2* (BS 3951 BSI – equates to ISO 668-1988 and ISO 1496/1-1990).
UIC (596-6) (1986) *Traffic of Road Vehicles on Wagons – Technical Organisation – Conditions for Coding Load Units in Combined Transport and Combined Transport Lines.*
HMSO, London (1986) *The Road Vehicles (Construction and Use) Regulations 1986* – as amended (SI 1078/1986).
UIC (592-4) (1985) *Technical Conditions for Swap Bodies Which Can Be Handled by Grabs.*

Books, Booklets and Conference Papers

Lowe D *The Transport Manager's and Operator's Handbook.* Kogan Page Ltd., London (published annually).
Railtrack plc. (April 1999) *A Guide to Rail Freight.*
Department for Transport. International Road Freight Office. Cambridge. *A Guide to Taking Your Lorry Abroad.* Free explanatory booklet.
International Union of Rail–Road Transport Companies (UIRR), Brussels (1995) *Focus on Combined Transport.* Free booklet.
UIRR, Brussels. *Reports of the International Union of Rail–Road Transport Companies* (published annually).
Fret SNCF, France (1995) *Intermodal Transport.*
Piggyback Consortium (1995) *Working to put Trucks on Trains.* Brochure.

IIR Publications Ltd., London (1995) *Intermodal '95.* Amsterdam Conference Papers.

Muller, Gerhardt, USA (1995) *Intermodal Freight Transportation* (3rd edition) Eno Transportation Foundation – jointly published with the Intermodal Association of North America (IANA).

Institute of Logistics (1994) *Understanding European Intermodal Transport – A User's Guide.* Guideline Series No. 7.

SP Tyres UK Limited (1994) *International Freighting via the Channel Tunnel.* Usercare leaflet No. 26.

Lowe D (1994) *The European Road Freighting Handbook.* Kogan Page, London.

Gibbins E (1994) *Intermodal Transport.* Road Transport Operation bulletin. Croner Publications.

Dawson RJ (1994) *EC Transport Policy.* Road Transport Operation bulletin, Croner Publications.

Cargo Systems (1993) *Intermodal '93.* Hamburg Conference Papers.

Rex Walden & Associates (1993) *Freightconnection '93* Manchester Conference Papers.

Department of Transport (1993) *Heavier Lorries for Combined Road–Rail Transport.* Consultation Paper, London.

Rail Freight Handbook. Clarke & Company, Oxford.

TRRL (1992) *Review of European Intermodal Freight Transport.*

European Conference of Ministers of Transport (1992) *Improvements in Main International Piggyback Links*, Report.

Lowe D (1992) *The Road to Europe.* Headlight Publications, London.

Lowe D (1992) *1992 – The Long Haul into Europe.* Iveco Ford Truck Ltd., Watford.

Smith RI (1992) *Combined Transport: The Way Forward.* Scottish Transport Studies Group.

Rex Walden & Associates (1992) *Freightconnection '92.* Birmingham Conference Papers.

Federal Department of Transport, Communications and Energy General Secretariat/Bureau for Transport Studies (1991) *Transport – Yesterday, Today, Tomorrow.* Switzerland's transport system and transport policy. Berne, Switzerland.

Glass DA and Cashmore C (1989) *Introduction to the Law of Carriage of Goods.* Sweet & Maxwell, London.

Lowe D (1989) *The Transport and Distribution Manager's Guide to 1992.* Kogan Page, London.

Stack JG (1988) *The Application of Freight Containers and Their Effects on Surface and Sea Transport Modes.* London.

Cargo Systems (1988) *Intermodal Europe '88.* Conference Papers.

Transport Tutorial Association (1988) *Intermodal Through Freight Transport.*

Hayuth Y (1987) *Intermodality: Concept and Practice.* London.

Baker DRW (1986) *The Rail Trailer* Bath.

Temple SR (1986) *Modal Split in the British Inland Container Market.* London.

Donald A (1981) *The CMR.* Derek Beattie Publishing, London.

Lowe D (1973) *Transport and Delivery to EC Customers.* Kogan Page, London.

Hadfield C (1968) *The Canal Age.* Pan Books

Press Cuttings

ILT Focus (October 2003) *Achieving Real Dividends – Intermodal Solutions.*

ILT Focus (October 2003) *Full Ahead for Waterborne Freight.*

MT Logistica (1994) *Calculating the Cost (of Transferring Freight from Road to Rail).*

MT Logistica (1994) *By Piggyback to the Future.*

MT Logistica (1994) *Now Take to the Tracks.*

MT Logistica (1994) *TDG Boosts Bimodal Transport.*

The Times (August 1993) *Piggy-back Train Takes Strain off Road.*

Department of Transport (1993) *MacGregor Moves to Switch Freight from Road to Rail.* Press Release, London.
The Independent (August 1993) *Bigger, Better Lorries and a Rail Piggy-back.*
Commercial Motor (July 1993) *Back on the Rails.*
Rail International (June/July 1993) *Road–Rail Transport Claims: Is Combined Transport a Long-Term Solution for the European Transport Market?*
ILDM Focus. Issue No 4 (1993) *Logistics by Combined Transport: Barriers to Market Entry and Strategies of Main Suppliers.*
Commercial Motor (April 1993) *The Continentals Are Ahead of Us in This Game: We Need to Catch Up.*
Transport Management (February/March 1993) *Handling European Freight Traffic.*
Rail International (February 1993) *Combined Transport: A Key Asset or a Challenge.*
Financial Times (August 1992) *Rail Failure to Affect Roads.*
Seaway Review (July/September 1992) *The St Lawrence to Europe: Trends in Intermodal Transportation.*
The Times (August 1992) *Road–Rail Freight Venture Collapses.*
ILDM Focus (July/August 1992) *Combined Road–Rail Transport: Shippers and the Market.*
Freight Management (June/July 1992) *Intermodal Europe Won't Be Built in a Day.*
Cargo Systems (June 1992) *Chartering a New Course.*
Cargo Systems (June 1992) *Heading for a Rail Clash.*
Transport Innovation, Supplement (Summer 1992). *Intermodal Freight.*
Cargoware International (May 1992) *EC Unveils Combined Transport Initiative.*
Transport (March/April 1992) *Expanding the Options.*
Railway Gazette International (March 1992) *Intermodal Holds the Key to European Freight Revival.*
Cargo Systems (February 1992) *Count the Bodies.*
Cargo Systems (February 1992) *Tunnel Vision: A Long Haul.*
Transport Engineer (January 1992) *Implications of Intermodal.*

Statistics

Department for Transport (2003) *Focus on Freight – 2003 Edition.*
Department for Transport (2004) *Focus on Freight – 2004 Edition.*

Index